The Culture of Vegetables and Flowers from Seeds and Roots

SIXTEENTH EDITION

Sutton and Sons

The Culture of Vegetables and Flowers from Seeds and Roots

The Culture of Vegetables and Flowers from Seeds and Roots

CONTENTS

THE CULTURE OF VEGETABLES

Horticulture has a full share in the progressive character of the age. Changes have been effected in the Kitchen Garden which are quite as remarkable as the altered methods of locomotion, lighting and sanitation. Vegetables are grown in greater variety, of higher quality, and are sent to table both earlier and later in the season than was considered possible by gardeners of former generations.

When Parkinson directed his readers to prepare Melons for eating by mixing with the pulp 'salt and pepper and good store of wine,' he must have been familiar with fruit differing widely from the superb varieties which are now in favour. A kindred plant, the Cucumber, is more prolific than ever, and the fruits win admiration for their symmetrical form.

The Tomato has ceased to be a summer luxury for the few, and is now prized as a delicacy throughout the year by all classes of the community.

As a result of the hybridiser's skill modern Potatoes produce heavier crops, less liable to succumb to the attacks of disease, than the old varieties, and the finest table quality has been maintained.

Peas are not what they were because they are so immensely better. While the powers of the plant have been concentrated, with the result that it occupies less room and occasions less trouble, its productiveness has been augmented and the quality improved. All the pulse tribe have shared in the advance, and a comparison of any dozen or score of the favourite sorts of Peas or Beans grown to-day with the same number of favourites of half or even a quarter of a century since will at once prove that progress in horticulture is no dream of the enthusiast.

Among the Brassicas, such as Broccoli, Brussels Sprouts, Cabbage and Cauliflower, a series of remarkable examples might be mentioned; and roots such as Beet, Carrot, Onion, Radish and Turnip afford other striking instances of improvement. Salads also, including Celery, Chicory, Endive and Lettuce, have participated in the beneficial change and offer a large choice of dainties, adapted to various periods of the year. Indeed it may be truly said that none of the occupants of the vegetable garden have refused to be improved by scientific crossing and selection.

The vegetables which are available for daily use offer a wide and most interesting field to the expert in selecting and hybridising. For past achievements we are indebted to the untiring labours of specialists, and to their continued efforts we look for further results. Whether the future may have in store greater changes than have already been witnessed none can tell. One thing only is certain, that finality is unattainable, and the knowledge of this fact adds to the charm of a fascinating pursuit. Happily, innovations are no longer received with the suspicion or hostility they formerly encountered. In gardens conducted with a spirit of enterprise novelties are welcome and have an impartial trial. The prudent gardener will regard these sowings as purely experimental, made for the express purpose of ascertaining whether better crops can be secured in future years. For his principal supplies he will rely on those varieties which experience has proved to be suitable for the soil and adapted to the requirements of the household he has to serve. By growing the best of everything, and growing everything well, not only is the finest produce insured in abundance, but every year the garden presents new features of interest.

In considering the general order of work in the Kitchen Garden, the first principle is that its productive powers shall be taxed to the utmost. There need be no fallowing—no resting of the ground; and if it should so happen that by hard cropping perplexity arises about the disposal of produce, the proverbial three courses are open—to sell, to give, or to dig the stuff in as manure. The last-named course will pay well, especially in the disposal of the remains of Cabbage, Kale, Turnips, and other vegetables that have stood through the winter and occupy ground required for spring seeds. Bury them in trenches, and sow Peas, Beans, &c., over them, and in due time full value will be obtained for the buried crops and the labour bestowed upon them. But hard cropping implies abundant manuring and incessant stirring of the soil. To take much off and put little on is like burning the candle at both ends, or expecting the whip to be an efficient substitute for corn when the horse has extra work to do. Dig deep always: if the soil be shallow it is advisable to turn the top spit in the usual manner, and break up the subsoil thoroughly for another twelve or fifteen inches. Where the soil is deep and the staple good, trench a piece every year two spits deep, the autumn being the best time for this work, because of the immense benefit which results from the exposure of newly turned soil to rain, snow, frost, and the rest of Nature's great army of fertilising agencies.

In practical work there is nothing like method. Crop the ground systematically, as if an account of the procedure had to be laid before a committee of severe critics. Constantly forecast future work and the disposition of the ground for various crops, keeping in mind the proportions they should bear to each other. Be particular to have a sufficiency of the flavouring and garnishing herbs always ready and near at hand. These are sometimes wanted suddenly, and in a well-ordered garden it should not be difficult to gather a tuft of Parsley in the dark. Change crops from place to place, so as to avoid growing the same things on the same plots in two successive seasons. This rule, though of great importance, cannot be strictly followed, and may be disregarded to a certain extent where the land is constantly and heavily manured. It is, however, of more consequence in connection with the Potato than with aught else, and this valuable root should, if possible, be grown on a different plot every year, so that it shall be three or four years in travelling round the garden. Lastly, sow everything in drills at the proper distances apart. Broadcasting is a slovenly mode of sowing, and necessitates slovenly cultivation afterwards. When crops are in drills they can be efficiently thinned, weeded and hoed—in other words, they can be cultivated. But broadcasting pretty well excludes the cultivator from the land, and can only be commended to the idle man, who will be content with half a crop of poor quality, while the land may be capable of producing a crop at once the heaviest and the best.

GLOBE ARTICHOKE

Cynara Scolymus

The Globe Artichoke is grown mainly for the sake of its flower-heads which make a delightful dish when cooked while immature. The plant is easily raised from seed, although not quite hardy in some districts. It will grow on almost any soil, but for the production of large fleshy heads, deep rich ground is requisite. The preparation of the soil should be liberal, and apart from the use of animal manure the plant may be greatly aided by wood-ashes and seaweed, for it is partial to saline manures, its home being the sandy seashores of Northern Africa.

The simplest routine of cultivation consists in sowing annually, and allowing each plantation to stand to the close of the second season. Seed may be sown in February in boxes of light soil, or in the open ground in March or April. In the former case, put in the seeds one inch deep and four inches apart, and start them in gentle heat. Grow on the seedlings steadily, and thoroughly harden off preparatory to planting out at the end of April, giving each a space of three to four feet apart each way. Under favourable conditions the plants from the February sowing will produce heads in the following August, September, and October. In the second year, the heads will be formed during June and July. This arrangement not only insures a supply of heads from June to October, but admits of a more effective rotation of crops in the garden.

Sowings in the open ground should be made in March or April, in drills one foot apart. Thin out the plants to six inches apart in the rows and allow them to stand until the following spring, when they may be transplanted to permanent beds.

Globe Artichokes may also be grown from suckers planted out in April when about nine inches high. Put them in rather deep, tread in firmly, and lay on any rough mulch that may be handy. Should the weather be dry they will require watering, and during a hot dry spell water and liquid manure should be given freely to insure a good supply of large heads. Seedlings that are started well in a suitable bed take better care of themselves than do plants from suckers, especially in a dry season. Vigorous seedlings send down their roots to a great depth.

To advise on weeding and hoeing for the promotion of a clean and strong growth should be needless, because all crops require such attention. But as to the production of large heads, a few words of advice may be useful. It is the practice with some growers to twist a piece of wire round the stem about three inches below the head. This certainly does tend to increase the size, but the same end may be accomplished by other means. In the first place, a rich deep bed and abundant supplies of water will encourage the growth of fine heads. Further aid in the same direction will be derived from the removal of all the lateral heads that appear when they are about as large as an egg. Up to this stage they do not tax the energies of the plants in any great degree; but as the flowers are forming within them their demands increase rapidly. Their removal, therefore, has an immediate effect on the main heads, and these attain to large dimensions without the aid of wire. The small heads will be valued at many tables for eating raw, as they are eaten in Italy, or cooked as 'artichauts frits.' The larger main heads are the best for serving boiled in the usual way. After the heads are used the plants should be cut down.

Chards are the blanched summer growth of Globe Artichokes, and are by many preferred to blanched Cardoons. In the early part of July the plants selected for Chards must be cut over about six inches above the ground. In a few days after this operation they will need a copious watering, which should be repeated weekly, except when heavy rains occur. By the end of September the plants will have made much growth and be ready for blanching. Draw them together, put a band of hay or straw around them, and earth them up, finishing the work neatly. The blanching will take fully six weeks, during which time there will be but little growth made—hence the necessity for promoting free growth before earthing up. Any Chards not used before winter sets in may be lifted and preserved by packing in sand in a dry shed.

The Artichoke is hardy on dry soils when the winter is of only average severity. But on retentive soils, which are most favourable to the production of fine heads, a severe winter will destroy the plantations unless they have some kind of protection. The usual course of procedure is to cut down the stems and large leaves without touching the smaller central leaves, and, when severe frost appears probable, partially earth up the rows with soil taken from between; this protection is strengthened by the addition of light dry litter loosely thrown over. With the return of spring the litter is removed, the earth is dug back, and all the suckers but about three removed: then a liberal dressing of manure is dug in, care being taken to do as little injury to the plants above and below ground as possible. At the end of five years a plantation will be quite worn out; in somewhat poor soil it will be exhausted in three years. But on any kind of soil the cultivation of this elegant vegetable is greatly simplified by sowing annually, and allowing the plants to stand for two years only, as already advised.

JERUSALEM ARTICHOKE

Helianthus tuberosus

The Jerusalem Artichoke is a member of the Sunflower tribe, quite hardy, and productive of wholesome roots that are in favour with many as a delicacy, and by others are regarded as worthless. It is said that wise men learn to eat every good thing the earth produces, and this root is a good thing when properly served; but when cooked in the same way as a Potato it certainly is a very poor vegetable indeed. It is a matter of some interest, however, that in respect of nutritive value it is about equal to the Potato; therefore, in growing it for domestic use nothing is lost in the way of food, though it needs to be cooked in a different way.

The Jerusalem Artichoke will grow anywhere; indeed, it will often yield a profitable return on land which is unsuitable for any other crop, but to insure a fine sample it requires a deep friable loam and an open situation. We have grown immense crops on a strong deep clay, but it is not a clay plant, because it soon suffers from any excess of moisture. To prepare the ground well for this crop is a matter of importance, for it roots freely and makes an immense top-growth, reaching, when very vigorous, a height of ten or twelve feet. Trench and manure in autumn, and leave the land rough for the winter. Plant in February or March, using whole or cut sets with about three eyes to each, and put them in trenches six inches deep and three feet apart, the sets being one foot apart in the trenches. When the plants appear, hoe the ground between, draw a little fine earth to the stems, and leave the rest to Nature. Take up a portion of the crop in November and store in sand and dig the remainder when wanted, as recommended in the case of Parsnips. The tubers must be dug with a fork by opening trenches and cleaning out every scrap of the roots, for whatever remains will grow and become troublesome in the following season.

ASPARAGUS

Asparagus officinalis

Asparagus is a liliaceous plant of perennial duration, and it demands more generous treatment than the majority of Kitchen Garden crops.

Under favourable conditions it improves with age to such an extent as to justify the best possible cultivation. Plantations that have stood and prospered for twenty or even thirty years are not uncommon, but a fair average term is ten years, after which it is generally advisable to break up a bed, the precaution being first taken to secure a succession bed on fresh soil well prepared for the purpose. Plantations are made either by sowing seeds or from transplanted roots; and although roots are extremely sensitive when moved, success can, as a rule, be insured by special care and prompt action, assuming that the proper time of year is chosen for the operation. The advantage of using roots is the saving of time, and in most gardens this is an important consideration. Fortunately roots may be planted almost as safely when two or three years old as at one year.

Soil.—Asparagus will grow in any soil that is well cultivated; a deep rich sandy loam being especially suitable. Calcareous soil is by no means unfavourable to Asparagus; still, a sand rich in humus is not the less to be desired, as the finest samples of European growth are the produce of the districts around Paris and Brussels. The London Asparagus, which is prized by many for its full flavour and tenderness, is for the most part grown near at hand, in deep alluvial soils enriched with abundance of manure. Nature gives us the key to every secret that concerns our happiness, and on the cultivation of Asparagus she is liberal in her teaching. The plant is found growing wild on the sandy coasts of the British Islands—a proof that it loves sand and salt.

Preparation of Ground.—The routine cultivation must begin with a thorough preparation of the ground. Efficient drainage is imperative, for stagnant water in the subsoil is fatal to the plant. But a rich loam does not need the extravagant manuring that has been recommended and practised. Deep digging and, where the subsoil is good, trenching may be recommended, but an average manuring will suffice, because Asparagus can be effectually aided by annual top-dressings, and proper surface culture is of great importance in the subsequent stages. It is necessary to choose an open spot for the plantation. Preparation of the ground should commence in the autumn and be continued through the winter, a heavy dressing of half-rotten stable manure being put on in the first instance, and trenched in two feet deep. In the course of a month the whole piece should be trenched back. If labour is at command a third trenching may be done with advantage, and the surface may be left ridged up until the time arrives to level it for seeding. It will be obvious that this routine is of a somewhat costly character, but we are supposing the plantation is to remain for many years, making an abundant return for the first investment. Still we are bound to say that a capital supply for a moderate table may be obtained by preparing a piece of good ground in an open situation in a quite ordinary manner with one deep digging in winter, adding at the time some six inches or so of fat stable manure, and leaving it thus until the time arrives for sowing the seed. Then it will be well to level down and point in, half a spade deep, a thin coat of decayed manure to make a nice kindly seed-bed.

Where soil known to be unsuitable, such as a damp clay or pasty loam, has to be prepared for Asparagus, it will be found an economical practice to remove the top spit, which we will suppose to be turf or old cultivated soil, and on the space so cleared make up a bed of the best possible materials at command. Towards this mixture there is the top spit just referred to. Add any available lime rubbish from destroyed buildings, sand, peat, leaf-mould, surface soil raked from the rear of the shrubberies, &c., and the result should be a good compost obtained at an almost nominal cost.

Size of Bed, and Sowing Seed.—At this juncture several questions of considerable importance arise. And first, whether the crop shall be grown on the flat or in raised beds. Where the soil is sufficiently deep, and the drainage perfect, the flat system answers well. The advantages of raised beds are that they deepen the soil, assist the drainage, promote warmth, and thus aid the growth of an early crop. In fact, raised beds render it possible to grow Asparagus on soils from which this vegetable could not otherwise be obtained. The preparation is the same in either case, and therefore we shall make no further allusion to flat beds, but leave those to adopt them who find their soil and requirements suitable. Now comes the question of distance, on which depends the width of the beds. The first point may be settled by the measure of the plant, and the second by the measure of the man. Monster sticks are valued at some tables, and we shall refer to these later on, but an abundant crop of handsome, though not abnormal, Asparagus meets the requirements of most households. After many experiments, we have come to the conclusion that the best mode of insuring a full return of really good sticks, with the least amount of labour, is to lay out the land in three-feet beds, with two-feet alleys between. In some instances, no doubt, five-feet beds, containing three rows of roots, one down the middle and one on each side at a distance of eighteen inches, are preferable. For the majority of gardens, however, the three-feet bed is a distinct advantage, were it only for the fact that all excuse for putting a foot on the bed is avoided. On this narrow bed only two rows of plants will be necessary. Put down the line at nine inches from the edge on both sides, and at intervals of fifteen inches in the rows dibble holes two inches deep, dropping two or three seeds in each. This will give a distance between the rows of eighteen inches. In very strong land, heavily manured, the holes may be eighteen inches apart instead of fifteen. April is the right month for sowing.

Thinning.—When the 'grass' from seeds has grown about six inches high, only the strongest plant must be left at each station, and they should finally stand at a distance of fifteen or eighteen inches in the row. Much of the injury reported to follow from close planting has been the result of carelessness in thinning. The young plant is such a slender, delicate thing, that, to the thoughtless operator, it seems folly to thin down to one only. The consequence is that two or three, or perhaps half a dozen, plants are left at each station to 'fight it out,' and these become so intermixed as to appear to be one, though really many, and of course amongst them they produce more shoots than can be fed properly by the limited range of their roots. Severe, or may we say mathematical, thinning is a *sine quâ non*, and it requires sharp eyes and careful fingers; but it must be done if the Asparagus beds are to become, as they should be, the pride of the Kitchen Garden.

Blanching.—The grave question of white *versus* green Asparagus we cannot entertain, except so far as concerns the cultivator only. On the point of taste, therefore, we say nothing; and it is a mere matter of management whether the sticks are blanched to the very tip, or allowed to become green for some few inches. Blanching is effected in various ways. The heaping up of soft soil, such as leaf-mould, will accomplish it. On the Continent many contrivances are resorted to, such as covering the heads with wooden or earthen pipes. In a few districts in France champagne-bottles with the bottoms cut away are employed. But a strong growth being secured, the cultivator will find it an easy matter to regulate the degree of colour according to the requirements of the table he has to serve. As a rule, a moderately stout growth, with a fair show of purple colour, is everywhere appreciated, and is the easiest to produce, because the most natural.

There is, however, an interesting point in connection with the production of green Asparagus, and it is that if wintry weather prevails when the heads are rising (as unfortunately is often the case) the tender green tops may be melted by frost and become worthless, or may be rendered so tough as to place the quality below that of blanched Asparagus; for the blanching is also a protective process, and quickly grown white Asparagus is often more tender and tasty than that which is green, but has been grown slowly. As the season advances and the heads rise rapidly the green Asparagus acquires its proper flavour and tenderness, and thus practical considerations should more or less influence final decisions on matters of taste. The business of the cultivator is to produce the kind of growth that is required, whether white or green, or of a quality intermediate between the two. This is easily done, making allowance for conditions. When green Asparagus is alone in demand, the cultivator may be advised to have in readiness, as the heads are making their first show, a sufficient supply of some rough and cheap protecting material, such as grass and coarse weeds, cut with a sickle from odd corners of the shrubbery and meadow land, or clean hay and straw perfectly free from mildew; but for obvious reasons stable litter should not be used. A very light sprinkling of material over an Asparagus bed that is making a first show of produce will ward off the morning frosts, and amply compensate for the little trouble in saving many tender green sticks that the frosts would melt to a jelly and render worthless. After the second or third week in May the litter may be removed if needful; but if appearances are of secondary importance, it may be left to shrink away on the spot.

Cutting.—Asparagus as supplied by market growers is needlessly long in the stem. The bundles have an imposing appearance, no doubt, but the useless length adds nothing to the comfort of those at table, and is a wasteful tax on the energy of the plant. For home consumption it will generally suffice if the white portion is about four inches long, and this determines the depth at which the sticks should be cut. Here it may be useful to remark that deeply buried roots do not thrive so well as those which are nearer the surface, nor do they produce such early crops. The sticks are usually cut by thrusting down a stiff narrow-pointed knife, or specially made saw, close to each shoot; and it is necessary to do this with judgment, or adjacent shoots, which are not sufficiently advanced to reveal their presence by lifting the soil, may be damaged. To avoid this risk of injury by the knife it is possible from some beds to obtain the sticks without the aid of any implement by a twist and pull combined, but the process needs a dexterous hand and is impracticable in tenacious soils. The sticks of a handsome sample will be white four or five inches of their length; the tops close, plump, of a purplish-green colour, and the colour extending two or at most three inches down the stems. Both size and degree of colouring are, however, so entirely questions of taste that no definite rule can be stated. It is more to the purpose to say that, if liberally grown, the plant may be cut from in the third year; and that cutting should cease about the middle of June, or early in July, according to the district. For the good of the plant the sooner cutting ceases the better, as the next year's buds have to be formed in the roots by the aid of the top-growth of the current season.

Weeding and Staking.—Two other points relating to the general management are worthy of attention. Some crops get on fairly well when neglected and crowded with weeds. Not so with Asparagus. The plant appears to have been designed to enjoy life in solitude, being unfit for competition; and if weeds make way in an Asparagus bed, the cultivator will pay a heavy penalty for his neglect of duty. The limitation of the beds to a width of three feet, therefore, is of consequence, because it facilitates weeding without putting a foot on them. The other point arises out of the necessity of affording support to the frail plant in places where it may happen to be exposed to wind. When Asparagus in high summer is rudely shaken, the stems snap off at the base, and the roots lose the service of the top-growth in maturing buds for the next season. To prevent this injury is easy enough, but the precautions must be adopted in good time. A free use of light, feathery stakes, such as are employed for the support of Peas, thrust in firmly all over the bed, will insure all needful support when gales are blowing. In the absence of pea-sticks, stout stakes, placed at suitable distances and connected with lengths of thick tarred twine, will answer equally well. In sheltered gardens the protection of the young growth with litter, and of the mature growth with stakes, need not be resorted to, but in exposed situations these precautions should not be neglected.

Manuring Permanent Beds.—The management of Asparagus includes a careful clean-up of the beds in autumn. The plants should not be cut down until they change colour; then all the top-growth may be cleared away and the surface raked clean. Give the beds a liberal dressing of half-decayed manure, and carefully touch up the sides to make them neat and tidy. It is usual at the same time to dig and manure the alleys, but this practice we object to *in toto*, because it tends directly to the production of lean sticks where fat ones are possible; for the roots run freely in the alleys, and to dig is to destroy them. In the spring clear the beds of the autumn dressing by raking any remnant of manure into the alleys, and the beds and the alleys should then be carefully pricked over with a fork two or three inches deep only, and with great care not to wound any roots.

The application of salt requires judgment. For a time it renders the bed cold, and when followed by snow the two combine to make a freezing mixture which arrests the growth of established plants. On a newly made bed salt is unnecessary, and may prove destructive to the roots. The proper time for applying salt must be determined by the district and the character of the season; but in no case should the mineral be used until active growth has commenced, although it is not needful to wait until the growth is visible above the surface. In the southern counties a suitable opportunity may generally be found from the beginning to the middle of April. Second and third dressings may follow at intervals of three weeks, which not only stimulate the roots but keep down weeds.

Planting Roots.—In many gardens where there is space for two or three beds only there will be the very natural desire to secure Asparagus in a shorter time than is possible from seed, and we therefore proceed to indicate the best method of planting roots. Asparagus roots do not take kindly to removal, especially old and established plants. The mere drying of the roots by exposure to the atmosphere is distinctly injurious to them. They will travel safely a long distance when well packed, but the critical time is between the unpacking and getting them safely into their final home. Everything should be made ready for the transfer before the package is opened, and the actual task of planting should be accomplished in the shortest time possible.

A three-feet bed should be prepared by taking out the soil in such a manner as to leave two ridges for the roots. The space between ridges to be eighteen inches, and the tops of the ridges to be so far below the level of the bed that when the soil is returned, and the bed made to its normal level, the crowns will be about five inches beneath the surface. This may be understood from the following illustration of a section cut across the bed.

A, A represent the alleys between the beds, and B the top of one bed. The dotted lines show the ridges on which the roots are to rest at C, C. When the bed is ready, open the package and place the Asparagus on the ridges at fifteen or eighteen inches apart, allowing about half the roots of each plant to fall down on either side of the ridge. As a rule it will be wise to have two pairs of hands engaged in the task. The soil should be filled in expeditiously, and a finishing touch be given to the bed. Very rarely will it be safe to transplant Asparagus until the end of March or beginning of April, for although established roots will pass unharmed through a very severe winter, those which have recently been removed are often killed outright by a lengthened period of cold wet weather, and especially by thawed snow followed by frost.

Giant Asparagus.—Some of the most critical judges of Asparagus in the country are extremely partial to giant sticks. Their preference is not based on mere superiority in size, but on the special flavour which is the peculiar merit of these extra-large Asparagus when they are properly grown. Although there is no difficulty whatever in producing them, it must be admitted that to insure specimens weighing nearly or quite half a pound, plenty of space must be allowed for the full development of each plant and a prodigal use of manure is imperative. Where drainage is effectual, the soil of any well-tilled garden can be made suitable. The roots may be grown in clumps or in rows. Clumps are planted in triangular form, two feet being allowed between the three plants of each group, with a distance of five feet between the groups. The more usual method, however, is to plant in rows. In both cases the cultural details are almost identical, and to obtain the finest results it is wise to get the preparatory work done at convenient times in advance of the planting season. Assuming that rows are decided on, commence operations by digging a broad deep trench, throwing out the soil to the right and left to form sloping sides until there is a perpendicular depth of twenty-seven inches from the top of the ridge. About one foot of prepared soil should be placed in the bottom of the trench. This may be composed of such material as the trimmings of hedges, sweepings of shrubberies, twigs from a faggot pile, wood ashes and leaf-mould. The constituents must to some extent depend on the materials at command. What is wanted is a light compost, consisting almost wholly of vegetable matter in a more or less advanced state of decomposition. Add three or four inches of rich loam, and on this, at the beginning of April, plant strong one-year roots of a robust-growing variety. Between the plants it is customary to allow a space of at least two feet, and some growers put them a full yard apart. Cover the crowns with three inches of rich soil, previously mixed with manure and laid up for the purpose. The second and following rows are to be treated in the same way, and the work must be so managed that an equal distance of four and a half or five feet is left between the rows. When the foliage dies down in autumn, a layer of fertile loam mixed with rotten manure should be spread over the surface. In the succeeding spring remove just the top crust of soil and give a thick dressing of decayed manure alone, upon which the soil can be restored. During the autumn of the second year the furrow must be filled with horse manure for the winter. Remove this manure in March, and substitute good loam containing a liberal admixture of decayed manure previously incorporated with the soil. The slight ridges that remain can then be levelled down. By this treatment large handsome sticks of Asparagus may be cut in the third year. To maintain the plants in a high state of efficiency, it must be clearly understood that forcing with horse manure will be necessary every subsequent year. Blanching may be carried out by any of the usual methods, and Sea Kale pots are both convenient and effectual. Not a weed should be visible on the beds at any time.

Forcing is variously practised, and the best possible system, doubtless, is to force in the beds, and thereby train the plants to their work so that they become used to it. The growers who supply Paris with forced Asparagus produce the white sample in the beds, and the green by removal of the roots to frames. Forcing in beds may be accomplished by means of trenches filled with fermenting material or by hot-water pipes, the beds in either case being covered with frames. Where the demand for forced Asparagus is constant, there can be no doubt the hot-water system is the cheapest as well as the cleanest and most reliable; for a casual supply forcing in frames answers very well, but it is attended with the disadvantage that when the crop has been secured the roots are worthless. The practice of forcing may be said to commence with the formation of the seed-bed, for if it is to be carried on in a systematic and profitable manner, every detail must be provided for in the original arrangements. The width of the beds and of the alleys, and the disposition of the plants, will have to be carefully considered, so as to insure the best results of a costly procedure, and it will be waste of time to begin forcing until the plants have attained their fourth year. The rough method of market growers consists in the employment of hot manure in trenches, and also on the beds, after the frames are put on. The beds are usually four feet wide, the alleys two feet wide and twenty inches deep, and the plants not more than nine inches apart in the row, there being three or four rows of plants in the bed. The frames are put on when forcing commences, but the lights are withheld until the shoots begin to appear. Then the fermenting material is removed from the beds, the lights are put on, and no air is given, mats being added in cold weather, both to retain warmth and promote blanching. This method produces a fair market sample, but a much better growth may be obtained by a good hot-water system, as will be understood from a momentary consideration of details. By the employment of fermenting material the temperature runs up rapidly, sometimes extravagantly, so that it is no uncommon event for the growth to commence at 70° to 80° Fahr., which may produce a handsome sample, but it will be flavourless. The hot-water system allows of perfect control, and the prudent grower will begin at 50°, rise slowly to 60°, and take care not to exceed 65°; the result will be a sample full of flavour, with a finer appearance than the best obtainable by the rougher method.

Forcing in frames is systematically practised in many gardens, and as it exhausts the roots there must be a corresponding production of roots for the purpose. The first requisite is a good lasting hot-bed, covered with about four inches of light soil of any kind, but preferably leaf-mould. The roots are carefully lifted and planted as closely as possible on this bed, and covered with fine soil to a depth of six inches. The sashes are then put on and kept close; but a little air may be given as the heads rise, to promote colour and flavour. The heat will generally run to 70°, and that figure should be the maximum allowed. Experienced growers prefer to force at 60° or 65°, and to take a little more time for the advantage of a finer sample.

BROAD BEAN

Faba vulgaris

The Broad Bean is a thrifty plant, as hardy as any in the garden, and very accommodating as to soil. It is quite at home on heavy land, but in common with nearly all other vegetables it thrives on a deep sandy loam. Considering the productive nature of the plant and its comparatively brief occupation of the ground, the common Bean must be regarded as one of our most profitable garden crops. Both the Longpod and Windsor classes should be grown. For general work the Longpods are invaluable; they are early, thoroughly hardy, produce heavy crops, and in appearance and flavour satisfy the world at large, as may be proved by appeal to the markets. The Windsor Beans are especially prized for their superior quality, being tender, full of flavour, and, if well managed, most tempting in colour when put upon the table.

For early crops the Longpods claim attention, and sowings may be made towards the end of October or during November on a dry soil in a warm situation, sheltered from the north. Choose a dry day for the operation. On no account should the attempt be made while the soil conditions are unfavourable, even if the sowing is thereby deferred for some time. The distance must depend upon the sorts, but two feet will answer generally as the distance between the double rows; the two lines forming the double rows may be nine inches apart, and the seed two inches deep. On strong ground a distance of three feet can be allowed between the double rows, but it is not well to give overmuch space, because the plants protect each other somewhat, and earliness of production is the matter of chief moment. Thoroughly consolidate the soil to encourage sturdy hard growth which will successfully withstand the excessive moisture and cold of winter. It is an excellent practice to prepare a piece of good ground sloping to the south, and on this to make a plantation in February of plants carefully lifted from the seed rows, wherever they can be spared as proper thinnings. These should be put in double rows, three feet apart. If transplanted with care they will receive but a slight check, and will give a successional supply.

Main Crops.—Another sowing may be made towards the end of January, but for the main crop wait until February or March. For succession crops sowings may be made until mid-April, after which time there is risk of failure, especially on hot soils. A strong soil is suitable, and generally speaking a heavy crop of Beans may be taken from a well-managed clay. But any deep cool soil will answer, and where there is a regular demand for Beans the cultivator may be advised to grow both Longpods and Windsors—the first for earliness and bulk, the second for quality. The double rows of maincrop Beans should be fully three feet apart, and the plants quite nine inches apart in the rows. The preparation of the seed-bed must be of a generous nature. Where grass land or land of questionable quality is broken up and trenched, it will be tolerably safe to crop it with Beans as a first start; and to prepare it for the crop a good body of fat stable manure should be laid in between the first and second spits, as this will carry the crop through, while insuring to the subsoil that has been brought up a time of seasoning with the least risk of any consequent loss.

There is not much more to be said about growing Beans; the ground must be kept clean, and the hoe will have its work here as elsewhere. The pinching out of the tops as soon as there is a fair show of blossom is a good plan, whether fly is visible or not, and it is also advisable to root out all plants as fast as they finish their work, for if left they throw up suckers and exhaust the soil. The gathering of the crop is often so carelessly performed that the supply is suddenly arrested.

Sowings under Glass.—In an emergency, Beans may be started in pots in the greenhouse, or on turf sods in frames for planting out, in precisely the same way as Peas for early crops. This practice is convenient in cases where heavy water-logged ground precludes outdoor sowing in autumn and early spring. In all such cases care must be taken that the forcing is of the most moderate character, or the crop will be poor and late, instead of being plentiful and early. When pushed on under glass for planting out, the young stock must have as much light and air as possible consistent with safety, and a slow healthy growth will better answer the purpose than a rapid growth producing long legs and pale leaves, because the physique of infancy determines in a great degree that of maturity, not less in plants than in animals.

DWARF FRENCH BEAN

Phaseolus vulgaris

Among summer vegetables Dwarf French Beans are deservedly in high favour, and are everywhere sown at the earliest moment consistent with reasonable expectations of their safety. This early sowing is altogether laudable, for although it occasionally entails the loss of a plantation, the aggregate result is advantageous, and a very little protection suffices to carry the early plant through the late spring frosts. But those who supply our tables with green delicacies do not all recognise the importance of late sowings of Dwarf Beans. Here, again, a risk must be incurred, but the cost is trifling, and when the summer is prolonged to October the late-sown Beans are highly prized. Even if they produce plentifully through September there is a great point gained, but that cannot be secured from the earliest sowings; it is impossible. After July it is useless to sow Beans, but where the demand is constant, two or three sowings may be made in this month, choosing the most sheltered nooks that can be found for them. For late sowings the earliest sorts should have preference.

Dwarf Beans for main crops require a good though somewhat light soil; but any fairly productive loam will answer the purpose, and the crop will yield an ample return for such reasonable digging and dressing as a careful cultivator will not fail to bestow. At the same time, it is a matter of some practical importance that the poorest land ever put under tillage will, in an average season, yield serviceable crops of these legumes, and on a rich soil of some depth the Dwarf Bean will endure summer drought better than any other crop in the Kitchen Garden. Earliness of production is of the highest importance up to a certain point; but an early crop being provided for, abundance of production next claims consideration, the heaviest bearers being of course best adapted for main-crop sowing. As regards the sowing and general culture, it is too often true that Dwarf Beans are crowded injuriously, even in gardens that are usually well managed. Nothing is gained by crowding. On the contrary, loss always ensues when the individual plant, through deficiency of space, is hindered in its full development.

For early crops which are eventually to come to maturity in the open ground, the first sowings may be made in the month of April, either in boxes in a gentle heat, or better still in a frame on a sunny border without artificial heat. In districts where frost frequently prevails in May, and on heavy soils where early sowings outdoors are impracticable in a wet spring, the forwarding of plants under glass is very desirable, but the actual date for sowing must depend on local conditions. The tender growth that is produced by a forcing process is not well adapted for planting out in May; but a plant produced slowly, with plenty of light and air, will be stout and strong, and if put out with care as soon as mild weather occurs in May, will make good progress and yield an early crop. The seed for this purpose should be sown in rather light turfy soil, as the plants may then be lifted without injury to their fleshy roots. Careful treatment will be desirable for some time after they are planted, such as protection from sun and frost, and watering, if necessary, although the less watering the better, provided the plants can hold their ground. The plot to which these early sowings are to be transplanted should be light and rich, and lying towards the sun; open the lines with the spade or hoe in preference to using the dibber, and as fast as the roots are dropped into their places with their balls of earth unbroken, carefully restore the fine soil from the surface. Rough handling will seriously interfere with the ultimate result, but ordinary care will insure abundant gatherings of first-class produce at a time when there are but few in the market. On dry soils a small sowing may be made about the second week of April on a sheltered south border. Sow in double rows six inches apart, and allow a distance of two feet between the double rows. When the seedlings appear give protection if necessary, and in due course thin the plants to six inches apart in the rows.

Main crops are sown from the last week in April to the middle of June. The distance for the rows may be from one and a half to two feet apart, according to the vigour of the variety, the strongest growers requiring fully two feet, and the distance between the plants may be eight to twelve inches; therefore it is well to sow the seed two to three inches apart, and thin out as soon as the rough leaves appear. The ground being in fairly good condition, it will only be necessary to chop over the surface, if at all lumpy, and with the hoe draw drills about two inches deep, which is far better than dibbling, except on very light soil, when dibbling about three inches deep is quite allowable. Generally speaking, if the plot be kept clean, the Beans will take care of themselves; but in droughty weather a heavy watering now and then will be visibly beneficial, for although the plant bears drought well, it is like other good things in requiring something to live upon. In exposed situations and where storms are prevalent, it is an excellent practice to support the plants with bushy twigs.

Late Crops.—To extend the outdoor supply sowings may be made early in July. When the ground has become dry and hard, it is advisable to soak the seed in water for five or six hours; the drills should also be watered, and, if possible, the ground should be covered with rotten dung, spent hops, or some other mulchy stuff to promote and sustain vegetation.

The gathering of the crop should be a matter of discipline. Where it is done carelessly, there will very soon be none to gather, for the swelling of a few seeds in neglected pods will cause the plants to cease bearing. Therefore all the Beans should be gathered when of a proper size, whether they are wanted or not; this is the only way to insure a long-continued supply of good quality both as to colour and tenderness.

Autumn, Winter and Spring Supplies.—By successional sowings under glass a continuous supply of Beans may be obtained through autumn, winter, and spring. The earliest sowings should be made at fortnightly intervals, from mid-July to mid-September, in cold frames filled with well-manured soil. Put in the seeds two inches deep and six inches apart, in rows one foot apart. Water copiously during the hot months and give protection when the nights become cold. After mid-September crops of dwarf-growing varieties should be raised in heated pits, or in pots placed in a warm temperature. In pits the beds should be one foot deep, the drills one foot apart, and the plants six inches asunder in the rows. When pots are used the ten-inch size will be found most convenient. Only three-parts fill the pots with a good compost, and insure perfect drainage. Place eight or nine beans one and a half inches deep in each pot, eventually reducing the number of plants to five. As the plants progress soil may be added to within an inch and a half of the rims. Air-giving and watering will need careful management, for the most robust growth possible is required, but there must be no chill, and any excess of either moisture or dryness will be immediately injurious. When a few pods are formed feed the plants with alternate applications of soot water and liquid manure, commencing with highly diluted doses. Thoroughly syringe the plants twice daily to combat Red Spider. At night a temperature of from 55° to 60° must be maintained. In mid-February sowings may be made in frames in which six inches of fertile soil has been placed over a good layer of litter or leaves. From these sowings heavy crops may be secured in spring and early summer before the outdoor supplies are ready.

Flageolets is the name given to the seeds of certain types of Dwarf and Climbing Beans when used in a state intermediate between the green pods *(Haricots verts)* and the fully ripe seeds *(Haricots secs)*, and they are strongly to be recommended for culinary purposes. The use of Bean seeds as *Flageolets*, although so little known in this country, is very largely practised abroad, and in the vegetable markets of many French towns the shelling of the beans from the semi-ripe pods by women, in readiness for cooking in the manner of green peas, is a very familiar sight. The seeds of almost all varieties are suitable for use in this way, irrespective of colour, as this is not developed as would be the case if the seeds were quite ripe.

CLIMBING FRENCH BEAN

The Climbing French Bean has all the merits of the Dwarf French Bean, and the climbing habit not only extends the period of bearing but results in a yield such as cannot be obtained from the most prolific strains in the Dwarf section. Although the modern Climbing Bean is less vigorous in growth than the ordinary Runner, the former may generally be had in bearing before the most forward crop of Runners is ready. For an early supply out of doors seed should be sown under glass in April, in the manner advised for early crops of the Dwarf class. Gradually harden off the plants and transfer to permanent quarters on the first favourable opportunity. In the open ground successive sowings may be made from the end of April to June. The outdoor culture of Climbing French Beans is practically the same as for the Dwarf varieties, except that the former are usually grown in double rows about four to five feet apart. Allow the plants to stand finally at nine to twelve inches each way, and support them with bushy sticks such as are used for Peas, for Climbing Beans will run far more readily on these than on single sticks.

The Climbing French Bean is especially useful for producing crops under glass in spring and autumn, and the plants do well when grown in narrow borders with the vines trained close to the roof-glass by means of wire or string to which the growth readily clings. The general treatment may be much the same as that recommended for the Dwarf varieties, special care being taken with regard to watering and the giving of air. During the autumn months atmospheric moisture must be cautiously regulated or much of the foliage will damp off, while in spring a humid atmosphere should be maintained and systematic watering practised. Cucumber, Melon, and Tomato beds from which the crops have been cleared may often be used to advantage for raising a crop of Climbing Beans, and generally these beds are in excellent condition for the plants without the addition of manure.

HARICOT BEAN

Although in France the term *Haricot* is given to all types of Beans, except those of the English Broad Bean, in this country the word *Haricot* is generally applied only to the dried seeds of certain Dwarf and Climbing Beans, notably those which are white. Almost any variety, however, may be used as *Haricots*, but the most popular are those which produce self-coloured seeds, such as white, green, and the various shades of brown. Seed should be sown early in May and the plants treated as advised for French Beans. The pods should not be removed from the plants until the seeds are thoroughly ripe. If ripening cannot be completed in the open, pull up the plants and hang them in a shed until the seeds are quite dry.

RUNNER BEAN

Phaseolus multiflorus

Runner beans need generous cultivation and will amply repay for the most liberal treatment. The main point to be borne in mind is that the plant possesses the most extensive root-system of any garden vegetable. Deep digging and liberal manuring are therefore essential where the production of the finest crops is aimed at. If possible the whole of the ground to be allotted to Runners should be deeply tilled and well manured in autumn or winter. But where this is inconvenient, trenching must be carried out in March or early April. Remove the soil to a depth of two feet, and the trench may be two feet wide for a double row of Beans. Thoroughly break up the subsoil, half-fill the trench with well-rotted manure, and restore the surface soil to within a few inches of the level.

Time of Sowing.—It is seldom advisable to sow Runners in the open before the month of May is fairly in, for they are less hardy than Dwarf Beans, but as late supplies are everywhere valued it is important to sow again in June. Of course these late crops are subject to the caprices of autumnal weather, although they often continue in bearing until quite late in the season. In districts where spring frosts are destructive, and on cold soils or in very exposed situations, plants may be raised in boxes for transferring to the open ground, as advised for Dwarf Beans, but in the case of Runners allow a space of three inches between the seeds.

Distances for Rows, &c.—Frequently the rows of Runner Beans are injuriously close, and the total crop is thereby diminished. On deep, well-prepared soils, single rows generally prove most productive, and they should be not less than five feet apart. But where the soil is shallow and generous preparation is not possible, and in wind-swept positions, double rows, set nine inches apart, are more satisfactory. Between the double rows allow a space of from six to eight feet, on which Cauliflower, Lettuce, or other small-growing subjects may be planted out. Two inches is the proper depth for putting in the seed, and it is a wise policy to sow liberally and eventually to thin the plants to a distance of from nine to twelve inches apart in the rows.

Staking.—It will always pay to give support by stakes, but where these are not available wire netting or strands of stout string make efficient substitutes. Immediately the plants are a few inches high, insert the sticks on either side of the rows and tie them firmly to the horizontal stakes placed in the fork near to the top. The means of support should be decided upon and erected in advance of planting out Runners which have been raised in boxes, thus avoiding any risk of injury to the roots.

But Runners make a good return when kept low by topping, and without any support whatever, a system adopted by many market gardeners. For this method of culture space the plants one foot apart in single rows set three feet apart. Pinch out the tips when the plants are eighteen inches high and repeat the operation when a further eighteen inches of growth has formed.

General Cultivation.—As slugs and snails are particularly partial to the young plants, an occasional dusting of old soot, slaked lime, or any gritty substance should be given to render the leaves unpalatable to these pests. During drought copious watering of the rows is essential, especially on shallow soils; spraying the plants in the evening with soft water is also freely practised and this assists the setting of flowers in dry weather. A mulch of decayed manure will prove of great benefit to the plants and will prolong the period of bearing.

In some gardens Runners are grown in groups running up rods tied together at the top, and when these groups are arranged at regular intervals on each side of a path, the result is extremely pleasing. This mode of culture interferes to a very trifling extent with other crops, and the ornamental effect may be enhanced by growing varieties which have white, red, and bicolor flowers.

Preserving the roots of Runners is sometimes recommended. We can only say that it is a ridiculous proceeding. The utmost care is required to keep the roots through the winter, and they are comparatively worthless in the end. A pint of seed will give a better crop than a number of roots that have cost great pains for their preservation.

Runner Beans for Exhibition.—Although fine specimens fit for exhibition may frequently be gathered from the general garden crop, a little extra attention to the cultivation of Runner Beans for show work will be well repaid. When staged the pods must possess not only the merit of mere size, but they should be perfect in shape and quite young. Rapid as well as robust growth is therefore essential to success. Select the strongest-growing plants in the rows, and for a few weeks before the pods are wanted give alternate applications of liquid manure and clear water. Pinch out all side growths, and limit the number of pods to two in each cluster.

WAXPOD BEAN

Many visitors to the Continent have learned to appreciate the fine qualities of the Waxpod Beans, sometimes known as Butter Beans, the pods of which are usually cooked whole. There are two types, the dwarf and the runner, for which respectively the culture usual for Dwarf French Beans and Runner Beans will be quite suitable.

GARDEN BEET

Beta vulgaris

As a food plant the Beet scarcely obtains the attention it deserves. There is no lack of appreciation of its beauty for purposes of garnishing, or of its flavour as the component of a salad; but other uses to which it is amenable for the comfort and sustenance of man are sometimes neglected. As a simple dish to accompany cold meats the Beet is most acceptable. Dressed with vinegar and white pepper, it is at once appetising, nutritive, and digestible. Served as fritters, it is by some people preferred to Mushrooms, as it then resembles them in flavour, and is more easily digested. It makes a first-rate pickle, and as an agent in colouring it has a recognised value, because of the perfect wholesomeness of the rich crimson hue it imparts to any article of food requiring it.

Frame Culture.—Where the demand for Beet exists the whole year through, early sowings in heat are indispensable. For this method of cultivation the Globe variety should be employed, and two sowings, the first in February and another in March, will generally provide a good supply of roots in advance of the outdoor crops. Sow in drills on a gentle hot-bed and thin the plants from six to nine inches apart in the rows. As soon as the plants are large enough, give air at every suitable opportunity. Fresh young Beets grown in this way find far more favour at table than those which have been stored for several months. They are also of great service for exhibition, especially in collections of early vegetables.

Preparation of Ground.—The cultivation of Beet is of the most simple nature, but a certain amount of care is requisite for the production of a handsome and profitable crop. Beet will make a fair return on any soil that is properly prepared for it; but to grow this root to perfection a rich light loam is necessary, free from any trace of recent or strong manure. A rank soil, or one to which manure has been added shortly before sowing the seed, will produce ugly roots, some coarse with overgrowth, others forked and therefore of little value, and others, perhaps, cankered and worthless. The soil should be well prepared by deep digging some time before making up the seed-bed, and it is sound practice to grow Beet on plots that have been heavily manured in the previous year for Cauliflower, Celery, or any other crop requiring good cultivation. If the soil from an old Melon or Cucumber bed can be spared, it may be spread over the land and dug in, and the piece should be broken up in good time to become mellow before the seed is sown. Seaweed is a capital manure for Beet, especially if laid at the bottom of the trench when preparing the ground. A moderate dressing of salt may be added with advantage, as the Beet is a seaside plant.

Early Crops.—Where frames are not available for providing early supplies of Beetroot, forward crops may often be obtained from the open ground by making sowings of the Globe variety from the end of March to mid-April, in a sheltered position. Of course, the earlier the sowing the greater the risk of destruction by frost, and birds may take the seedlings. A double thickness of fish netting, however, stretched over stakes about one foot above the soil, will afford protection from the former and prevent the depredations of the latter. Set the drills about twelve inches apart and sow the seed one and a half to two inches deep. Thin the plants early and allow them to stand finally at nine inches in the rows.

Main Crop.—The most important crop is that required for salading, for which a deep-coloured Beet of rich flavour is to be preferred, and the aim of the cultivator should be to obtain roots of moderate size and of perfect shape and finish. The ground having been trenched two spades deep early in the year, may be made up into four-and-a-half-feet beds some time in March, preparatory to sowing the seed. The main sowing should never be made until quite the end of April or beginning of May. For a neat crop, sow in drills one and a half to two inches deep, and spaced from twelve to fifteen inches apart. When finally thinned the plants should stand about nine inches apart in the rows. Hand weeding will have to follow soon after sowing, and perhaps the hoe may be required to supplement the hand. The thinning should be commenced as early as possible, but it is waste of time to plant the thinnings, and it is equally waste of time to water the crop. In fact, if the ground is well prepared, weeding and thinning comprise the whole remainder of the cultivation.

Some of the smaller and more delicate Beets, of a very dark colour, may be sown in drills a foot or fifteen inches apart and thinned to six inches distance in the drills. We have, indeed, lifted pretty crops of the smaller Beets at four inches, but it is not prudent to crowd the plants, as the result will be thin roots with long necks.

On stony shallow soils, where it is difficult to grow handsome long Beets, the Globe and Intermediate varieties may be tried with the prospect of a satisfactory result. We have in hot seasons found these most useful on a damp clay where fine specimens of long Beet were rarely obtainable. From this same unkind clay it is possible to secure good crops of long Beets, by making deep holes with a dibber a foot apart and filling these with sandy stuff from the compost yard and sowing the seed over them. It is a tedious process, but it benefits the land for the next crop, and the Beets pay for it in the first instance.

Late Crops.—By sowing the Globe or Turnip-rooted varieties in July, useful roots may be obtained during the autumn and winter. Space the drills as advised for early crops. Seed may also with advantage be thinly sown broadcast; the young plants will thus protect one another, and the roots may be pulled as they mature.

Lifting and Storing.—A Beet crop may be left in the ground during the winter if aided by a covering of litter during severe frost. But it is safer out of the ground than in it, and the proper time to lift is when a touch of autumn frost has been experienced. Dry earth or sand, in sufficient quantity, should be ready for the storing, and a clamp in a sheltered corner will answer if shed room is scarce. In any case, a dry and cool spot is required, for damp will beget mildew, and warmth will cause growth. In cutting off the tops before storing, take care not to cut too near the crown, or injurious bleeding will follow. On the other hand, the long fang-like roots may be shortened without harm, for the slight bleeding that will occur at that end will not affect more than the half-inch or so next to the cut part. A little experience will teach anyone that Beets must be handled with care, or the goodness will run out of them. Many cooks bake Beets because boiling so often spoils them; but if they are in no way cut or bruised, and are plunged into boiling water and kept boiling for a sufficient length of time—half an hour to two hours, according to size—there will be but a trifling difference between boiling and baking.

The Silver, or Sea Kale, Beet is grown principally for the stalk and the midrib of the leaf, considered by some to be equal to Asparagus. In a rank soil, with plenty of liquid manure, the growth is quick, robust, and the plant of good quality, without the necessity of earthing up. Sow in April and May, thinly in drills, and allow the plants eventually to stand at about fifteen inches apart each way. The leaves should be pulled, not cut. As the stalks often turn black in cooking, it is advisable to add a few drops of lemon-juice to the water in which they are boiled, and, of course, soda should never be used. They should be served up in the same manner as Asparagus. The remainder of the leaf is dressed as Spinach.

BORECOLE, or KALE

Brassica oleracea acephala

The Borecoles or Kales are indispensable for the supply of winter vegetables, and their importance becomes especially manifest when severe frost has made general havoc in the Kitchen Garden. Then it is seen that the hardier Borecoles are proof against the lowest temperature experienced in these islands; and, while frost leaves the plants unharmed, it improves the tops and side sprouts that are required for table purposes.

As regards soil, the Borecoles are the least particular of the whole race of Brassicas. They appear to be capable of supplying the table with winter greens even when grown on hard rocky soil, but good loam suits them admirably, and a strong clay, well tilled, will produce a grand sample. Granting, then, that a good soil is better than a bad one, we urge the sowing of seed as early as possible for insuring to the plant a long season of growth. But early sowing should be followed by early planting, for it is bad practice to leave the plants crowded in the seed-bed until the summer is far advanced. This, however, is often unavoidable, and it is well to consider in time where the plants are to go, and when, according to averages, the ground will be vacant to receive them. The first sowing may be made early in March, and another in the middle of April. These two sowings will suffice for almost all the purposes that can be imagined. A good seed-bed in an open spot is absolutely necessary. It is usual to draw direct from the seed-bed for planting out as opportunities occur, and this method answers fairly well. But when large enough it is better practice to prick out as a preparation for the final planting, because a stouter and handsomer plant is thereby secured. If it is intended to follow the rough and ready plan, the seed drills should be nine inches apart; but for pricking out six inches will answer, and thus a very small bed will provide a lot of plants. When pricked out, the plants should be six inches apart each way, and they should go to final quarters as soon as the leaves touch one another. On the flat, a fair distance between Borecoles is two feet apart each way, but some vigorous kinds in good ground will pay for another foot of space, and will yield enormous crops when their time arrives. Transplanting is usually done in June and July, and in many gardens Kales are planted between the rows of second-early or maincrop Potatoes. The work should be done during showery weather if possible, but these Brassicas have an astonishing degree of vitality. If put out during drought very little water is required to start them, and as the cool weather returns they will grow with vigour. But good cultivation saves a plant from extreme conditions; and it is an excellent practice to dig in green manure when preparing ground for Kales, because a free summer growth is needful to the formation of a stout productive plant.

We have suggested that two sowings may be regarded as generally sufficient, but we are bound to take notice of the fact that the late supplies of these vegetables are sometimes disappointing. In a mild winter the Kales reserved for use in spring will be likely to grow when they should stand still, and at the first break of pleasant spring weather they will bolt, very much to the vexation of those who expected many a basket of sprouts from them. A May sowing planted out in a cold place may stand without bolting until spring is somewhat advanced. Kale of the

'Asparagus' type, such as Sutton's Favourite, will often prove successful when sown as late as July.

As regards the varieties, they agree pretty nearly in constitution, although they differ much in appearance and in the power of resisting the excitement of spring weather. But in this section of vegetables there are a few very interesting subjects. The Variegated and Crested Kales are extremely ornamental and eminently useful in large places for decorative purposes. These do not require so rich a soil as Sutton's A1 or Curled Scotch, and they must have the fullest exposure to bring out their peculiarities. It is found that in somewhat dry calcareous soils these plants acquire their highest colour and most elegant proportions. When planted by the sides of carriage drives and in other places where their colours may be suitably displayed, it is a good plan to cut off the heads soon after the turn of the year, as this promotes the production of side shoots of the most beautiful fresh colours. A crop of Kale may be advantageously followed by Celery.

BROCCOLI

Brassica oleracea botrytis asparagoides

The great importance of this crop is indicated by the long list of varieties and the still longer list of synonyms. As a vegetable it needs no praise, and our sole business will be to treat of the cultivation.

Of necessity we begin with generalities. Any good soil will grow Broccoli, but it is a strong-land plant, and a well-tilled clay should yield first-class crops. But there are so many kinds coming into use at various seasons, that the cultivation may be regarded as a somewhat complex subject. We will therefore premise that the best must be made of the soil at command, whatever it may be. The Cornish growers owe their success in great part to their climate, which carries their crops through the winter unhurt; but they grow Broccoli only on rich soil, and keep it in good heart by means of seaweed and other fertilisers. All the details of Broccoli culture require a liberal spirit and careful attention, and the value of a well-grown crop justifies first-class treatment. On the other hand, a badly-grown crop will not pay rent for the space it covers, to say nothing of the labour that has been devoted to it.

The Seed-bed.—Broccoli should always be sown on good seed-beds and be planted out; the seed-beds should be narrow, say three or three and a half feet wide, and the seed must be sown in drills half an inch deep at the utmost—less if possible; and where sparrows haunt the garden it will be well to cover the beds with netting, or protect the rows with wire pea guards. A quick way of protecting all round seeds against small birds is to put a little red lead in a saucer, then lightly sprinkle the seed with water and shake it about in the red lead. Not a bird or mouse will touch seed so treated.

The seed-beds must be tended with scrupulous care to keep down weeds and avert other dangers. It is of great importance to secure a robust plant, short, full of colour, and free from club at the root. Now, cleanliness is in itself a safeguard. It promotes a short sturdy growth, because where there are no weeds or other rubbish the young plant has ample light and air. Early thinning and planting is another important matter. If the land is not ready for planting, thin the seed-bed and prick out the seedlings. A good crop of Broccoli is worth any amount of trouble, although trouble ought to be an unknown word in the dictionary of a gardener.

Manuring Ground.—As a rule, Broccoli should be planted in fresh ground, and, in mild districts, if the soil is in some degree rank with green manure the crop will be none the worse for it. But rank manure is not needful; a deep, well-dug, sweet loam will produce a healthy growth and neat handsome heads. However, it is proper to remark, that if any rank manure is in the way, or if the ground is poor and wants it, the Broccoli will take to it kindly, and all the rankness will be gone long before they produce their creamy heads. Still, it must be clearly understood that the more generous the treatment, the more succulent will be the growth, and in cold climates a succulent condition may endanger the crop when hard weather sets in.

Method of Planting.—Broccoli follows well upon Peas, early Potatoes, early French Beans, and Strawberries that are dug in when gathered from for the last time. But it does not follow well upon Cabbage, Turnip, or Cauliflower; if Broccoli must follow any of these, dig deeply, manure heavily, and in planting, dust a little freshly slaked lime in the holes. The times of planting will depend on the state of the plants and the proper season of their heading in. But everywhere and always the plants should be got out of the seed-bed into their permanent quarters as soon as possible, for the longer they stay in the seed-bed the more likely are they to become drawn above and clubbed below. As regards distances, too, the soil, the variety, and the season must be considered. For all sorts the distances range from two to two and a half feet; and for most of the medium-sized sorts that have to stand out through the winter for use in spring, a distance of eighteen to twenty-four inches is usually enough, because if they are rather close they protect one another. But with strong sorts in strong soils and kind climates, two feet and a half every way is none too much even for safe wintering. Plant firmly, water if needful, and do not stint it; but, if possible, plant in showery weather, and give no water at all. Watering may save the crop, but the finest pieces of Broccoli are those that are secured without any watering whatever.

Autumn Broccoli.—To grow Autumn Broccoli profitably, sow in February, March, and April, the early sowings in a frame to insure vigorous growth, and the later sowings in the open ground. Plant out as soon as possible in fresh land that has been deeply tilled. If the soil is poor, draw deep drills, fill them with fat manure, and plant by hand, taking care to press round each root crumbs from the surface soil. This will give them a good start, and they will take care of themselves afterwards. When they show signs of heading in, run in shallow drills of Prickly Spinach between them, and as this comes up the Broccoli will be drawn, leaving the Spinach a fair chance of making a good stolen crop, needing no special preparation whatever. Another sowing of Broccoli may be made in May, but the early sowings, if a little nursed in the first instance, will pay best, because early heads are scarce, whereas late Broccoli are plentiful.

Winter Broccoli should not be sown before the end of March and thence to the end of April. As a rule, the April sowing will make the best crop, although much depends on season, soil, and climate. Begin to plant out early, and continue planting until a sufficient breadth of ground is covered. Within reasonable limits it will be found that the time of planting does not much affect the date when the heads turn in, and only in a moderate degree influences the size of them.

Spring Broccoli are capricious, no matter what the world may say. It will occasionally happen that sorts planted for cutting late in spring will turn in earlier than they are wanted, and the sun rather than the seedsman must be blamed for their precocity. In average seasons the late sorts turn in late; but the Broccoli is a sensitive plant, and unseasonable warmth results in premature development. Sow the Spring Broccoli in April and May, the April sowing being the more important. It will not do, however, to follow a strict rule save to this effect, that early and late sowings are the least likely to succeed, while mid-season sowings—say from the middle of April to the middle of May—will, as a rule, make the best crops. Where there is a constant demand for Broccoli in the early months of the year, two or three small sowings will be better than one large sowing.

Summer Broccoli are useful when Peas are late, and they are always over in time to make way for the glut of the Pea crop. Late Queen may, in average seasons, be cut at the end of May and sometimes in June, if sown about the middle of May in the previous year, and carefully managed. This excellent variety can, as a rule, be relied on, both to withstand a severe winter in an exposed situation and to keep up the supplies of first-class vegetables until the first crop of Cauliflower is ready, and Peas are coming in freely. Generally speaking, smallish heads, neat in shape and pure in colour, are preferred. They are the most profitable as a crop and the most acceptable for the table. An open, breezy place should be selected for a plantation of late Broccoli, the land well drained, and it need not be made particularly rich with manure. But good land is required, with plenty of light and air to promote a dwarf sturdy growth and late turning in.

Protection in Winter.—Various plans are adopted for the protection of Broccoli during winter. Much is to be said in favour of leaving them to the risk of all events, for certain it is that finer heads are obtained from undisturbed plants than by any interference with them, provided they escape the assaults of winter frost. But in such a matter it is wise to be guided by the light of experience. In cold districts, and on wet soils where Broccoli do not winter well, heeling over may be adopted. There are several ways of accomplishing the task, the most successful method being managed thus. Open a trench at the northern end, and gently push over each plant in the first row so that the heads incline to the north. Put a little mould over each stem to settle it, but do not earth it up any more than is needful to render it secure. Push over the next row, and the next, and so on, finishing off between them neatly and leaving the plants nearly as they were before, save that they now all look northward, and their sloping stems are a little deeper in the earth than they were in the first instance. This should be done during fine weather in November, and if the plants flag a little they should have one good watering at the roots. In the course of about ten days it will be scarcely perceptible that they have been operated on. They may be lifted and replanted with their heads to the north, but this is apt to check them too much. In exceptionally cold seasons cover the plot with straw or bracken, but this must be removed in wet weather. When it is seen that the heads are forming and hard weather is apprehended, some growers take them up with good balls of earth and plant them in a frame, or even pack them neatly in a cellar, and the heads finish fairly well, but not so well as undisturbed plants. It is impossible, however, to cut good heads in a very severe winter without some such protective measures. In many gardens glass is employed for protecting Winter Broccoli, in which case the plantations are so shaped that the frames will be easily adapted to them without any disturbance of the plants whatever. There must be allowed a good space between the beds to be covered, and the plants must be fifteen to eighteen inches apart, with the object of protecting the largest number by means of a given stock of frames.

Sprouting Broccoli, both white and purple, are invaluable to supply a large bulk of a most acceptable vegetable in winter and early spring. Sow in April and the plants may be treated in the same way as other hardy winter greens. They should have the most liberal culture possible, for which they will not fail to make an ample return. The Purple Sprouting Broccoli is a favourite vegetable in the kitchen, because of its freedom from the attacks of all kinds of vermin.

BRUSSELS SPROUTS

Brassica oleracea bullata gemmifera

Brussels Sprouts are everywhere regarded as the finest autumnal vegetable of the strictly green class. They are, however, often very poorly grown, because the first principle of success—a long growing season—is not recognised. It is in the power of the cultivator to secure this by sowing seed at the end of February, or early in March, on a bed of light rich soil made in a frame, and from the frame the plants should be pricked out into an open bed of similar light fresh soil as soon as they have made half a dozen leaves. From this bed they should be transferred to their permanent quarters before they crowd one another, the object being at each stage to obtain free growth with a sturdy habit, for mere length of stem is no advantage; it is a disadvantage when the plant is deficient of corresponding substance. The ground should be made quite firm, in order to encourage robust growth which in turn will produce shapely solid buttons. This crop is often grown on Potato land, the plants being put out between the rows in the course of the summer. It is better practice, however, to plant Kales or Broccoli in Potato ground, because of the comparative slowness of their growth, and to put the Sprouts on an open plot freely dressed with somewhat fresh manure. If a first-class strain, such as Sutton's Exhibition, is grown, it will not only pay for this little extra care, but will pay also for plenty of room, say two and a half feet apart every way at the least; and one lot, made up of the strongest plants drawn separately, may be in rows three feet apart, and the plants two and a half feet asunder. For the compact-growing varieties two feet apart each way will generally suffice. Maintain a good tilth by the frequent use of the hoe during summer, and as autumn approaches regularly remove all decaying leaves. Those who have been accustomed to treat Sprouts and Kales on one uniform rough plan will be surprised at the result of the routine we now recommend. The plants will button from the ground line to the top, and the buttons will set so closely that, once taken off, it will be impossible to replace them. Moderate-sized, spherical, close, grass-green Sprouts are everywhere esteemed, and there is nothing in the season more attractive in the markets.

Crops treated as advised will give early supplies of the very finest Sprouts. For successional crops it will be sufficient to sow in the open ground in the latter part of March, or early in April, and plant out in the usual manner; in other words, to treat in the commonplace way of the ordinary run of Borecoles. With a good season and in suitable ground there will be an average crop, which will probably hold out far into the winter. It is important to gather the crop systematically. The Sprouts are perfect when round and close, with not a leaf unfolded. They can be snapped off rapidly, and where the quantity is considerable they should be sorted into sizes. The season of use will be greatly prolonged, and the tendency of the Sprouts to burst be lessened, if the head is cut last of all.

CABBAGE

Brassica oleracea capitata

The Cabbage is a great subject, and competes with the Potato for preeminence in the cottage garden, in the market garden, and on the farm, sometimes with such success as to prove the better paying crop of the two. It may be said in a general way that a Cabbage may be grown almost anywhere and anyhow; that it will thrive on any soil, and that the seed may be sown any day in the year. All this is nearly possible, and proves that we have a wonderful plant to deal with; but it is too good a friend of man to be treated, even in a book, in an off-hand manner. The Cabbage may be called a lime plant, and a clay plant; but, like almost every other plant that is worth growing, a deep well-tilled loam will suit it better than any other soil under the sun. It has one persistent plague only. Not the Cabbage butterfly; for although that is occasionally a troublesome scourge, it is not persistent, and may be almost invisible for years together. Nor is it the aphis, although in a hot dry season that pest is a fell destroyer of the crop. The great plague is club or anbury, for which there is no direct remedy or preventive known. But indirectly the foe may be fought successfully. The crop should be moved about, and wherever Cabbage has been grown, whether in a mere seed-bed or planted out, it should be grown no more until the ground has been well tilled and put to other uses for one year at least, and better if for two or three years. There are happy lands whereon club has never been seen, and the way to keep these clear of the pest is to practise deep digging, liberal manuring, and changing the crops to different ground as much as possible. A mild outbreak of club may generally be met by first removing the warts from the young plants, and then dipping them in a puddle made of soot, lime, and clay. But when it appears badly amongst the forward plants, their growth is arrested, the plot becomes offensive, and the only course left is to draw the bad plants, burn them, and give up Cabbage growing on those quarters for several years. The question as to why the roots of brassicaceous plants are subject to this scourge on some soils, while plants from the same seed-bed remain healthy when transferred to different land, is deeply interesting, and the subject is discussed later on in the chapter on 'The Fungus Pests of certain Garden Plants.' Here it is sufficient to say that the presence of the disease is generally an indication that the soil is deficient in lime. A dressing at the rate of from 14 to 28 or even 56 pounds per square pole may be necessary to restore healthy conditions. The outlay will not be wasted, for lime is not merely a preventive, it has often an almost magical influence on the fertility of land.

For general purposes Cabbages may be classified as early and late. The early kinds are extremely valuable for their earliness, but only a sufficient quantity should be grown, because, as compared with mid-season and late sorts, they are less profitable. In the scheme of cropping it may be reckoned that a paying crop of Cabbage will occupy the ground through a whole year; for although this may not be an exact statement, the growing time will be pretty well gone before the ground is clear. After Cabbage, none of the Brassica tribe should be put on the land, and, if possible, the crop to follow should be one requiring less of sulphur and alkalies, for of these the Cabbage is a great consumer, hence the need for abundant manuring in preparation for it. The presence of sulphur explains the offensiveness of the exhalations from Cabbage when in a state of decay.

Spring-sown Cabbage for Summer and Autumn use.—To insure the best succession of Cabbage it will be necessary to recognise four distinct sowings, any of which, save the autumnal sowing, may be omitted. Begin with a sowing of the earliest kinds in the month of February. For this, pans or boxes must be used, and the seed should be started in a pit or frame, or in a cool greenhouse. When forward enough, prick out in a bed of light rich soil in a cold frame, and give plenty of air. Before the seedlings become crowded harden them off and plant out, taking care to lift them tenderly with earth attached to their roots to minimise the check.

These will heart quickly and be valued as summer Cabbages. The second sowing is to be made in the last week of March, and to consist of early kinds, including a few of the best type of Coleworts. As these advance to a planting size, they may be put out a few at a time as plots become vacant, and they will be useful in various ways from July to November or later. A third sowing may be made in the first or second week of May of small sorts and Coleworts; and these again may be planted out as opportunities occur, both in vacant plots for hearting late in the year, and as stolen crops in odd places to draw while young. The second and third sowings need not be pricked out from the seed-bed, but may be taken direct therefrom to the places where they are to finish their course.

In planting out, the spacing must be regulated according to the size of the variety grown. If put out in beds, the plants may be placed from one to two feet apart, and the rows one and a half to two feet asunder. All planting should be done in showery weather if possible, or with a falling barometer. It may not always be convenient to wait for rain, and happily it is a peculiarity of Brassicas, and of Cabbage in particular, that the plants will endure, after removal, heat and drought for some time with but little harm, and again grow freely after rain has fallen. But good cultivation has in view the prevention of any such check. At the best it is a serious loss of time in the brief growing season. Therefore in droughty weather it will be advisable to draw shallow furrows and water these a day in advance of the planting, and if labour and stuff can be found it will be well to lay in the furrows a sprinkling of short mulchy manure to follow instantly upon the watering; then plant with the dibber, and the work is done. If the mulch cannot be afforded, water must be given, and to water the furrows in advance is better than watering after the planting, as a few observations will effectually prove. If drought continues, water should be given again and again. The trouble must be counted as nothing compared with the certain loss of time while the plant stands still, to become, perhaps, infested with blue aphis, and utterly ruined. As a matter of fact, a little water may be made to go a long way, and every drop judiciously administered will more than repay its cost. The use of the hoe will greatly help the growth, and a little earth may be drawn towards the stems, not to the extent of 'moulding-up,' for that is injurious, but to 'firm' the plants in some degree against the gales that are to be expected as the days decline.

Autumn-sown Cabbage for Spring and Summer use.—The fourth, or autumn, sowing is by far the most important of the year, and the exact time when seed should be put in deserves careful consideration. A strong plant is wanted before winter, but the growth must not be so far advanced as to stand in peril from severe and prolonged frost. There is also the risk that plants which are too forward may bolt when spring arrives. In some districts it is the practice to sow in July, and to those who find the results entirely satisfactory we have nothing to say. Our own experiments have convinced us that, for the southern counties, August is preferable, and it is wise to make two sowings in that month, the first quite early and the second about a fortnight later. Here it is necessary to observe that the selection of suitable varieties is of even greater consequence than the date of sowing. A considerable number of the Cabbages which possess a recognised value for spring sowing are comparatively useless when sown in August. Success depends on the capability of the plant to form a heart when the winter is past instead of starting a seed-stem, and this reduces the choice to very narrow limits. Among the few Cabbages which are specially adapted for August sowing, Sutton's Harbinger, April, Flower of Spring, Favourite, and Imperial may be favourably mentioned, and even in small gardens at least two varieties should be sown. Where Spring Cabbages manifest an unusual tendency to bolt, sowing late in August, followed by late planting, will generally prove a remedy, always assuming that suitable varieties have been sown.

The planting of autumn-sown Cabbages should be on well-made ground, following Peas, Beans, or Potatoes, and as much manure should be dug in as can be spared, for Cabbage will take all it can get in the way of nourishment. If the entire crop is to be left for hearting, a minimum of fifteen inches each way will be a safe distance for the smallest varieties. Supposing every alternate plant is to be drawn young for consumption as Coleworts, a foot apart will suffice, but in this case the surplus plants must be cleared off by the time spring growth commences. This procedure will leave a crop for hearting two feet apart, and when the heads are cut the stumps will yield a supply of Sprouts. As these Sprouts appear when vegetables are none too plentiful, they are welcome in many households, and make a really delicate dish of greens.

By sowing quick-growing varieties of Cabbage in drills during July and August, and thinning the plants early, thus avoiding the check of transplanting, heads may often be had fit for cutting in October and November.

The Red Cabbage is grown for pickling and also for stewing, being in demand at many tables as an accompaniment to roasted partridges. The plant requires the best ground that can be provided for it, with double digging and plenty of manure. Two sowings may be made, the first in April for a supply in autumn for cooking, and the second in August for a crop to stand the winter and to supply large heads for pickling.

SAVOY CABBAGE

Brassica oleracea bullata

The Savoy Cabbage is directly related to Brussels Sprouts, though differing immensely in appearance. It is of great value for the bulk of food it produces, as well as for its quality as a table vegetable during the autumn and winter. In all the essential points the Savoy may be grown in the same way as any other Cabbage, but it is the general practice to sow the seed in spring only, the time being determined by requirements. For an early supply, sow in February in a frame, and in an open bed in March, April, and May for succession. This vegetable needs a rich deep soil to produce fine heads, but it will pay better on poor soil than most other kinds of Cabbage, more especially if the smaller sorts are selected. Savoys are not profitable in the form of Collards; hence it is advisable to plant in the first instance at the proper distances, say twelve inches for the small sorts, eighteen for those of medium growth, and twenty to twenty-four where the ground is strong and large heads are required. In private gardens the smaller kinds are much the best, but the market grower must give preference to those that make large, showy heads.

CAPSICUM and CHILI

Capsicum annuum, C. baccatum

Capsicums and Chilis are so interesting and ornamental that it is surprising they are grown in comparatively few gardens. Sometimes there is reason to lament that Cayenne pepper is coloured with drugs, but the remedy is within reach of those who find the culture of Capsicums easy, and to compound the pepper is not a difficult task. The large-fruited varieties may also be prepared in various ways for the table, if gathered while quite young and before the fruits change colour.

The cultivation of Capsicums is a fairly simple matter. The best course of procedure is to sow seed thinly in February or March in pots or pans of fine soil placed on a gentle hot-bed or in a house where the temperature is maintained at about 55°. Pot on the young plants as they develop and keep them growing without a check. Spray twice daily, for Capsicums require atmospheric moisture and the Red Spider is partial to the plant. Nice specimens may be grown in pots five to eight inches in diameter, beyond which it is not desirable to go, and as the summer advances these may be taken to the conservatory. Plants intended for fruiting in warm positions out of doors should be hardened off in readiness for transfer at the end of May. In gardens favourably situated, as are many in the South of England, it is sufficient to sow a pinch of seed on an open border in the middle of May, and put a hand glass over the spot. The plants from this sowing may be transferred to any sunny position, and will yield an abundant crop of peppers.

The Bird Pepper or Chili is grown in precisely the same way as advised for Capsicum.

To prepare the pods for pepper, put the required number into a wire basket, and consign them to a mild oven for about twelve hours. They are not to be cooked, but desiccated, and in most cases an ordinary oven, with the door kept open to prevent the heat rising too high, will answer perfectly. Being thus prepared, the next proceeding is to pound them in a mortar with one-fourth their weight of salt, which also should be dried in the oven, and used while hot. When finely pounded, bottle securely, and there will be a perfect sample of Cayenne pepper without any poisonous colouring. One hundred Chilis will make about two ounces of pepper, which will be sufficient in most houses for one year's supply. The large ornamental Capsicums may be put on strings, and hung up in a dry store-room, for use as required, to flavour soups, make Chili vinegar, Cayenne essence, &c. The last-named condiment is prepared by steeping Capsicums in pure spirits of wine. A few drops of the essence may be used in any soup, or indeed wherever the flavour of Cayenne pepper is required.

CARDOON

Cynara Cardunculus

This plant is nearly related to the Globe Artichoke, and it makes a stately appearance when allowed to flower. Although the Cardoon is not widely cultivated in this country, it is found in some of our best gardens, and is undoubtedly a wholesome esculent from which a skilful cook will present an excellent dish. The stalks of the inner leaves are stewed, and are also used in soups, as well as for salads, during autumn and winter.

The flowers, after being dried, possess the property of coagulating milk, for which purpose they are used in France.

In a retentive soil Cardoons should be grown on the flat, but the plant is a tolerably thirsty subject, and must have sufficient water. Hence on very dry soils it may be necessary to put it in trenches after the manner of Celery, and then it will obtain the full benefit of all the water that may be administered. In any case the soil must be rich and well pulverised if a satisfactory growth is to be obtained.

Towards the end of April rows are marked out three or four feet apart, and groups of seed sown at intervals of eighteen inches in the rows. The plants are thinned to one at each station, and in due time secured to stakes. Full growth is attained in August, when blanching is commenced by gathering the leaves together, wrapping them round with bands of hay, and earthing up. It requires from eight to ten weeks to accomplish the object fully. The French method is quicker. Seed is sown in pots under glass, and in May the plants are put out three feet apart. When fully grown the Cardoons are firmly secured to stakes by three small straw bands. A covering of straw, three inches thick, is thatched round every plant from bottom to top, and each top is tied and turned over like a nightcap. A little soil is then drawn to the foot, but earthing up is needless. In about a month blanching is completed.

CARROT

Daucus Carota

The Carrot is a somewhat fastidious root, for although it is grown in every garden, it is not everywhere produced in the best style possible. The handsome long roots that are seen in the leading markets are the growth of deep sandy soils well tilled. On heavy lumpy land long clean roots cannot be secured by any kind of tillage. But for these unsuitable soils there are Sutton's Early Gem, the Champion Horn, and Intermediate, which require no great depth of earth; while for deep loams the New Red Intermediate answers admirably.

Forcing.—Carrots are forced in frames on very gentle hot-beds. They cannot be well grown in houses, and they must be grown slowly to be palatable. It is usual to begin in November, and to sow down a bed every three or four weeks until February. A lasting hot-bed is of the first importance, and it is therefore necessary to have a good supply of stable manure and leaves. The material should be thoroughly mixed and allowed to ferment for a few days. Then turn the heap again, and a few days later the bed may be made up. In order to conserve the heat the material will need to be three to four feet deep, and if a box frame is used the bed should be at least two feet wider than the frame. Build up the material in even, well-consolidated layers, to prevent unequal and undue sinking, and make the corners of the bed perfectly sound. Put on the bed about one foot depth of fine, rich soil; if there is any difficulty about this, eight inches must suffice, but twelve is to be preferred. As the season advances less fermenting material will be needed, and a simple but effective hot-bed may be made by digging out a hole of the required size and filling it with the manure. The latter will in due time sink, when the soil may be added and the frame placed in position. The bed should always be near the glass, and a great point is gained if the crop can be carried through without once giving water, for watering tends to damage the shape of the roots. No seed should be sown until the temperature has declined to 80°. Sow broadcast, cover with siftings just deep enough to hide the seed, and close the frame. If after an interval the heat rises above 70°, give air to keep it down to that figure or to 65°. It will probably decline to 60° by the time the plant appears, but if the bed is a good one it will stand at that figure long enough to make the crop. Thin betimes to two or three inches, give air at every opportunity, let the plant have all the light possible, and cover up when hard weather is expected. Should the heat go down too soon, linings must be used to finish the crop. Radishes and other small things can be grown on the same bed. In cold frames seed may be sown in February.

Warm Borders.—In March the first sowings on warm borders in the open garden may be made. These may need the shelter of mats or old lights until the plant has made a good start, but it is not often the plant suffers in any serious degree from spring frosts, as the seed will not germinate until the soil acquires a safe temperature. All the early crops of Carrot can be grown on a prepared soil, or a light sandy loam, free from recent manure. The drills may be spaced from six to nine inches apart.

For the main crops double digging should be practised, and if the staple is poor a dressing of half-rotten dung may be put in with the bottom spit. But a general manuring as for a surface-rooting crop is not to be thought of, the sure effect being to cause the roots to fork and fang most injuriously. It is sound practice to select for Carrots a deep soil that was heavily manured the year before, and to prepare this by double digging without manure in the autumn or winter, so as to have the ground well pulverised by the time the seed is sown. Then dig it over one spit deep, break the lumps, and make seed-beds four feet wide. Sow in April and onwards in drills, mixing the seed with dry earth, the distance between rows to be eight to twelve inches according to the sort; cover the seed with a sprinkling of fine earth and finish the bed neatly. As soon as possible thin the crop, but not to the full distance in the first instance. The final spacing for main crops may be from six to nine inches, determined by the variety. By a little management it will be an easy matter during showery weather to draw delicate young Carrots for the final thinning, and these will admirably succeed the latest of the sowings in frames and warm borders.

Late Crops.—Sowings of early varieties made in July will give delicate little roots during the autumn and winter. The rows may be placed nine inches apart, and it is essential to thin the plants early to about three inches apart in the rows. In the event of very severe weather protect with dry litter. For providing young Carrots throughout the winter it is also an excellent plan to broadcast seed thinly. When grown in this way the plants afford each other protection, and the roots may be drawn immediately they are large enough.

In July the culture of the smaller sorts may also be undertaken in frames, but hot-beds may be dispensed with, and lights will not be wanted until there is a crop needing protection, when the lights may be put on, or the frames may be covered with shutters or mats.

Storing.—Before autumn frosts set in the main crop should be lifted and stored in dry earth or sand, the tops being removed and the earth rubbed off, but without any attempt to clean them thoroughly until they are wanted for use.

Carrots for Exhibition.—It will be found well worth while to give a little extra attention to the preparation of the ground when growing Carrots for exhibition. As in the case of Beet and Parsnip, holes should be bored to the requisite depth and about one foot apart in the rows. Where the soil is at all unfavourable to the growth of clean symmetrical roots the adoption of this practice will be essential to success. Any light soil of good quality will be suitable for filling the holes. Well firm the material in and sow about half a dozen seeds at a station, eventually thinning out to one plant at each. The tendency of Carrots to become green at the tops in the later stages of growth, thus spoiling them for show work, may be prevented by lightly covering the protruding portion of the root with sifted fine earth.

Destructive Enemies.—The Carrot maggot and the wire-worm are destructive enemies of this crop. In a later chapter on 'The Pests of Garden Plants,' both these foes are referred to. Here it is only necessary to say that sound judgment as to the choice of ground, deep digging, and the preparation of the beds in good time, are the preventives of these as of many other garden plagues. It is often observed that main crops sown early in April suffer more than those sown late, and the lesson is plain. It has also been noticed that where the crops have suffered most severely the land was made ready in haste, and the wild birds had no time to purge it of the insects which they daily seek for food.

CAULIFLOWER

Brassica oleracea botrytis cauliflora

This fine vegetable is managed in much the same way as Broccoli, and it requires similar conditions. But it is less hardy in constitution, more elegant in appearance, more delicate on the table, and needs greater care in cultivation to insure satisfactory results. As regards soil, the Cauliflower thrives best on very rich ground of medium texture. It will also do well on light land, if heavily manured, and quick growth is promoted by abundant watering. In Holland, Cauliflowers are grown in sand with water at the depth of a foot only below the surface, and the ground is prepared by liberal dressings of cow-manure, which, with the moisture rising from below, promotes a quick growth and a fine quality. In any case, good cultivation is necessary or the crop will be worthless; and whatever may be the nature of the soil, it must be well broken up and liberally manured.

In gardens where Cauliflower are in great demand, an unbroken supply of heads from May to November may be obtained by selecting suitable varieties and with careful management of the crop. But in arranging for a succession it should be borne in mind that some varieties are specially adapted for producing heads in spring and summer, while others are only suitable for use in late summer and autumn.

For Spring and Early Summer use.—To have Cauliflower in perfection in spring and early summer, seed should be sown in autumn. The exact time is a question of climate. In the northern counties the middle of August is none too early, but for the south seed may be got in during August and September, according to local conditions. The most satisfactory course is to sow in boxes, placed in a cool greenhouse or a cold frame, or even in a sheltered spot out of doors. For these sowings it is desirable to use poor soil of a calcareous nature, as at this period of the year the seedlings are liable to damp off in rich earth. From the commencement every endeavour must be made to keep the growth sturdy and to avoid a check of any kind. When the plants have made some progress, prick them off three inches apart each way into frames for the winter. No elaborate appliances are necessary. A suitable frame may be easily constructed by erecting wooden sides around a prepared bed of soil, over which lights, window frames, or even a canvas covering may be placed. Brick pits, or frames made with turf walls, will also answer well. The soil should not be rich, or undesirable fleshy growth will result, especially in a mild winter. It is important to ventilate freely at all times, except during severe weather when the structures should have the protection of mats or straw, and excessive moisture must be guarded against. As soon as conditions are favourable in February or March, transfer the plants to open quarters on the best land at command, and give them every possible care. For these early-maturing varieties a space of eighteen inches apart each way will generally suffice. With liberal treatment, vigorous healthy growth should be made and heads of the finest quality be ready for table from May onwards.

As we have already said, the best results with early Cauliflower are obtained from an autumn sowing, but there are many growers who prefer to sow in January or February. At this season the seed should be started in pans or boxes placed in a house just sufficiently heated to exclude frost. Prick out the plants early, in a frame or on a protected border made up with light rich soil, and when strong enough plant out on good ground. Spring sowings put out on poor land, or in dry seasons, are sometimes disappointing, because the heads are too small to please the majority of growers. Where, however, the soil is rich and the district suitable there is this advantage in quick cultivation, that while time is shortened and the worry of wintering is avoided, the crop is safer against buttoning and bolting, which will occasionally occur if the plants become too forward under glass and receive a check when planted out.

In well-prepared sheltered ground seed may also be sown in March and April, from which the plants should be pricked out once before being transferred to permanent positions. Occasional hoeing between the plants and heavy watering in dry weather will materially tend to their well-doing, the object being to maintain growth from the first without a check. If the plants turn in during very hot weather, snap one of the inner leaves without breaking it off, and bend it over to protect the head.

For use in Late Summer and Autumn.—Seed may be sown in April or very early in May, and where only one sowing is made the first week of April should be selected. A fine seed-bed in a sheltered spot is desirable, and as soon as the seedlings are large enough they should be pricked out, three inches or so apart. Shift to final quarters while in a smallish state. If the plants are allowed to become somewhat large in the seed-bed they are liable to 'button,' which means that small, worthless heads will be produced as the result of an untimely check. The distances between the plants may vary from one and a half to two feet or more, and between the rows from two to two and a half feet, according to the size of the variety. If put out on good ground, the crop will almost take care of itself, but should the plants need water it must be copiously given.

Cutting and Preserving.—The management of the crop has been treated so far as to growth, but we must now say a word about its appropriation. The two points for practical consideration are, how to economise a glut, and how to avoid destruction by frost. Cauliflowers should be cut at daybreak, or as soon after as possible, and be taken from the ground with the dew upon them. If cut after the dew has evaporated, the heads will be inferior by several degrees as compared with those cut at the dawn of the day. When the heads appear at too rapid a rate for immediate consumption, draw the plants, allowing the earth to remain attached to the roots, and suspend them head downwards in a cool, dark, dry place, and every evening give them a light shower of water from a syringe. The deterioration will be but trifling, and the gain may be considerable, but if left to battle with a burning sun the Cauliflowers will certainly be the worse for it. After being kept in this way for a week, they will still be good, although, like other preserved vegetables, they will not be so good as those freshly cut and in their prime. It often happens that frost occurs before the crop is finished. A similar plan of preserving those that are turning in may be adopted, but it is better to bury them in sand in a shed or under a wall, and, if kept dry, they may remain sound for a month or more.

Cauliflower for Exhibition.—On the exhibition stage few vegetables win greater admiration than well-grown heads of Cauliflower. Indeed, Cauliflower and Broccoli, in their respective seasons, are indispensable items in the composition of any first-class collection. By closely following the cultural directions contained in the foregoing pages no difficulty should be experienced in obtaining heads of the finest texture and spotless purity during many months of the year. The degree of success achieved is generally in proportion to the amount of attention devoted to minor details. Select the most robust plants and treat them generously. As soon as the heads are formed, examine them frequently to prevent disfiguration by vermin. The best period of the day for cutting has already been discussed. Do not allow the heads to stand a day longer than is necessary, and if not wanted immediately the plants should be lifted and preserved in the manner described in the preceding paragraph.

CELERY

Apium graveolens

Celery is everywhere esteemed, not only as a salad, but as a wholesome and delicious vegetable. The crop requires the very best of cultivation, and care should be taken not to push the growth too far, for the gigantic Celery occasionally seen at Shows has, generally speaking, the quality of size only, being tough and tasteless. Nevertheless, the sorts that are held in high favour by growers of prize Celery are good in themselves when grown to a moderate size; it is the forcing system alone that deprives them of flavour. Yet another precaution may be needful to prevent a mishap. In a hot summer, Celery will sometimes 'bolt' or run up to flower, in which case it is worthless. This may be the fault of the cultivator more than of the seed or the weather, for a check in many cases hastens the flowering of plants, and it is not unusual for Celery to receive a check through mismanagement. If sown too early, it may be impossible to plant out when of suitable size, and the consequent arrest of growth at a most important stage may result in a disposition to flower the first year, instead of waiting for the second. It should be understood, therefore, that early sowing necessitates early planting, and the cultivator should see his way clearly from the commencement.

Sowing and Transplanting.—The 1st of March is early enough for a first sowing anywhere of a small variety, and this will require a mild hot-bed, or a place in the propagating house. Sow on rich fine soil in boxes, cover lightly, and place in a temperature of 60°. When forward enough prick out the plants on a rich bed close to the glass, in a temperature of 60° to 65°, keep liberally moist, and give air, at first with great caution, but increasing as the natural temperature rises until the lights can be removed during the day. The plant may thus be hardened for a first planting on a warm border in a bed consisting of one-half rotten hot-bed manure and one-half of turfy loam. The bed need not be deep, but it must be constantly moist, and old lights should be at hand to give shelter when needful. If well grown in trenches, this first crop will be of excellent quality, and will come in early.

For the general crop a second sowing may be made of the finest Red and White varieties, also on a mild hot-bed, in the second week of March, and have treatment similar to the first, but once pricking out into the open bed will be sufficient, the largest plants being put out first at six inches, and to have shelter if needful; other plantings in the same way to follow until the seed-bed is cleared. By good management this sowing may be made to serve the purpose of three sowings, the chief point being to prick out the most forward plants on another mild bed as soon as they are large enough to be lifted, and to make a succession from the same seed-bed as the plants advance to a suitable size.

The third and last sowing may be made in the second week of April, in an open border, on rich light soil, and should have the shelter of mats or old lights during cold weather. From this, also, there should be two or three prickings out, the first to be transferred to a bit of hard ground, covered with about three inches of rich mulchy stuff, in the warmest spot that can be found, and the last to a similar bed on the coldest spot in the garden. In the final planting the same order should be followed. The result will be a prolonged supply from one sowing, and the first lot will come in early, though sown late, if the plants are kept growing without a check, and receive thoroughly generous culture.

The planting outis an important matter, and each lot will require separate treatment, subordinate to one general and very simple plan. Celery must have rich soil, abundant moisture, and must be blanched to make it fit for table. There are various ways of accomplishing these ends, although they differ but slightly, and common sense will guide us in the matter. For the earliest crops the ground must be laid out in trenches, with as much rich stable manure dug in as can be afforded. To overdo it in this respect seems impossible, for Celery, like Cauliflower, will grow freely in rotten manure alone, without any admixture of loam. The trenches should be eighteen inches wide at bottom, ten inches deep, and four feet from centre to centre, and should run north and south. The plants are to be carefully lifted with a trowel, and placed six to nine inches apart in single or double rows, and should have water as planted, that there may be no check. In a cold soil and a cold season the trenches may be less in depth by two or three inches with advantage. If dry weather ensues, water must be given ungrudgingly, but earthing up should not commence until the plant has made a full and profitable growth, for the earthing pretty well stops the growth, and is but a finishing process, requiring from five to seven weeks to bring the crop to perfection. The second lot can be put out in the same way, and other plantings may follow at discretion; but as the season advances the trenches must be less deep.

Earthing up is often performed in a rough way, as though the plant were made of wood instead of the most delicate tissue. The first earthing should be done with a hand-fork, and quite loosely, to allow the heart of the plant room to expand. The result should be a little ring of light earth scarcely pressing the outside leaves, and leaving the whole plant as free as it was before. A fortnight or so later the earthing must be carried a stage further by means of the spade. Chop the earth over, and lay it in heaps on each side of the plant. Then gather a plant together with both hands, liberate one hand, and with it bring the earth to the plant half round the base, and, changing hands, pack up the earth on the other side. Be careful not to press the soil very close; also avoid putting any crumbs into the heart of the plant; and do not earth higher than the base of the leaves. As soon as may be necessary repeat this process, carrying the earth a stage higher; and about a week from this finish the operation.

The top of the plant must now be closed, and the earth carefully packed so high that only the very tops of the leaves are visible. Finish to a proper slope with the spade, but do not press the plants unduly, the object being simply to obtain a final growth of the innermost leaves in darkness, but otherwise free from restraint.

The Bed System answers particularly well for producing a large supply of Celery with the least amount of labour. This method of cultivation is also especially suitable for raising Celery intended to be served when boiled, or for soups. Celery beds are made four and a half feet wide and ten inches deep, the soil which is taken out being laid up in a slope round the outside of the bed, and the bank thus formed may be planted with any quick crop, such as Dwarf Beans. The ground will need to be heavily manured in the same manner as for the trench system. Space the plants six inches apart in single or double lines, as may be preferred, and allow not less than twelve inches between the rows. Water must be given to each row as planted; afterwards the surface to be several times chopped over with the hoe or a small fork, and watering repeated until the plants have made a start. An easy means of blanching is by the use of stiff paper collars as described below; another simple method is to place mats over the tops of the plants when nearly full grown. The bed system is not only economical, but convenient for sheltering in winter, and should have the attention of gardeners who are expected to supply abundance of Celery throughout the winter and spring, for in such cases a large sample is not required, but quality and continuance are of importance.

It is a great point to keep Celery unhurt by frost far on in the winter, and the advantage of growing the late crops on dry light soil, and on the bed system, will be seen in the ease with which the plants can be preserved. On heavy soil Celery soon suffers from frost, but not so readily on a soil naturally light and dry. Moreover, the bed system allows of many methods of protection, with whatever materials are at command. In heavy soil fine crops of Celery for autumn use may be grown, but in consequence of the liability of the plant to suffer by winter damp, it is advisable to plant late crops on the level, and earth up from the adjoining plots in order to keep the roots dry in winter. Another step towards securing a late supply consists in bending the tops on one side at the final earthing, which prevents the trickling of water into the heart of the plant during heavy rain or snow.

Celery for Exhibition.—From the opening paragraph it will be gathered that to produce extra fine specimens of Celery for exhibition very generous treatment of the plants is necessary. Apart from the choice of varieties—and only the finest strains should be considered—four points are of especial importance to the cultivator. The ground must be liberally enriched; at no period should the plant receive a check or suffer for want of water; there must be the closest inspection at frequent intervals to prevent disfiguration of the stalks or leaves by slugs, snails, or the Celery fly; and finally the operation of blanching will need great care and discretion. These points have already been dealt with at some length. But on the question of blanching it may be well to add that in order to insure perfect specimens, free from blemish, artificial means of some kind must be adopted in place of earthing up in the ordinary way. The use of strips of good quality brown paper will prove both simple and effectual. These strips need not exceed a width of five or six inches, fresh bands being added as growth develops. Tie them securely with raffia or twine, making due allowance for expansion of the plant, and when in position carefully draw the soil towards the base.

The numerous enemies of Celery, such as slugs, snails, the mole-cricket, and the maggot, do not seriously interfere with the crop where good cultivation prevails, but the Celery fly appears to be indifferent to good cultivation, and therefore must be dealt with directly. Dusting the leaves occasionally with soot has been found to operate beneficially. It should be done during the month of June on the mornings of days that promise to be sunny. If the soot is put on carelessly it will do more harm than good; a very fine dusting will suffice to render the plant distasteful to the fly. Syringing the leaves with water impregnated with tar has also saved plants from attack. Where the eggs are lodged the leaves will soon appear blistered, and the maggot within must be crushed by pinching the blister between the thumb and finger. Leaves that are much blistered should be removed and burned, but to rob the plants of many leaves will seriously reduce the vigour of growth.

Celeriac, or Turnip-rooted Celery, is much prized on the Continent as a cooked vegetable, and as a salad. In ordinary Celery the stem forms a mere basis to the leaves, but in Celeriac it is developed into a knob weighing from one to five pounds, and the root is more easily preserved than Celery. When cooked in the same manner as Sea Kale, Celery is well known as a delicacy at English tables, and the cooked Celeriac ranks in importance with it, though it affords quite a different dish. The stem or axis of the plant is used, and not the stalks. To grow fine Celeriac a long season is requisite; and therefore it is advisable to sow the seed in a gentle heat early in March, and afterwards prick out and treat as Celery; but after the first stage the treatment is altogether different. For the plantation a light and rich soil is required, and where the staple is heavy, a small bed can easily be prepared by spreading six inches depth of any sandy soil over the surface. The plants must be put out on the level a foot and a half apart each way, and be planted as shallow as possible. Before planting, trim carefully to remove lateral shoots that might divide the stems, and after planting water freely. The cultivation will consist in keeping the crop clean, and frequently drawing the soil away from the plants, for the more they stand out of the ground the better, provided they are not distressed. They must never stand still for want of water, or the roots will not attain to a proper size. The lateral shoots and fibres must be removed to keep the roots intact, but not to such an extent as to arrest progress. When a good growth has been made, and the season is declining, cover the bulbs or stems with a thin coat of fine soil, and in the first week of October lift a portion of the crop and store it in sand, all the leaves being first removed, except those in the centre, which must remain, or the roots may waste their energies in producing another set. The portion of the crop left in the ground will need protection from frost, and this can be accomplished by earthing them over with soil taken from between the rows.

Celeriac is cooked in the same manner as Beet, and requires about the same length of time. The stems, bulbs, or roots (for the knobs, which are true stems, are known by various names) are trimmed, washed, and put into boiling water without salt or any flavouring, and kept boiling until quite tender; they may then be pared, sliced, and served with white sauce, or left uncut to be sliced up for salads when cold.

CHICORY

Cichorium Intybus

A valuable addition to the supply of winter and spring roots. When stewed and served with melted butter, Chicory bears a slight resemblance to Sea Kale. More frequently, however, it is eaten in the same manner as Celery, with cheese, and it also makes an excellent and most wholesome salad. All the garden varieties have been obtained from the wild plant, and some of the stocks show a decided tendency to revert to the wild condition. It is therefore important to sow a carefully selected strain, or the roots may be worthless for producing heads.

Seed should be sown in May or June, in rows one foot apart, and the plants thinned out to about nine inches in the rows. The soil must be deep and rich, but free from recent manure, except at a depth of twelve inches, when the roots will attain the size of a good Parsnip.

In autumn the roots must be lifted uninjured with the aid of a fork, and only a few at a time, as required. After cutting off the tops just above the crown, they can at once be started into growth, and it is essential that this be made in absolute darkness. French growers plant in a warm bed of the temperature suited to Mushrooms, but this treatment ruins the flavour, and has the effect of making the fibre of the leaves woolly. It is far simpler and better to put the roots into a cellar or shed in which a temperature above the freezing point may be relied on, and from which every ray of light can be excluded. They can be closely packed in deep boxes, with light soil or leaf-mould between. If the soil be fairly moist, watering will not be necessary for a month, and had better not be resorted to until the plants show signs of flagging. Instead of boxes, a couple of long and very wide boards, stood on edge and supported from the outside, make a convenient and effective trough. The packing of the roots with soil can be commenced at one end, and be gradually extended through the entire length, until the part first used is ready for a fresh start. Breaking the leaves is better than cutting, and gathering may begin about three weeks after the roots are stored. From well-grown specimens, heads may be obtained equal to a compact Cos Lettuce, and by a little management it is easy to maintain a supply from October until the end of May. The quantity of salading to be obtained from a few roots is really astonishing.

CORN SALAD

Valerianella olitoria

Corn Salad, or Lamb's Lettuce, so often seen on Continental tables, is comparatively unknown in this country. The reason for this is, perhaps, to be found in the fact that, as a raw vegetable, it is not particularly palatable, although when dressed as a salad with oil and the usual condiments it is altogether delicious, and forms a most refreshing episode in the routine of a good dinner. Corn Salad is a plant of quick growth, and is valued for its early appearance in spring, when elegant salads are much in request. It may be mixed with other vegetables for the purpose, or served alone with a little suitable preparation.

The most important sowings are made in August and September. Seed may, however, be sown at any time from February to October, but only those who are accustomed to the plant should trouble to secure summer crops; when Lettuces are plentiful Corn Salad is seldom required. Any good soil will grow it, but the situation should be dry and open. Sow in drills six inches apart, and thin to six inches in the rows. The crop is taken in the same way as Spinach, either by the removal of separate leaves or cutting over in tufts.

COUVE TRONCHUDA

Brassica oleracea costata

Couve Tronchuda, or Portugal Cabbage, is a fine vegetable that should be grown in every garden, including those in which Cabbages generally are not regarded as of much importance. The plant is of noble growth, and in rich ground requires abundant room for the spread of its great leaves, the midribs of which are thick, white, tender, and when cooked in the same manner as Sea Kale quite superb in quality. When a fair crop of these midribs has been taken there remains the top Cabbage, which is excellent.

Two or three sowings may be made in February, March, and April, and the early ones must be in heat. Transfer to rich soil as early as possible, giving the plants ample room, from two to three feet each way, and aid with plentiful supplies of water in dry weather.

CRESS

Lepidium sativum

Cress is best grown in small lots from frequent sowings, and the sorts should be kept separate, and, if possible, on the same border. Fresh fine soil is requisite, and there is no occasion for manuring, in fact it is objectionable, but a change of soil must be made occasionally to insure a good growth. The seed is usually sown too thick, yet thin sowing is not to be recommended. It is important to cut Cress when it is just ready—tender, green, short, and plump. This it will never be if sown too thick, or allowed to stand too long. Immediately the plant grows beyond salad size it becomes worthless, and should be dug in. From small sowings at frequent intervals under glass a constant supply of Cress may be kept up through the cold months of the year, for which purpose shallow boxes or pans will be found most convenient. Cress generally requires rather more time than Mustard.

American or Land Cress *(Barbarea præcox)* is of excellent quality when grown on a good border, and two or three sowings should be made in the spring and autumn in shady spots. If the site is not naturally moist, water must be copiously given.

Water Cress *(Nasturtium officinale)* is so highly prized that many who are out of the reach of ordinary sources of supply would gladly cultivate it were there a reasonable prospect of success. Assertions have been made that it can be grown in any garden without water, but we have never yet seen a sample fit to eat which has been grown without assistance from the water can. A running stream is not necessary. Make a trench in a shady spot, and well enrich the soil at the bottom of it. In this sow the seed in March, and when the plants are established keep the soil well moistened. The more freely this is done the better will be the result. Other sowings may be made in April, August, and September. We have seen Water Cress successfully cultivated in pots and pans immersed in saucers of water placed in shady positions.

CUCUMBER

Cucumis sativus

The Cucumber is everywhere valued. Its exceeding usefulness explains its popularity, and happily the plant is of an accommodating character. In large establishments, Cucumbers are grown at all seasons of the year; in medium-sized gardens, summer Cucumbers are generally deemed sufficient, and there is no difficulty in growing an abundant and continuous supply of the finest quality. The winter cultivation demands suitable appliances and skilful management; but a very small house, with an efficient heating apparatus, will suffice to produce a large and constant supply, and therefore winter Cucumbers need not be regarded as beyond the range of practice of any ordinary well-kept garden.

Frame Cucumbers are the most in demand, and the easiest to grow. The very first point for the cultivator is to determine when to begin, for the rule is to begin too early, and to waste time and opportunity in consequence. We will suppose the Cucumbers are to be grown in a two-light frame, for which will be required four good cartloads of stable manure. This should be put in a heap three weeks before the bed is made up, and the bed will have to last until the season is sufficiently advanced to sustain the heat without any further fermentation. Considering these points, it will be understood that it is a far safer proceeding to begin the first week in April than the first week in March, and unless the way is clearly seen, the later date is certainly preferable, for it reduces to a minimum the conflict with time in the matter of bottom heat. Make up the heap; then, early in March, turn it twice, and at the end of the month prepare the bed, firming the stuff with a fork as the work proceeds, but taking care not to tread on the bed. Put on the lights and leave the affair for five or six days; then lay down a bed of rich loamy soil of a somewhat light and turfy texture, about nine inches deep. It is now optional to sow or plant as may be most convenient. Strong plants in pots, put out at once, will fruit earlier than plants from seeds sown on the bed. But sowing on the bed is good practice for all that, and if this plan is adopted a few more seeds must be sown than the number of plants required, to provide a margin for enemies; any surplus plants will generally prove useful, for Cucumber plants seldom go begging. If it is preferred to begin with plants, the question of providing them must be considered in good time. The seed should be sown at least a month in advance, and should be brought forward on a hot-bed or in a cool part of a stove. Many a successful Cucumber grower has no better means of raising plants than by sowing the seeds in a box or pan of light rich earth, kept in a sunny corner of a common greenhouse, with a slate or tile laid over until the seeds start, and by a little careful management nice thrifty plants are secured in the course of about four weeks. In some books on horticulture a great deal is said as to the soil in which Cucumber seed should be sown. We advise the reader not to make too much of that question. Any turfy loam, or even peat, will answer; but a rank soil is certainly unfit. The object should be to obtain short, stout plants of a healthy green colour; not the long-drawn, pallid things that are often to be seen on sale, and which by their evident weakness seem destined to illustrate the problems of Cucumber disease.

Having made a beginning with strong plants on a good bed, the two matters of importance are to regulate the temperature and the watering. In the first instance, it will be necessary to shade the plants a little, but as they acquire strength they should have more light and more air than are usually allowed to Cucumbers. A temperature averaging 60° by night and 80° by day will be found safe and profitable, as promoting a healthy growth and lasting fruitfulness. But the rule must be elastic. You may shut up at 90° without harm, and during sunshine the glass may rise to 95° without injury, provided the plants have air and are not dry at the roots. But it is of great moment that the night temperature should be kept near 60° and not go below it. If the thermometer shows that the night temperature has been above the proper point owing to the heat of the bed, wedge up the lights about half an inch in the evening, and as the season advances increase this supply of night air, for it keeps the plants in health, provided there is no chill accompanying it. As regards watering, the important point is to employ soft water of the same temperature as the frame, and therefore a spare can, filled with water, must be always kept in the frame ready for use, and when emptied should be filled again and left for the next watering. Twice a day at least the plants and the sides of the frame should receive a shower from the syringe. It is better to syringe three times than twice, but this must be in some degree determined by the temperature. The greater the heat, the more freely should air and water be supplied; on the other hand, if the heat runs down, give water with caution, or disaster may follow. In case of emergency the plants will go through a bad time without serious damage if kept almost dry, and then it will be prudent to give but little air. Sometimes the heat of the bed runs out before there is sufficient sun heat to keep the plants growing, but if they can be maintained in health for a week or so, hot weather may set in, and all will come right. But to carry Cucumbers through at such a time demands particular care as to watering and air-giving.

As regards stopping and training, we may as well say at once, that the less of both the better. Free healthy natural growth will result in an abundant production of fruit, and stopping and training will do very little to promote the end in view. But there is something to be done to secure an even growth and the exposure of every leaf to light. When the young plant has made three rough leaves, nip out the point to encourage the production of shoots from the base. When the shoots have made four leaves, nip out the points to promote a further growth of side shoots, and after this there must be no more stopping until there is a show of fruit. The growth should be pegged out to cover the bed in the most regular manner possible, and wherever superfluous shoots appear they must be removed. Any crowding will have to be paid for, because crowded shoots are not fruitful. If a great show of fruit appears suddenly, remove a large portion of it, as over-cropping makes a troublesome glut for a short time, and then there is an end of the business; but by keeping the crop down to a reasonable limit, the plants will bear freely to the end of the season. Every fruiting shoot should be stopped at two leaves beyond the fruit, and as the crop progresses there must be occasional pruning out of old shoots to make room for young ones. An error of management likely to occur with a beginner is allowing the bed to become dry below while it is kept quite moist above by means of the syringe. Many cultivators drive sticks into the bed here and there, and from time to time they draw these out and judge by their appearance whether or not the bed needs a heavy watering. To be dry at the root is deadly to the Cucumber plant, and to be in a swamp is not less deadly. It must have abundance of moisture above and below, but stagnation of either air or water will bring disease, ending in a waste of labour.

The greenhouse cultivation of the Cucumber for a summer crop only is the most profitable and simple as well as the most interesting of all the methods practised. In many gardens the houses that have been filled during the winter with Geraniums and other plants are very poorly furnished during the summer, and present a most unsightly appearance. Now, it is a very easy matter to render them at once profitable and beautiful, for when clothed with green vines bearing handsome Cucumbers, such houses are attractive and pay their way amazingly well. To carry out the routine properly, the house should be cleared at the end of April, the plants being removed to pits and frames. If possible, make up the beds on slates laid close over the hot-water pipes, and use a bushel or more of soil under each light to begin with. First lay on the slate a large seed-pan, bottom upwards, and on that a few flat tiles, and then heap up a shallow cone of nice light turfy loam. Start the fire and shut up, and raise the heat of the empty house to 80° or 90° for one whole day. The next day plant on each hillock a short stout Cucumber plant, or sow three seeds. Proceed as advised for frame culture, keeping a temperature of 60° by night and 80° by day, with a rise of 5° to 10° during sunshine. Ply the syringe freely, give air carefully, and use the least amount of shading possible. It will very soon be found that by judicious management in shutting up and air-giving, the firing may be dispensed with, and then it remains only to syringe freely and train with care. The plants should not be stopped at all, but be taken up direct to the roof and be trained out on a few wires or tarred string, in the first instance right and left, and afterwards along the rafters to meet at the ridge, and form a rich leafy arcade. The fruits will appear in quantity, and must be thinned to prevent over-cropping. As the plants grow, earth must be added to the hillocks until there is a continuous bed, on which a certain number of shoots may be trained where there is sufficient light for them. It is best to begin as advised above, with the aid of fire heat to start the crop for the sake of gaining time; but if this is not convenient begin without fire heat in the last week of May, and the plants will produce fruit until the chill of autumn makes an end of them, and the house is again required for the greenhouse plants.

Winter Cucumbers thrive best in lean-to houses with somewhat steep roofs, as such houses are less liable to chill during cold windy weather, and they catch a maximum of the winter sunshine. In a mild winter, Cucumbers may be grown in any kind of house that can be maintained at a suitable temperature, and the markets are supplied from rough constructions that do duty for many purposes. But in hard weather, the steep lean-to, with bed along the front, and tank to give equable bottom heat, will prove the most serviceable, as it will neither allow snow to lodge on the glass, nor suffer any serious decline of temperature during the prevalence of sharp frost and keen winds. For late autumn supply any kind of house will suffice, but best of all an airy span. A brick pit will answer every purpose from October to March with good management, and fermenting materials will afford the needful heat. In such cases trenches should be provided for occasional renewal of the bottom heat. But a roomy house and a service of hot water justly stand in favour with experienced cultivators, as combining the necessary conditions with convenience of management.

For winter culture, plants are raised from seeds and from cuttings. Seedling plants are the most vigorous, but they require a little more time than cuttings to arrive at a fruiting state. For pot culture cuttings are preferable, as only a moderate crop is expected, and quickness of production is of great importance. It is usual to sow the first lot of seeds on the 1st of September, and to sow again on the 1st of October and the 1st of November; after which it is not advisable to sow again until the 1st of February for the spring crop. If the management is good, the first sowing will be in fruit by the time the third batch of seed is sown, say, by the first week of November, and thenceforward throughout the winter there should be no break in the supply.

The management of Winter Cucumbers turns upon details chiefly, and will be found in the end to depend rather upon care than skill. The general principles are the same as in growing Cucumbers in frames, the task for the cultivator being to carry them out successfully. Begin by sowing the seed singly in small pots in light turfy loam, or peat with which a fair proportion of sharp sand has been mixed. These pots to be placed in a heat of 70° to 75°, and for plants to last long the lower temperature is preferable. As regards the next stage, the plants may be trained up rafters, or spread out on beds, the first being always the better plan where it happens to be convenient. But the prudent cultivator will not be tied to rules; he will cut his coat according to his cloth, and while he has a house of Cucumbers trained to the roof, he will, perhaps, also have a pit filled with plants on beds. To stop severely is bad practice, for vigorous growth is wanted; but a certain amount of stopping must be done to promote an even growth, and to distribute the fruit fairly both in space and time. We have already admitted that in some books on gardening too much has been said about soil. In many places a suitable turfy loam, or a good fibrous peat, may be obtained, and the accidents that have befallen Cucumbers have usually been the result of bad management in respect of heat, water, and air, rather than the use of unsuitable soil. But it must not be supposed that we are careless about this matter. Neither a pasty clay, a sour sticky loam, nor a poor sandy or chalky soil will produce fine Cucumbers. On the other hand, rank manure and poor leaf-mould are both unfavourable materials. There is nothing like mellow loam, which can be enriched and modified at discretion, without going to extremes.

Ridge Cucumbers are grown in much the same way as recommended for Vegetable Marrows. They may be put on hillocks or beds, and in either case a foundation of fermenting material is required to insure a crop in the early part of the summer. For a late crop, the natural heat of the soil will be sufficient should the summer prove to be fine, but in a cold season Ridge Cucumbers are disappointing. Of the many methods of growing them, one of the best is to lay out the ground in four-feet beds by taking out the soil to a depth of fifteen inches, and spreading about that depth or more of half-rotted manure, to which may be added any leaves and other litter that may be handy. Cover with a foot depth of good loam. About mid-April sow the seeds in three-inch pots or in boxes and place in a cool greenhouse. After careful hardening, plant out about the third week of May. If preferred, seeds may be sown on the bed early in May. Give the plants the protection of a hand-light should the weather prove unfavourable, and some care will be needed to keep them moving fairly until the season is so far advanced as to allow for the removal of the lights. Put the plants at thirty inches apart down the middle of the bed, and when growing freely, nip out the points *once only*. A crop of Lettuce may be taken from the beds while the plants are advancing.

DANDELION

Taraxacum officinale

As a salad Dandelion has won general esteem for its wholesome medicinal qualities. Nature teaches the way to grow this plant, for she sows the seed in early summer, and we find the finest plants on dry ground, while there are none to be found in bogs and swamps. Any gravelly or chalky soil will grow good Dandelion, one fair digging without manure being a sufficient preparation for it. Sow in May or June, and thin to one foot apart every way, keeping the crop scrupulously clean by flat hoeing. Any time in the winter the roots may be lifted and forced in the same way as Sea Kale, or they may be covered with pots in spring to blanch where grown. In any case the spring growth must be made in darkness, for when green the flavour is bitter. Invalids who require this salutary salad may obtain early supplies by planting the roots in boxes in a cellar, and covering with empty boxes. Only as much water should be given as will keep the roots reasonably moist.

EGG PLANT (AUBERGINE)

Solatium Melongena, S. esculentum

In this country the Egg Plant is generally grown merely as an ornament, but it is a delicious vegetable when sliced and fried in oil, the purple- and black-fruited kinds being especially serviceable for the table. The common white, which is best known, is fairly good when cooked young, though less rich in flavour than the purple. The cultivation recommended for Capsicum will suit the Egg Plant, but little atmospheric moisture is needed or the seedlings may damp off. They are not well adapted for planting out, although in a warm season they will fruit freely under a sunny wall, and will grow in a gravel walk if helped at first with a little good soil round the roots. If required in quantity for the table, the purple variety may be grown in a frame from plants raised on a hot-bed. Generally speaking, a few plants in pots are all that are required where the fruit is not valued as an esculent.

ENDIVE

Cichorium Endivia

As a result of the growing taste for wholesome salads Endive has considerably advanced in public esteem. The flavour of well-blanched Endive suits most palates that have had experience of salads, and of the salutary properties of the plant we have a hint in its close relation to the Chicory.

The selection of sorts is a question of importance, because the handsome curled varieties that make the best appearance on the table, and might be regarded as ornaments if they were not edible, are the very finest for salads, being tender, with a fresh nutty flavour. The broad-leaved sorts are not so well adapted for salads as for stews, and they take the place of Lettuces when the latter are not available for soups and ragoûts. However, when an emergency occurs, the curled varieties will be found suitable for cooking, and the broad-leaved for salading, and therefore there need be no waste where one sort predominates.

Soil.—A difficulty common to Endive culture may be got over in the way advised for Celeriac. The plant requires a light, dry, sandy soil; and a portion, at least, of the crop is expected to stand through the winter. Thus on a heavy soil there is a prospect of failure in respect of the late crop, but that is obviated by adopting a made bed—one of smallish dimensions being sufficient to accommodate a large stock of plants. Select an open spot, make a foundation of any hard rubbish that is at hand, and on this put one to two feet of sandy soil. This will form a raised bed of a kind exactly suited to the plant, and will cost but little as compared with its ultimate value. If regularly dressed with manure, and otherwise well managed, the bed will supply Endive in winter and other salads in summer, or it may be cropped with Dwarf Beans, which can be removed in August to make way for the usual planting of Endive. Where the soil is naturally light and dry no such preparation is needed, but Endive does not come to perfection without food, and therefore the soil should be rich and deeply dug.

Sowing and Transplanting.—The seed may be sown as early as March, in a moderate heat, but the latter part of April is early enough for most purposes, and the main sowings are made in June. Later sowings may follow in July and August. But the June sowing is the most important, as by a little careful management it will supply a few early heads and many late ones. Sow in shallow drills six inches apart, and when the plants are an inch high draw the most forward, and prick them out on a bed of rich light soil in the same way as Celery, and with a little nursing these will make a first plantation. The plants in the seed-bed should be thinned to three inches, and must have water in dry weather. All the thinnings should be pricked out in the first instance to make them strong for planting, but the last lot may go direct to the beds to finish.

The final planting must be on rich, light, dry soil, and water given to encourage growth. The distance for the curled varieties is a foot each way, and for the broad-leaved fifteen inches. In taking the last lot from the seed-bed, a crop should be left untouched to mature at twelve to fifteen inches apart. These plants will give a first and most excellent supply if carefully blanched.

If more convenient, seed may be sown where the crop is intended to stand, the plants being thinned to the distances already given.

The blanching is an important business, and is variously performed. The customary mode is to tie the leaves together in the manner usual with Lettuce and mould them up. This method answers perfectly, except in wet seasons, when, if the plants stand for some time, the outer leaves begin to rot, and the decay proceeds inwards, to the deterioration or destruction of the plant. A clean and effective process is to cover the heart of the plant with a flower-pot. The hole is darkened with part of a tile or slate, on which should be laid a piece of turf or a handful of mould. A plate or clean tile placed over the centre of the plant will also blanch Endives satisfactorily in autumn. For winter supplies, the plants may be lifted as wanted and placed in boxes or pots of soil, these being covered with other boxes or pots to exclude light. A Mushroom-house, cellar, or under a greenhouse stage, will serve for storing the lifted plants. The blanching must be carried on in such a way as to insure a succession without a glut at any time, for when sufficiently blanched Endive should be used, or decay will soon set in.

GARLIC

Allium sativura

The mode of culture advised for Shallots will suit Garlic also, except that the latter should be planted in February about two inches beneath the surface of the soil, and the bulbs may be grown closer together, about eight or nine inches apart each way.

When large bulbs are required for exhibition or other purposes, the cloves—as the divisions of each root are called—should be planted separately; but for general use moderate-sized bulbs, planted whole, will produce a heavier crop.

GOURD and PUMPKIN

(Cucurbita)

Gourds and Pumpkins may be grown to perfection by precisely the same method recommended for Ridge Cucumbers; but as the plants occupy more space, room must be left for them to extend south wards beyond the limits of the ridge. It is well to put out strong plants from seeds sown in pots in April or May, and protect them until established. If these are not obtainable, the seed may be sown where the plants are intended to stand, and there will in time be plenty of produce, but of course somewhat later in the season than if strong plants had been put out in the first instance. Keep a sharp look-out for slugs, which will flock in from all quarters to feast upon them, but will scarcely touch them after they have been planted a week or so. Any rough fermenting material, such as grass mowings, may be used in making the hills, to give them the aid of a warm bed for a brief space of time, and it is a great gain if they grow freely from the first. Later on the natural heat will be enough for them.

The edible Gourds are useful in all their stages and ages; and if the cultivator has a fancy to grow large, handsome fruits, he can make the business answer by hanging them up for use in winter, when they may be employed in soups in place of Carrots, or in addition to the usual vegetables, and may indeed be cooked in half a dozen different ways. There remains yet one more purpose to which the plants may be applied: supposing you have a great plantation of edible Gourds and Marrows, and would like a peculiarly elegant and delicious dish of Spinach, pinch off a sufficiency of the tops of the advancing shoots, and cook them Spinach fashion. If properly done, it is one of the finest vegetables ever eaten. As pinching off the tender tops of the shoots lessens the fruitfulness of the vines, we only recommend this procedure where there is a large plantation.

Gourds may be trained to trellises, fences, and walls. In all such cases, a good bed should be prepared of any light, rich loam, and it will be none the less effective if made on a mound of fermenting material.

HERBS

With certain exceptions, the growing of Sweet Herbs from seeds is altogether advantageous. The plants come perfectly true, and are so vigorous that it is easier to raise them from seed than to secure a succession from slips or cuttings. To meet a large and continuous demand in the kitchen there must be a proportionate plantation in the border; but in gardens of medium size we do not advocate the culture of Herbs on an extensive scale, unless there be a special object in view. A moderate number of Herbs will meet the necessities of most families. Still it is a fact that the tendency is always in the direction of increased variety, and gardeners are called on to provide frequent changes of flavouring Herbs, some of which are quite as highly prized in salads as they are for culinary purposes.

In the smallest gardens, Mint, Parsley, Sage, and both Common and Lemon Thyme, must find a place. In gardens which have any pretension to supply the needs of a luxurious table there should be added Basil, Chives, Pot and Sweet Marjoram, Summer and Winter Savory, Sorrel, Tarragon, and others that may be in especial favour. Large gardens generally contain a plot, proportioned to demands, of all the varieties which follow.

Several of the most popular Herbs, such as Chives, Mint, Tarragon, and Lemon Thyme, are not grown from seed—at all events, those who venture on the pastime might employ their labour to greater advantage. But others, such as Basil, Borage, Chervil, Fennel, Marjoram, Marigold, Parsley, Savory, &c, are grown from seed, in some cases of necessity, and in others because it is the quicker and easier way of securing a crop.

Angelica and Mint flourish in moist soil, but the majority of aromatic Herbs succeed on land that is dry, poor, and somewhat sandy, rather than in the rich borders that usually prevail in the Kitchen Garden. Happily they are not very particular, but sunshine they must have for the secretion of their fragrant essences. A narrow border marked off in drills, and, if possible, sloping to the south, will answer admirably. Thin the plants in good time, and the thinnings of those wanted in quantity may, if necessary, be transplanted. The soil must be kept free from weeds, and every variety be allowed sufficient space for full development.

Angelica *(A. Archangelica)*.—A native biennial which is not easily raised from seed treated in the ordinary way. Germination is always capricious, slow and irregular. It may be several months before the plants begin to appear. The best results are obtained by placing the seed in sand, kept moist for several weeks before sowing. The leaves and stalks are sometimes blanched and eaten as Celery, and are also boiled with meat and fish. Occasionally the tender stems and midribs are coated with candied sugar as a confection. Angelica was formerly supposed to possess great medicinal virtues, but its reputation as a remedy for poison and as a preventive of infectious diseases is not supported by the disciples of modern chemistry. The seeds are still used for flavouring liqueurs.

Balm *(Melissa officinalis)*.—A perennial herb, which can be propagated by cuttings or grown as an annual from seed. An essential oil is distilled from the leaves, but they are chiefly used, when dried, for making tea for invalids, especially those suffering from fever. The plant has also been used for making Balm wine. Sow in May.

Basil, Bush *(Ocymum minimum)*.—A dwarf-growing variety, used for the same purposes as the Sweet Basil. Sow in April.

Basil, Sweet *(Ocymum Basilicum)*.—A tender annual, originally obtained from India, and one of the most popular of the flavouring Herbs. Seeds should be sown in February or March in gentle heat. When large enough the seedlings must be pricked off into boxes until they are ready for transferring to a rich border in June, or seed may be sown in the open ground during April and May. A space of eight inches between the plants in the rows will suffice, but the rows should be at least a foot apart. The flower-stems must be cut as they rise, and be tied in bundles for winter use. This practice will prolong the life of the plant until late in the season. Many gardeners lift plants in September, pot them, and so maintain a supply of fresh green leaves until winter is far advanced.

Borage *(Borago officinalis)*.—A native hardy plant, which thrives in poor, stony soil. The flowers are used for flavouring purposes, especially for claret-cup. Borage is also a great favourite with bee-masters. Sow in

April or May in good loam, and thin to fifteen or eighteen inches apart. The rows should be from eighteen to twenty-four inches asunder, for the plant is tall, and strong in growth.

Chervil, Curled *(Anthriscus Cerefolium)*.—Used for salads, garnishing, and culinary purposes. To secure a regular supply of leaves small successional sowings are necessary from spring to autumn, and frequent watering in dry weather will prevent the plants from being spoiled by throwing up seed-stems. For winter use, sow in boxes kept in a warm temperature.

Chives *(Allium Schænoprasum)*.—A mild substitute for the Onion in salads and soups. The plant is a native of Britain, and will grow freely in any ordinary garden soil. Propagation is effected by division of the roots either in spring or autumn. The clumps should be cut regularly in succession whether wanted or not, with the object of maintaining a continuous growth of young and tender shoots. At intervals of four years it will be necessary to lift, divide, and replant the roots on fresh ground.

Fennel *(Fæniculum officinale)*.—A hardy perennial which has been naturalised in some parts of this country. It is grown in gardens to furnish a supply of its elegant feathery foliage for garnishing and for use in fish sauces. Occasionally the stems are blanched and eaten in the same way as Celery, and in the natural state they are boiled as a vegetable. The seeds are also employed for flavouring. Sow in drills in April and May, and thin the plants to fifteen inches apart.

Finocchio, or Florence Fennel *(Fæniculum dulce*, DC).—A sweet-tasting herb, very largely grown in the south of Italy, where it is eaten both in the natural state and when boiled. Sow in the open ground during spring or early summer, in rows about eighteen inches apart, and thin or transplant to six or nine inches. When the base begins to swell, earth up the plants in the same manner as Celery. If transplanted, pinch off the tips of the roots.

Horehound *(Marrubium vulgare)*.—A well-known medicinal herb, from which an extract is obtained for subduing irritating coughs. Sow in April or May, and thin the plants until they stand fifteen inches apart.

Hyssop *(Hyssopus officinalis)*.—The leaves and young shoots are used as a pot-herb, and the leafy tops and flowers, when dried, are employed for medicinal purposes. Hyssop is also occasionally used as an edging plant. A dry soil and warm situation suit it. Sow in April, and thin the plants to a foot apart in the rows.

Lavender *(Lavandula)*.—Universally known and valued for its perfume. Although the plant is generally propagated from cuttings, it can easily be grown from seed sown in April or May. The plants attain a height of one or two feet, and the stems should not be cut until the flowers are expanded.

Marigold, Pot *(Calendula officinalis)*.—Employed both in flower and vegetable gardens: in the former as a bedding annual, and in the latter that the flowers may be dried and stored for colouring and flavouring soups; also for distilling. In April or May sow the seed in drills one foot apart, and thin the plants to the same distance in the rows.

Marjoram, Pot *(Origanum Onites)*.—One of the most familiar Herbs in British gardens. The aromatic leaves are used both green and when dried for flavouring. Strictly the plant is a perennial, but it is readily grown as an annual. Sow in February or March in gentle heat, and in the open ground a month later. The plants should be allowed a space often inches or a foot each way.

Marjoram, Sweet Knotted *(Origanum Majorana)*.—This plant is used for culinary purposes in the same way as the Pot Marjoram, and it is also regarded as a tonic and stomachic. The most satisfactory mode of cultivation is that of a half-hardy annual. Sow in March or April and allow each plant a square foot of ground.

Mint *(Mentha viridis)*.—Known also as Spearmint. It must be grown from divisions. Between the delicacy of fresh young green leaves and those which have been dried with the utmost care there is so wide a difference that the practice of forcing from November to May is fully justified. This is easily accomplished by packing roots in a box and keeping them moist in a temperature of 60°. Where this is impossible, stems must be cut, bunched, and hung in a cool store for use during winter and spring. Mint grows vigorously in damp soil, and the bed should have occasional attention, to prevent plants from extending beyond their proper boundary. To secure young and luxuriant growth a fresh plantation should be made annually in February or March. If allowed to occupy the same plot of land year after year the leaves become small and the stems wiry.

Parsley *(Carum Petroselinum)* will teach those who have eyes exactly how it should be grown. There will appear here and there in a garden stray or rogue Parsley plants. No matter how regularly the hoeing and weeding may be done, a stray Parsley plant will occasionally appear alone, perhaps in the midst of Lettuces, or Cauliflowers, or Onions. When these rogues escape destruction they become superb plants, and the gardener sometimes leaves them to enjoy the conditions they have selected, and in which they evidently prosper. The lesson for the cultivator is, that Parsley should have plenty of room from the very first; and this lesson, we feel bound to say, cannot be too often enforced upon young gardeners, for they are apt to sow Parsley far more thickly than is wise, and to be injuriously slow and timid in thinning the crop when the plants are crowding one another.

Parsley, like many other good things, will grow almost anywhere and anyhow, but to make a handsome crop a deep, rich, moist soil is required. It attains to fine quality on a well-tilled clay, but the kindly loam that suits almost every vegetable is adapted to produce perfect Parsley, and every good garden should show a handsome sample, for beauty is the first required qualification. To keep the house fairly well supplied sowings should be made in February, May, and July. The first of these will be in gentle heat. When large enough prick out the plants into boxes, or on to a mild hot-bed, and transfer to the open ground at the end of April, allowing each plant a space of one foot each way. In the open, it is best to sow in lines one foot apart, and thin out first to three inches, and finally to six inches, the strongest of the seedlings being put out one foot apart. By following this plan sufficient supplies for a small household may be obtained from one annual sowing made in April. It should not be overlooked that Parsley is indispensable to exhibitors of vegetables, especially as a groundwork for collections, and due allowance for such calls must be made in fixing the number and extent of the sowings. When the plant pushes for seed it becomes useless, and had best be got rid of; but by planting at various times in different places a sufficiency may be expected to go through a second season without bolting, after which it will be necessary to root them out and consign them to the rubbish-heap. Parsley is often grown as an edging, but it is only in large gardens that this can be done advantageously, and then a very handsome edging is secured. In small gardens it is best to sow on a bed in lines one foot apart, and thin out first to three inches, and finally to six inches, the strongest of the thinnings being planted a foot apart, to last over as proposed above. When Parsley has stood some time it becomes coarse, but the young growth may be renewed by cutting over; this operation being also useful to defer the flowering, which is surely hastened by leaving the plants alone. For the winter supply a late plantation made in a sheltered spot will usually suffice, for the plant is very hardy; but it may be expedient sometimes to put old frames over a piece worth keeping, or to protect during hard weather with dry litter. A few plants lifted into five-inch pots and placed in a cool house will often tide over a difficult period. In gathering, care should be taken to pick separately the young leaves that are nearly full grown, and to take only one or two from each plant. It costs no more time to fill a basket by taking a leaf or two here and there from a whole row than to strip two or three plants, and the difference in the end will be considerable as regards the total produce and quality of the crop.

Pennyroyal *(Mentha Pulegium)* is a native perennial which must be propagated by divisions, and this can be done either in spring or autumn.

The rows may be twelve or fifteen inches apart, but in the rows the plants do well at a distance of eight inches. The taste for Pennyroyal is by no means universal, but some persons like the tender tops in culinary preparations. The belief in its supposed medicinal virtues is slowly dying.

Purslane *(Portulaca oleracea)*.—This annual plant thrives best in a sunny position. Seed should be sown from mid-April onwards to insure a succession of young leaves and shoots which may be cooked as a vegetable or eaten raw as a salad. Space the rows nine inches apart and thin the plants to a distance of six inches.

Rampion *(Campanula Rapunculus)*.—Both leaves and roots are used in winter salads; the roots are also boiled. If the seed be sown earlier than the end of May the plants are liable to bolt. Choose a shady situation where the soil is rich and light, and do not stint water. The rows need not exceed six inches apart, and four inches in the rows will be a sufficient space between plants.

Rosemary *(Rosmarinus officinalis)*.—A hardy evergreen shrub easily grown from seed, the leaves of which are used for making Rosemary tea for relieving headache. An essential oil is also obtained by distillation. A dry, warm, sunny border suits the plant. Sow in April and May.

Rue *(Ruta graveolens)*.—A hardy evergreen shrub, chiefly cultivated for its medicinal qualities. The leaves are acrid, and emit a pungent odour when handled. The plant is shrubby, and as it attains a height of two or three feet it occupies a considerable space. Sow in April.

Sage *(Salvia officinalis)*.—Although Sage can be raised from seed with a minimum of trouble, yet this is one of the few instances where it is an advantage to propagate plants from a good stock. The difference will be obvious to any gardener who will grow seedlings by the side of propagated plants. Still, seedlings are often raised, and as annuals the plants are quite satisfactory. Sow under glass in February and March, and in open ground during April and May. Prick off the seedlings into a nursery bed before transferring to final positions, in which each plant should be allowed a space of fifteen inches.

Savory, Summer *(Satureia hortensis)*.—An aromatic seasoning and flavouring herb, which must be raised annually from seed. Sow early in April in drills one foot apart, and thin the plants to six or eight inches in the rows. Cut the stems when in full flower, and tie in bunches for winter use.

Savory, Winter *(Satureia montana)*.—A hardy dwarf evergreen which can be propagated by cuttings; but it is more economically grown from seed sown at the same time, and treated in the same manner, as Summer Savory.

Sorrel *(Rumex scutalus)*.—The large-leaved or French Sorrel is not only served as a separate dish, but is mingled with Spinach, and is also used as an ingredient in soups, sauces, and salads. Leaves of the finest quality are obtainable from plants a year old, and when the crop has been gathered the ground may with advantage be utilised for some other purpose. Light soil in fairly good heart suits the plant. The seed should be sown in March or early April, in shallow drills six or eight inches apart, and the seedlings must be thinned early, leaving three or four inches between them in the rows. To keep the bed free from weeds is the only attention necessary, unless an occasional watering becomes imperative. In September the entire crop may be transferred to fresh ground, allowing eighteen inches between the plants, or part may be drawn and the remainder left at that distance. In the following spring the flower-stems will begin to rise, and if these are allowed to develop they reduce the size of the leaves and seriously impair their quality; hence the heads should be pinched out as fast as they are presented.

Tarragon *(Artemisia Dracunculus)*.—This aromatic herb is used for a variety of purposes, but is most commonly employed for imparting its powerful flavour to vinegar. The plant is a perennial, and must be propagated by divisions in March or April, or by cuttings placed in gentle heat in spring. Later in the year they will succeed under a hand-glass in the open. Green leaves are preferable to those which have been dried, and by a little management a succession of plants is easily arranged. For winter use roots may be lifted in autumn and placed in heat. Those who have no facilities for maintaining a supply of green leaves rely on foliage cut in autumn and dried.

Thyme, Common *(Thymus vulgaris)*.—An aromatic herb, well known in every garden, and in constant demand for the house. Seedlings are easily raised from a sowing in April, or the plant can be grown from division of the roots in spring. Thyme makes a very effective edging, and is frequently employed for this purpose on dry, well-kept borders.

Thyme, Lemon *(Thymus Serpyllum vulgaris)*.—This plant cannot be grown from seed; only by division of the roots in March or April. It is an aromatic herb, generally regarded as indispensable in a well-ordered garden.

Wormwood *(Artemisia Absinthium)*.—An intensely bitter herb, used for medicinal purposes. The plant is a hardy perennial, and is usually propagated in spring by taking cuttings or dividing the roots.

HORSE-RADISH

Cochlearia Armoracia

This vegetable is highly prized as a condiment to roast beef, but as a rule it is badly grown. The common practice is to consign it to some neglected corner of the garden, where it struggles for existence, and produces sticks which are almost worthless for the table. In the same space a plentiful supply of large handsome sticks may be grown with as little trouble as Carrots or Parsnips. Choose for the crop a piece of good open ground, and in preparing it place a heavy dressing of rotten manure quite at the bottom of each trench. Early in the year select young straight roots from eight to twelve inches long, each having a single crown, and plant them one foot apart each way. By the following autumn these will become large, succulent sticks, which will put to shame the ugly striplings grown under starving conditions. The roots may be dug as required; but we do not advocate that method. It is better practice to clear the whole bed at once, and store the produce in sand for use when wanted. This plan should be repeated each year, and a fresh piece of land ought always to be found for the crop.

KALE—*see* BORECOLE, *page 27*

KOHL RABI (KNOL KOHL)

Brassica oleracea Caulo-rapa

Kohl Rabi, or Knol Kohl, is comparatively little grown in this country, because we can almost always command tender and tasty Turnips. On the Continent it is otherwise. There Kohl Rabi may be seen in every market, and on many a good table, where it proves a most acceptable vegetable. For all ordinary purposes the green variety is better than the purple. A small crop of this root should be annually grown in every garden. In case of failure with Turnips, Kohl Rabi will take their place to tide over an emergency. When. served it has the flavour of a Turnip with a somewhat nutty tendency, and may be prepared for table in the same manner.

Kohl Rabi is cultivated in much the same way as Turnips. Seed may be sown at any time from March to August in rows one and a half to two feet apart. As soon as possible thin the seedlings to three inches apart in the rows, and, as the leaves develop, to six inches apart. By drawing every other plant some small roots may be obtained early, and the remainder will be left to mature at twelve inches in the rows. The seedlings may be transplanted, if desired. Keep the ground clean and the surface open, but care should be taken not to damage the leaves, or in the least degree to earth up the roots. Any animal that can eat a Turnip will prefer a Kohl Rabi, and when substituted for the Turnip in feeding cows, it does not affect the flavour of the milk. The plant is hardy, and as a rule may stand, to be drawn as wanted, until the spring is far advanced, when the remnant should be cleared off for the benefit of the animals on the home farm, or be dug in as manure.

LEEK

Allium Porrum

The leek is not so fully appreciated in the southern parts of England as it is in the North, and in Scotland and Wales. It is a fine vegetable where it is well understood, and when stewed in gravy there is nothing of its class that can surpass it in flavour and wholesomeness. One reason of its fame in Scotland and the colder parts of Wales is its exceeding hardiness. The severest winters do not harm the plant, and it may remain in the open ground until wanted, occasioning no trouble for storage.

Times of Sowing.—To obtain large handsome specimens of the finest quality a start must be made in January or early February, and this early sowing is imperative for the production of Leeks for exhibition, as the roots must be given a longer season of growth than is generally allowed for ordinary crops. It is usual to sow in pans or boxes of moistened soil, placed in a temperature of about 55°. The seeds need only a very light covering of fine soil. When the seedlings are about two inches high transfer to shallow boxes of rich soil, spacing them three inches apart each way, or the finest may be placed in pots of the 32-size, taking care not to break the one slender root on which the plant depends at this stage. Grow on in the same temperature until mid-March, when they may be transferred to a cold frame to undergo progressive hardening in readiness for planting out at a favourable opportunity in April.

There may be three sowings of Leek made in the open ground in February, March, and April, to insure a succession, and also to make good any failures. But for most gardens one sowing about the middle of March will be sufficient. From this sowing it will be an easy matter to secure an early supply, a main crop, and a late crop, for they may be transplanted from the seed-bed at a very early stage, and successive thinnings will make several plantations; and finally, as many can be left in the seed-bed to mature as will form a proper plantation.

General Cultivation.—The Leek will grow in any soil, and when no thicker than the finger is useful; indeed, in many places where the soil is poor and the climate cold it rarely grows larger, but is, nevertheless, greatly valued. A rich dry soil suits the plant well, and when liberally grown it attains to a great size, and is very attractive, with its silvery root and brilliant green top. The economical course of management consists in thinning and planting as opportunities occur, beginning as soon as the plants are six inches high, and putting them in well-prepared ground, which should be thoroughly watered previously, unless already softened by rain. The distance for planting must depend upon the nature of the soil and the requirements of the cultivator. For an average crop, eighteen inches between the rows and six to nine inches between the plants is sufficient; but to grow large Leeks, they must be allowed a space of twelve to eighteen inches in the rows. In planting, first shorten the leaves a little (and very little), then drive down the dibber, and put the plant in as deep as the base of the leaves, and close in carefully without pressure. Water liberally, occasionally stir the ground between plants, and again cut off the tops of the leaves, when the roots will grow to a large size. If the ground is dangerously damp or pasty, make a bed for the crop with light rich soil, plant on the level and mould up as the growth advances. On light land, however, it is advisable to grow them in trenches, prepared as for Celery. The largest and whitest should not be left to battle with storms, but those left in the seed-bed will take no harm from winter weather, and will be useful when the grandees are eaten. The finest roots that remain when winter sets in may be taken up in good time and stored in dry sand, and will keep for at least a month. Any that remain over in spring can readily be turned to account. As the flower-stems rise nip them out; not one should be left. The result of this practice will be the formation on the roots of small roundish white bulbs, which make an excellent dish when stewed in gravy, and may be used for any purpose in cookery for which Onions or Shallots are employed. They are called 'Leek Bulbs,' and are obtainable only in early summer.

Blanching.—The edible part of the root should be blanched, and this may be effected in various ways. Drain-pipes not less than two and a half inches in diameter, and from twelve to fifteen inches in length, answer well for large stems. Tubes of stiff brown paper are also very serviceable. Drawing up the earth to the stem as growth develops is a simple method of blanching, and the edible portion may easily be increased according to the amount of earthing-up given. Perfect blanching is of first importance when specimens are wanted for the exhibition table, and a commencement must be made as soon as the plants may be said to have thoroughly recovered from the effects of transplanting.

LETTUCE

Lactuca sativa

The lettuce is the king of salads, and as a cooked vegetable it has its value; but as it does not compete with the Pea, the Asparagus, or the Cauliflower, we need not make comparisons, but may proceed to the consideration of its uses in the uncooked state. Scientific advisers on diet and health esteem the Lettuce highly for its anti-scorbutic properties, and especially for its wholesomeness as a corrective. It supplies the blood with vegetable juices that are needful to accompany flesh foods when cooked vegetables are unattainable. Our summers are usually too brief and too cool to permit us to acquire a knowledge of the real value of the Lettuce, but in Southern Europe and many parts of the East it becomes a necessary of life, and those large red Lettuces that are occasionally grown here as curiosities are prized above all others because of their crisp coolness and refreshing flavour under a burning sun.

The numerous varieties may, for practical purposes, be grouped in two classes—Cabbage and Cos Lettuces. They vary greatly in habit and are adapted for different purposes, the first group being invaluable for mixed salads at all seasons, but more especially in winter and early spring; the second group is most serviceable in the summer season, and is adapted for a simple kind of salad, the leaves being more crisp and juicy. A certain number of the two classes should be grown in every garden, both for their great value to appetite and health, and their elegance on the table, whether plain or dressed. In the selection of sorts, leading types should be kept in view. Some of the varieties which have been introduced have no claim to a place in a good list, because of their coarseness. Although they afford a great bulk of blanched material, it is too often destitute of flavour, or altogether objectionable. The best types are tender and delicately flavoured, representing centuries of cultivation, and the sub varieties of these types should retain their leading characteristics, though perhaps they are more hardy and stand longer, and are therefore much to be desired.

Preparation of the Soil.—The Lettuce requires a light, rich soil, but almost any kind of soil may be so prepared as to insure a fair supply, and in places where fine Cos Lettuces are not readily obtained, it may be possible to grow excellent Cabbage varieties in place of them. A tolerably good garden soil will answer for both classes, and fat stable manure should be liberally used. The best way to prepare ground for the summer crop is to select a piece that has been trenched, and go over it again, laying in a good body of rough green manure, one spade deep, so that the plant will be put on unmanured ground, but will reach the manure at the very period when it is needed, by which time contact with the earth will have rendered it sweet and mellow. By this mode of procedure the finest growth is secured, and the plants stand well without bolting, as they, are saved from the distress consequent on continued dry weather. As regards drought, it must be said that the red-leaved kinds stand remarkably well in a hot summer, and although they do not rank high as table Lettuces in this country, were we to experience a succession of roasting summers they would rise in repute and be in great demand. Cabbage Lettuces bear drought fairly well, more especially the diminutive section; but where water is available Lettuces have as good a claim to a share of it in a dry, hot season, as any crop in the garden.

Blanching.—A first-class strain of White Cos Lettuce will produce tender white hearts without being tied, and, as a rule, therefore, the labour of tying may be saved. The section of which Sutton's Superb White Cos is the type may be said to produce better samples without tying than with this imaginary aid to blanching. The market grower is still accustomed to tie Lettuces because they are more easily packed and travel better when tied, but when tying is practised it need not be done until one or two days before the Lettuces are cut. The coarser market kinds certainly are improved by tying, and in this case the operation must be performed when the plants are quite dry, and not more than ten days in advance of the day on which it is intended to pull them. The Bath Cos must be tied always, and when well managed the heart is white, with a pretty touch of pink in the centre.

Spring-sown Lettuces may be forwarded under glass from January to March, from which time sowings may be made successively in the open ground. In any and every case the finest Lettuces are obtained by sowing in the open ground, and leaving the plants to finish in the seed-bed without being transplanted. It will, of course, occur to the practical cultivator that the two systems may be combined, so as to vary the time of turning in, and thus from a single sowing insuring a longer succession than is possible by one system only. We will suppose small sowings made of three or four sorts in January or early in February, and put into a gentle heat to start them. A very little care will keep them going nicely, and of course they must have light and air to any extent commensurate with safety. When about three weeks old, it will be advisable to prick these out into a bed of light rich earth in frames; or if the season is backward, and they need a little more nursing, prick them into large shallow boxes, containing two or three inches of soil, which will be sufficient provided it consists in great part of decayed manure, kept always moist enough for healthy growing. The next step will be to plant them out about six inches apart, with a view to draw a certain number as soon as they are large enough to be useful, leaving the remainder at nine to twelve inches, taking care to thin out in time to prevent any leaves overlapping. If Peas are being grown under glass, a few plants of an early Cabbage variety may be put out between the rows, or they may be pricked out on the borders of a Peach-house, in either case spacing the plants nine inches apart. Successive sowings made in February and March will be treated in the same way, and will need less nursing. In planting out, it is important to have the seedlings well hardened, for they are naturally susceptible to wind and sunshine, and if suddenly exposed to either will be likely to perish. Again, when first planted out their delicate leaves will attract all the slugs and snails in the garden, and the discreet way of acting is to regard a plantation of Lettuce as an extensive vermin trap, and thus, knowing where the marauders are, to be ready to catch and kill, or to destroy them by sprinklings of lime, salt, or soot, in all cases being careful to keep these agents at a reasonable distance from the plants.

Sowings in the open ground from the end of March onwards should be made, not on an ordinary seed-bed, but on a plot loaded with rich manure at one spit deep, and the seed should be put in shallow drills one foot apart. From the time the young plants are two inches high they must be drawn freely for 'Cutting Lettuce,' or for planting out elsewhere; this thinning to proceed until a sufficient crop remains to finish off on the ground. The value of 'Cutting Lettuce' is better understood on the Continent than in this country. The small tender plants are in daily use, and appear in the salad bowl with Water Cress and Corn Salad, delicately dressed with delicious flavourings. After this brief digression it is necessary to add that a crowded Lettuce crop is an encumbrance to the ground; and one of the evils of the best system, that of sowing where the crop is to finish, is the tendency of the cultivator to be timid in the thinning, which should be done with a bold hand, and in good time.

July and August Sowing.—From sowings made during these months the supply of Lettuce from the open ground may be extended throughout the autumn, and even into December or January should the weather prove favourable. The main conditions essential to success are, the use of quick-growing varieties, sowing in good soil where the heads are to mature, and early and severe thinning. The thinnings may be transplanted if required.

Winter Lettuces are produced and provided for in various ways. In some places Lettuces stand out the winter without covering, and turn in early in the spring. But in other districts they seldom survive the winter without protection, even when the sparrows spare them. The summer sowings will afford supplies to a late season of the year, and the crop that remains when frost sets in may be preserved with slight and rough protection. But for the profitable production of Winter Lettuces frames are a necessity, and care must be taken not to promote a strong growth, for after a term of mild winter weather a sudden and severe frost will probably annihilate those that are in a too thriving condition. In the least likely places, however, it is well to have a small plantation of Winter Lettuces in the open, and to give some rough protection in bad times, as these often prove of great advantage, and even outlive frame crops which have been allowed to get too forward by the aid of warmth and a rich soil.

For winter and spring use sowings should commence in August and be continued, according to requirements, until the middle of October, after which it is waste of time and seed to sow any more. The August and September sowings may be made partly on an open border and partly in frames, but the October sowings must be in frames only, for winter may overtake them in the seed-leaf. The seedlings must in all cases be thinned and pricked out as soon as large enough, and should be planted in fine soil, free from recent manure, being carefully handled to avoid needless check. Some should be planted in frames on beds of light soil near the glass, at three inches apart, and when these meet they must be thinned for the house as may be necessary: the remainder of the thinnings may be put out on warm borders at six inches, and, if quite convenient, a crop should be left in the seed-bed at six inches. From the frames, the supplies will be ready in time to follow those from late summer sowings, and thus through the winter until the frames are cleared out for the work of the spring. The frame crop must have plenty of air, and be kept as hardy as possible, but with moisture enough to sustain a steady healthy growth. If roughly handled in the planting, or a little starved in respect of moisture, the plants will rise from the centre just when they ought to begin to turn in, and the first few days of warm sunshine will start them in the wrong way. As to those wintered out, there are many ways of protecting them, and when success has crowned the effort there will be a crowded plant. It will be necessary, therefore, to transplant at least half the crop by lifting every other one. This must be done with care, as though they were worth a guinea each. By transplanting early in March to a piece of rich light ground in a warm spot, and doing the work neatly and smartly, the result will be a valuable crop of early Summer Lettuce, while those that remain will help through the spring.

Forcing.—Lettuces do not force well; but as they are so constantly in demand, it is a matter of importance to grow them in every possible way. Nice promising plants from August and September sowings may be selected from the frames, and planted on gentle hot-beds from November to January, and will do well if tenderly lifted. The Commodore Nutt and Golden Ball are the best of the Cabbage varieties for forcing. The Cos varieties do not differ much as to forcing, none of them being well adapted for the purpose; but the Superb White Cos may be brought to fine condition by taking time enough, so as to make a very moderate warmth suffice. On sunny days the heat should not exceed 75°; but 65° is sufficient, with a night temperature of 45°to 50°.

One other method of providing small delicate salading may be adopted to meet emergencies. On the barrows of itinerant greengrocers in Paris the thinnings of Lettuce crops form part of the general stock, and in this country we do not sufficiently utilise this young tender stuff. But we have now in view the use of Lettuce in a still earlier stage of growth. By sowing rather thinly in boxes, kept under glass, a dense growth is produced in a short time which can be cut in the same manner as Mustard. For this purpose Sutton's Winter Gathering is especially valuable, or one of the best White Cos varieties should be sown.

MAIZE and SUGAR CORN

Zea Mays

Maize is a tender plant of great beauty that may be grown as a table vegetable, a forage plant, or a corn crop; but in the last-named capacity it is rarely profitable in this country, owing to the brevity of our summers. As an ornamental plant it is entitled to consideration, and the more so because, while adorning the garden with its noble outlines and splendid silken tufts, it will at the same time supply to the table the green cobs that are so much valued when cooked and served in the same manner as Asparagus.

There is a simple rough and ready way of growing Maize, the first step towards which is to prepare a deep rich soil, in a sunny and sheltered situation. Late in April or early in May dibble the seeds two inches deep, in rows two feet asunder and one foot apart in the rows. When the plants have made some progress, remove every other one, these thinnings to be destroyed or planted at discretion. Plants may also be started under glass by sowing seeds in gentle heat in April. Prick off into pots and gradually harden for transfer to the open. The crop will almost take care of itself when the weather is warm enough to suit it. But a deluge of water may be given during the hottest weather. In its native country, and indeed wherever Maize thoroughly thrives, it is dependent on frequent storms.

MELON

Cucumis Melo

The popularity of this cool and delicious fruit has in recent years been greatly enhanced by increased knowledge as to the best method of treating the plant, and also by the introduction of several varieties which are attractive in form and superb in flavour. It would shock a modern Melon eater to be advised to cook a Melon, and flavour it with vinegar and salt, as in the early days of English gardening. A good Melon of the present day does not even need the addition of sugar; the beauty, aroma, and flavour are such that it is not unusual for the epicure to push the luscious Pine aside in order to enjoy this cool, fresh, gratifying fruit that delights without cloying the palate. The newer varieties are remarkable alike for fruitfulness and high quality, and are somewhat hardier than the favourites of years gone by.

The Melon is grown in much the same way as the Cucumber, but it differs in requiring a firmer soil, a higher temperature, a much stronger light, less water, and more air. It may be said that no man should attempt to grow Melons until he has had some experience in growing Cucumbers. As regards this point, the hard and fast line is useless, but Cucumber-growing is certainly a good practical preparative for the higher walk wherein the Melon is found. But Cucumbers are grown advantageously all the winter through; Melons are not. The former are eaten green, and the latter are eaten ripe; this makes all the difference. Melons that are ripened between October and May are seldom worth the trouble bestowed upon them; therefore we shall say nothing about growing Melons in winter.

The Frame Culture may with advantage begin about the middle of March by the preparation of a good hot-bed. It is best to use a three-light frame, as the heat will be more constant than with one of smaller size. There should be six loads of stuff laid up for the bed, and the turning should be sufficient to take out the fire, without materially reducing the fermenting power. Begin a fortnight in advance of making up the bed, and be careful at every stage to do things well, as advised for the cultivation of frame Cucumbers. The best soil for Melons is a firm, turfy loam, nine inches of which should be placed on top of the manure. In a clay district, a certain amount of clay, disintegrated by frost, may be chopped over with turfy loam from an old pasture. If the soil is poor, decayed manure should be added, but the best possible Melons may be grown in a fertile loam without the aid of manures or stimulants of any kind. It is good practice to raise the plants in pots, and have them strong enough to plant out as soon as the newly-made beds have settled down to a steady temperature of about 80°, but below 70° will be unsafe. If plants cannot be prepared in advance, seed must be sown on the bed, and as a precaution against accidents and to permit of the removal of those which show any sign of weakness, a sufficient number of seeds should be sown to provide for contingencies.

As regards the bed, it may be made once and for all at the time of planting, a few days being allowed for warming the soil through. But we much prefer to begin with smallish hillocks, or with a thin sharp ridge raised so as almost to touch the lights, and to plant or sow on this ridge, which can be added to from time to time as the plants require more root room. The soil, coming fresh and fresh, sustains a vigorous and healthy root action. The high ridge favours the production of stout leaves, and the absorption by the soil of sun-heat is to the Melon of the first importance.

The practice of pruning Melons as if the plants were grown for fodder, and might be chopped at for supplies of herbage, must be heartily condemned. Melons should never be so crowded as to necessitate cutting out, except in a quite trivial manner. A free and vigorous plant is needed, and under skilful attention it will rarely happen that there is a single leaf anywhere that can be spared. We will propose a practical rule that we have followed in growing Melons for seed, of which a large crop of the most perfect fruits is absolutely needful to insure a fair return. The young plants are pinched when there are two rough leaves. The result is two side shoots. These are allowed to produce six or seven leaves, and are then pinched. After this, the plants are permitted to run, and there is no more pinching or pruning until the crop is visible. Then the fruits that are to remain must be selected, and the shoots be pinched to one eye above each fruit, and only one fruit should remain on a shoot; the others must be removed a few at a time. All overgrowth must be guarded against, for crowded plants will be comparatively worthless. It is not by rudely cutting out that crowding is to be prevented, but by timely pinching out every shoot that is likely to prove superfluous. From first to last there must be a regular plant, and not a shoot should be allowed to grow that is not wanted. Cutting out may produce canker, and crowding results in sterility.

As the Melon is required to ripen its fruits, and the Cucumber is not, the treatment varies in view of this difference. It is not necessary to fertilise the female flowers of the Cucumber, but it is certainly desirable, if not absolutely necessary, to operate on those of the Melon to insure a crop. The early morning, when the leaves are dry and the sun is shining, is the proper time for this task, which is described in a later paragraph. And the necessity for ripening the crop marks another difference of management, for Cucumbers may carry many fruits, and continue producing them until the plants are exhausted. But the production of Melons must be limited to about half a dozen on each plant, and good management requires that these should all ripen at the same time, or nearly so, fully exposed to the sun, and with plenty of ventilation.

The requisite supply of water is an important matter. The plant should never be dry at the root, and must have a light shower twice a day over the leafage, but the moisture which is necessary for Cucumbers would be excessive for Melons. It is a golden rule to grow Melons liberally, keeping them sturdy by judicious air-giving, and to give them a little extra watering just as they are coming into flower. Then, as the flowers open, the watering at the root should be discontinued, and the syringe should be used in the evening only at shutting up. If discontinued entirely, red spider will appear, and the crop will be in jeopardy, for that pest can be kept at a distance only by careful regulation of atmospheric moisture.

Melons in frames do better spread out on the beds than when trained on trellises. When so grown, each fruit must be supported with a flat tile or an inverted flower-pot, and means must be taken, by pegs or otherwise, to prevent it from rolling off, for the twist of stem that ensues may check the fruit or cause it to fall. When the fruits are as large as the top joint of a man's thumb, watering may be resumed, and the syringe used twice a day until the fruit begins to change colour, when there must be a return to the dry system, but with care to avoid carrying it to a dangerous extreme.

The Melon-house, heated by hot water, is adapted to supply fruit earlier than is obtainable by frame culture, and is entirely superior to any frame or pit. It appears, however, that in Melon-houses red spider is more frequently seen than in frames heated by fermenting material; but this point rests on management, and there can be nothing more certain than that a reasonable employment of atmospheric humidity may be made effectual for preventing and removing this pest. For the convenient cultivation of the crop a lean-to or half-span is to be preferred. The width should not exceed twelve feet, and ten to twelve feet should be the utmost height of the roof. A service of pipes under the bed will be required; but as Melons are not grown in winter, the heating of a Melon-house is a simple affair, and, indeed, very much of the cultivation as the summer advances will be carried on by the aid of sun-heat only. The treatment of the plants in a house differs from the frame management, because a trellis is employed, and the plants are taken up the trellis without stopping until they nearly reach the top, when the points are pinched out to promote the growth of side shoots. In setting the fruit, the same principles prevail as in frame culture, and it is advisable to 'set' the whole crop at once; if two or three fruits obtain a good start, others that are set later will drop off. As the fruits swell, support must be afforded to prevent any undue strain on the vine, and this should be accomplished by nets specially made for the purpose, or by suspending small flat boards of half-inch deal with copper wires, each fruit resting on its board, until the cracking round the stem gives warning that the fruit should be cut and placed in the fruit room for a few days to complete the ripening for the table. In houses of the kind described Melons and Cucumbers are occasionally grown together. But although this may be done, and there are many cultivators expert in the business, the practice cannot be recommended, for ships that sail near the wind will come to grief some day. The moisture and partial shade that suit the Cucumber do not suit the Melon, and it is a poor compromise to make one end of the house shady and moist, and the other end sunny and dry, to establish different conditions with one atmosphere. A glass partition pretty well disposes of the difficulty, because it is then possible to insure two atmospheres suitable for two different operations. *(See also pages 157, 175, and 184.)*

The Pollination of Melons is performed by plucking the mature male blooms, and after the removal of the petals, transferring the pollen of the male flower to the stigma of the female flower.

MERCURY

Chenopodium Bonus-Henricus

This perfectly hardy vegetable, known also by the name of Good King Henry, is much grown in Lincolnshire. The leaves are used in the same way as Spinach, and by earthing up the shoots they may be blanched as a substitute for Asparagus. Sow the seeds during April in drills twelve inches apart, and in due course thin the seedlings to one foot apart in the rows.

MUSHROOM

Agaricus campestris

The Mushroom has many friends among all classes, few benevolent neutrals, and fewer still who are absolutely hostile to it as an article of food. Those who find, or imagine they find, that this delicacy does not agree with them, might possibly arrive at another conclusion were a different mode of preparation adopted, or were the consumption of it accompanied with a full persuasion that the Mushroom is not merely delicious in flavour, but thoroughly wholesome, rich in flesh-forming constituents, and, for a vegetable, possessed of more than the average proportion of fat-formers and minerals. These facts have been clearly established by chemical analysis, and may dispose of timid misgivings, always supposing the true edible Mushroom, *Agaricus campestris*, to be in question.

Hitherto the artificial production of Mushrooms has never been equal to the demand. Notwithstanding the enormous quantities sent to Covent Garden by the growers around London, many tons are imported from France, although it is generally admitted that they are neither so fine nor so rich in flavour as those produced in this country. If, however, the large centres of population are inadequately supplied, the scarcity of Mushrooms is more keenly felt in the provinces, except, perhaps, in certain favoured districts, where, after a few warm days in autumn, an abundant crop may be gathered from the neighbouring pastures. Then there is a brave show in the greengrocers' windows for a brief period, followed by entire dearth for weeks, and perhaps months. Obviously, therefore, the demand, large as it already is, might be immensely augmented by a commensurate supply. Yet it is not only possible but quite easy to grow Mushrooms for the greater part of the year in very small gardens, even when such gardens are entirely destitute of the appliances usually considered necessary for the higher flights of horticulture. The idea that

Mushroom-growing is somewhat of a mystery, forbidden to all but the strictly initiated, has happily been dispelled. If we examine the conditions under which Mushrooms grow freely in pastures, it is surprising how few and simple are the elements of success. The crop generally appears in September, when temperature is genial and fairly equable, with sufficient but not superabundant moisture. The artificial production of Mushrooms in the garden needs only reliable spawn, a sweet fertile bed, and some means of maintaining a steady temperature under varying atmospheric conditions. When the principles of Mushroom culture are thoroughly mastered, they may be successfully applied in many different ways, and they render the practical work easy and tolerably certain.

The Spawn.—Although the Mushroom may be grown from seed, it is seldom done except for strictly scientific purposes. The seeds are, however, largely disseminated by Nature, and, having found a suitable home, they germinate and produce an underground growth which at a hasty glance resembles mildew. It really consists of white gossamer-like films, which increase in number and distinctness as they develop, until they push their way towards the surface, and give rise to the growth above ground of the Mushroom. It follows that if we do not begin the cultivation with seeds or spores, we must resort to the white films or 'mycelium,' that the growth of the plant may begin in Nature's own way below ground. What is called 'Mushroom Spawn' consists of certain materials from the stable and the field, mixed and prepared in such a manner as to favour the development of the mycelium of the Mushroom. When dried, the cakes have the appearance of an unburnt brick. The preparation of the spawn, though a very simple matter, demands the skill and care of experienced operators. If the work is not well done, the spawn will be of poor quality, and will yield a meagre crop, or perhaps fail to produce a single Mushroom. Whether the cakes or bricks are impregnated in the manner long practised in this country, or direct from the tissue of the Mushroom, the culture remains the same. Provided that the spawn is good, it has but to be broken into lumps of a suitable size, and inserted in the bed, to impregnate the entire mass with the necessary white films. These will take their time to collect from the soil the alkalies and phosphates of which Mushrooms principally consist, and this part of their work being done, the fruits of their labours will be displayed above ground in the elegant and sweet-smelling fungus that few human appetites can resist when it is placed upon the table in the way that it deserves. Experts can readily form an opinion as to whether a cake of Mushroom spawn is or is not in a fit state for planting, and it will be a safe proceeding for the amateur to buy from a Firm which has a large and constant sale; otherwise, spawn may be purchased which was originally well made and properly impregnated, but has lost its vitality through long keeping.

Soil.—As to soil, it is well known that in a favourable autumn Mushrooms abound in old rich pastures, and those who have command of turf cut from a field of this character have only to stack the sods grass side downwards for a year or two, and they will be in possession of first-class material for Mushroom beds either in the open or under cover. But small gardens, particularly in towns, have no such bank to honour their drafts, and for these it becomes a question of buying a load or two of turfy loam, or of making the soil of the garden answer, perhaps with a preliminary enrichment by artificial manure. In the general interests of the garden, the money for a limited quantity of good loam would probably be well spent, independently of the question of Mushrooms. No great bulk is necessary to cover a moderate-sized Mushroom bed, but the quality of the soil will certainly have an influence on the number and character of the Mushrooms. As a proof of the exhaustive nature of the fungus, it almost invariably happens that when the soil is used a second time it tends to diminish the size and lower the quality of the crop.

Manure.—In the management of the manure two essentials must be borne in mind. Not only is nourishment for the plant required, but warmth also. Probably a large proportion of the failures to grow Mushrooms might, if all the facts were known, be traced to some defect in the manure employed, or to some fault in its preparation. It must be rich in the properties which encourage and support the development of Mushrooms, absolutely free from the least objectionable odour, for the plant is most fastidious in its demand for sweetness, although it can dispense with light; and there must remain in the manure when made into a bed a sufficient reserve of fermentation to insure prolonged heat, no matter what the temperature of the atmosphere may be. Of course, the duration of the heat will depend very much on the care with which it is conserved by suitable covering and management. These requirements, formidable as they may seem, can be insured with extreme ease; indeed, the work is apparently far more difficult and complicated on paper than it proves to be in practice.

Preparation of the Bed.—The manure should come from stables occupied by horses in good health, fed exclusively on hard food. The most suitable store is the floor of a dry shed, or under some protection which will prevent the loss of vital forces. Ammonia, for example, is readily dissipated in the atmosphere or washed away by rain. The manure should neither be allowed to become dust dry, nor to waste its power in premature fermentation. Operations may be commenced with three or four loads. A smaller quantity increases the difficulty of maintaining the requisite temperature when fermentation begins to flag. The first procedure is to make the manure into a high oblong heap well trodden down. If the stuff be somewhat dry, a sprinkling of water over every layer will be necessary. In a few days fermentation will make the heap hot all through, and then it must be taken to pieces and remade, putting all the outside portions into the interior, with the object of insuring equal fermentation of the entire bulk. This process will have to be repeated several times at intervals of three or four days until the manure has not only been fermented but sweetened. When ready it will be of a dark colour, soft, damp enough to be cohesive under pressure, but not sufficiently damp to part with any of its moisture, and almost odourless; at all events the odour will not be objectionable, but may be suggestive of Mushrooms. Make a long bed, having a base about four feet wide, and sides sloping to a ridge like the roof of a house, with this difference—the narrow part of the ridge is useless, and the top should, therefore, be rounded off when about a foot across. Some growers prefer a circular bed of six or eight feet diameter at the bottom and tapering towards a point, after the shape of a military tent; but here again the point will be worthless, and the bed may terminate abruptly. Either the long bed or the round heap answers admirably. Tread the manure down compactly, and for the sake of appearances endeavour to finish it off in a workmanlike manner. During the first few days there will be a considerable rise in the temperature, which will gradually subside, and when the plunging thermometer shows that it has settled down to a comfortable condition of about 80° the bed must be spawned. Experienced men can determine by the sense of touch when the temperature is right, but the inexperienced should rely entirely on the thermometer. The question will arise as to the period of the year when operations should be commenced. Well, the experts who grow Mushrooms in the open ground for market gather crops almost the year round; but a beginner will do wisely to start under the most favourable natural conditions, and these will be found about midsummer, because the bed will commence bearing before winter creates difficulty as to temperature.

Spawning and After-management.—Break each cake of spawn into eight or ten pieces, and force every piece gently a little way into the manure at regular intervals of six to nine inches all over the bed, closing the manure over and round each piece of spawn. The practice of inserting spawn by means of the dibber is to be strongly condemned, for it leaves smooth, hollow spaces which arrest the mycelium; and very small pieces of spawn should be avoided because they generally result in small Mushrooms. Immediately the spawning is completed, a thick and even covering of clean straw or litter of some kind should be laid over the bed, secured from wind by canvas, mats, hurdles, or in some other way. From good spawn the films of mycelium will begin to extend within a week. In the contrary case an examination of the pieces will show that they have become darker than when put into the bed, which means that they have perished. Then the question will arise as to whether the bed or the spawn is at fault, and the former must either be spawned again or broken up. Supposing the spawn to show signs of vitality, the time has come for covering the bed with a layer of rather moist soil, pressed lightly but firmly on to the manure with the spade or fork, so that the earth will not slip down. At once restore the covering of litter, &c., and wait patiently for about seven or eight weeks for the crop. Meanwhile the plunging thermometer ought to be consulted daily. Until the Mushrooms appear the instrument should not indicate less than 60°, and while in bearing not less than 55°. Experience proves that the most violent alternations of temperature may be combated by regulating the thickness of the covering. Although it may possibly be necessary to resort to eighteen inches of litter or more during hard frost or the prevalence of a cutting east wind, a much thinner covering will suffice in milder weather.

Should the temperature of the bed, through inexperience in the management of it, sink below the point at which Mushrooms can grow, we advise the exercise of a little patience. We have known several instances of beds made in autumn producing no crop at the expected time, but which have borne fairly in the following spring or summer. But in the event of the first effort failing outright there is no great loss. The manure, which is the most costly item, will still be available for the garden, and an observant man will pretty well understand in what respect he must amend his course of procedure.

Water.—Moisture is of great consequence, for a dry Mushroom bed will soon be barren also; but whenever water is given it must be applied tepid and from a fine rose. To slop cold water over a Mushroom bed is about as reasonable a procedure as putting ice into hot soup. Water is best administered in the afternoon of a genial day, and should be sufficient to saturate the bed. Immediately it is done the covering of litter and canvas must be promptly restored to prevent the temperature from being seriously lowered by rapid evaporation. A couple of stakes driven from the crown to the bottom of the bed at the time of making up the heap are useful as indicators of moisture, and may occasionally be drawn out and examined.

In gathering the crop, only a small portion of the bed should be uncovered at a time. This should be the rule at all seasons, and the strict observance of it will prevent a mistake in cold weather, for then, if the bed is carelessly uncovered and much chilled, the crop will come to an end, when perhaps it would, if properly handled, be at high tide and full of profit. Another rule should be enforced, to this effect, that every Mushroom must be taken out complete, and if the root does not come with the stem, it must be dug out with a knife. Any trifling with this rule will prove a costly mistake. The stem of a Mushroom, if left in the ground, will produce nothing at all. But it may attract flies, and it certainly will interfere with the movements of the mycelium at that particular spot, and actually prevent the production of any more Mushrooms. The old practitioners were accustomed to leave the stem in the ground, and they were content with about one-third of the crop now produced on beds that are, perhaps, not better made than were theirs. But they had a notion about the powers of the root which increased knowledge of the subject has shown to be fallacious.

In Pastures.—As already indicated, Mushrooms are often to be found in abundance in well-stocked pastures during the late summer months, and where favourable conditions exist it is an excellent plan to insert pieces of spawn two inches deep in the turf in June and July.

Turf Pits.—The facility with which Mushrooms may be raised under simple methods is illustrated by the practice of growing them inside the turf walls of cool pits. In the country turf walls are common, and they offer the advantage of growing Mushrooms in addition to the purpose they usually serve. After determining the size of the pit, and accurately marking it on the ground, cut the turf into narrow strips, say three or four inches wide, and of exactly eighteen inches length. The strips should be closely laid, grass side downwards, across the width of the walls—not longitudinally—except at the corners, where the layers should cross each other. The front and back walls to be rather above the required height, because the turf always scales down a little, and the two ends must gradually rise from front to back. The top layer may be right side up, when it will keep green for a long time. As the work proceeds insert lumps of spawn at intervals in every layer, about three or four inches from the inside edge. A wooden frame will be requisite on the top to carry the glass lights. This structure makes a useful cool pit and a Mushroom bed from which supplies may sometimes be gathered for years. In the summer it will be necessary to keep the walls moist by means of the syringe, or they will cease bearing.

Indoor Beds.—Mushrooms may be grown almost anywhere, evenly in a cellar, or on the wall of a warm stable, provided only that the mode of procedure is in a reasonable degree adapted to the requirements of the fungus. Ordinary pits and frames are also serviceable, and many gardeners obtain good crops in autumn by the simple process of inserting a few lumps of spawn in a Cucumber or Melon bed while the plants are still in bearing. Between spawning and cropping a period of six or eight weeks usually elapses, so that if the plan just mentioned be adopted, the spawn should be introduced in the height of summer, both to insure it a warm bed and to allow time for the crop to mature before the season runs out. Sheds and outhouses not only afford shelter and space for beds on the floor, but the walls can be fitted with shelves on which Mushrooms may be plentifully grown. In all cases the shelves should be two feet apart vertically, and each shelf should have a ledge nine inches deep. The walls of a house may be quickly and cheaply fitted with woodwork for the purpose, but brick is so much better than wood that whenever it is possible to employ brick it should have the preference. As regards the ledges, they should be of stout planking in any case, and should not be fixed, because of the necessity for clearing the shelves and renewing the soil periodically. The details of cultivation are the same within doors as without, but the roof gives valuable protection, and helps to maintain the beds at a suitable temperature.

A proper Mushroom-house for production during winter should be heated with hot water, and have an opaque roof. There is nothing so good for the crop as a roof of thatch, but there are many objections to it, and usually slate is employed. A double roof will pay for its extra cost by promoting an equable temperature. A few side lights fitted with shutters are necessary, as there should be a good light for working purposes; but the crop does not need light, and a more steady temperature can be maintained in a dark house than in one which has several windows. The most convenient dimensions for a Mushroom-house are: length, twenty-five feet; width, twelve feet; height at sides, six feet, to allow of a bed on the floor, and a shelf four feet above it; the ridge rising sufficiently for head room, and to shoot off water. There will be room for a central path of four feet, and a bed of four feet on each side. An earth or tile floor and a slate or stone shelf will, with one four-inch flow and return pipe, complete the arrangements. The less wood and the less concrete the better; there is nothing like porous red tiles for the floor and stone for the shelves, with loose planks on edge to keep up the soil, a few uprights being sufficient to hold them in their places.

Temperatures at every point are of great importance. The bed should be near 80° when the spawn is inserted. The air temperature requisite to the rising crop is 60° to 65°, which is the usual temperature of the season when Mushrooms appear in pastures. While the bed is bearing a temperature of 55° will suffice, but at any point below this minimum production will be slow and may come to a stop. When giving water, take care that it is at a temperature rather above than below that of the bed.

MUSTARD

Sinapis alba, and S. nigra

Mustard is much valued as a pungent salad, and for mixing in the bowl it may take the place of Water Cress when the latter is not at command. Mustard is often sown with Cress, but it is bad practice, for the two plants do not grow at the same pace, and there is nothing gained by mixing them. The proper sort for salading is the common White Mustard, but Brown Mustard may be used for the purpose. Rape is employed for market work, but should be shunned in the garden. As the crop is cut in the seed leaf, it is necessary to sow often, but the frequency must be regulated by the demand. Supplies may be kept up through the winter by sowing in shallow boxes, which can be put into vineries, forcing pits, and other odd places. Boxes answer admirably, as they can be placed on the pipes if needful; they favour the complete cutting of a crop without remainders, and this is of importance in the case of a salad that runs out of use quickly and is so easily produced. From Lady Day to Michaelmas Mustard may be sown on the open border with other saladings, but as the summer advances a shady place must be found for it.

ONION

Allium Cepa

The onion has the good fortune to be generally appreciated and well grown almost everywhere. It enhances the flavour and digestibility of many important articles of food that would fail to nourish us without its aid, while to others it adds a zest that contributes alike to enjoyment and health. Although there are but few difficulties to be encountered in the cultivation of the Onion, there is a marked difference between a well-grown crop and one under poor management. There is, moreover, what may be termed a fine art department in Onion culture, one result being special exhibitions, in which handsome bulbs of great weight are brought forward in competition for the amusement and edification of the sight-seeing public. Thus, when the first principles have been mastered, there may be, for the earnest cultivator of this useful root, many more things to be learned, and that may be worth learning, alike for their interest and utility.

Treatment of Soil.—The Onion can be grown on any kind of soil, but poor land must be assisted by liberal manuring. A soil that will not produce large Onions may produce small ones, and the smallest are acceptable when no others are to be had. But for handsome bulbs and a heavy crop a deep rich loam of a somewhat light texture is required, although an adhesive loam, or even a clay, may be improved for the purpose; while on a sandy soil excellent results may be obtained by good management, especially in a wet season. In any case the soil must be well prepared by deep digging, breaking the lumps, and laying up in ridges to be disintegrated by the weather, and if needful its texture should be amended, as far as possible, at the same time. A coat of clay may be spread over a piece of sand, to be thoroughly incorporated with it; on the other hand, where the staple is clay, the addition of sand will be advantageous. All such corrective measures yield an adequate return if prudently carried out, because it is possible to grow Onions from year to year on the same ground; and thus in places where the soil is decidedly unsuitable a plot may be specially prepared for Onions, and if the first crop does not fully pay the cost, those that follow will do so. But the plant is not fastidious, and it is easy work almost anywhere to grow useful Onions. The first step in preparing land is to make it loose and fine throughout, and as far as possible to do this some time before the seed is sown. For sowing in spring, the beds should be prepared in the rough before winter, and when the time comes for levelling down and finishing, the top crust will be found well pulverised, and in a kindly state to receive the seed. Stagnant moisture is deadly to Onions, therefore swampy ground is most unfit; but a sufficient degree of dryness for a summer crop may often be secured by trenching, and leaving rather deep alleys between the beds to carry off surface water during heavy rains.

Manures.—As almost any soil will suit the Onion, so also will almost any kind of manure, provided that it be not rank or offensive. This strongly flavoured plant likes good but sweet living, and it is sheer folly to load the ground for it with coarse and stimulating manures. Yet it is often done, and the result is a stiff-necked generation of bulbs that refuse to ripen, or there may be complete failure of the crop through disease or plethora. But any fertiliser that is at hand, whether from the pigstye, or the sweepings of poultry yards or pigeon lofts, may be turned to account by the simple process of first making it into a compost with fresh soil, and then digging it in some time in advance of the season for sowing, and in reasonable but not excessive quantity. All such aids to plant growth as guano, charcoal, and well-rotted farmyard manure, may be used advantageously for the Onion crop; but there are two materials of especial value, and costing least of any, that are universally employed by large growers, both to help the growth and prevent maggot and canker. These are lime and soot, which are sown together when the ground is finally prepared for the seed, and in quantity only sufficient to colour the ground. They exercise a magical influence, and those who make money by growing Onions take care to employ them as a necessary part of their business routine.

Spring-sown Onions require to be put on rich, mellow ground, the top spit of which is of a somewhat fine texture, and at the time of sowing almost dry. Having been well dug and manured in good time, the top spit only should be dug over when it is finally made ready for the seed. The work must be done with care, and the beds should be marked off in breadths of four feet, with one-foot alleys between. Break all lumps with the spade, and work the surface to a regular and finely crumbled texture. Light soil should be trodden over to consolidate it, and then the surface may be carefully touched with the rake to prepare it for the seed. March and April are the usual months for spring sowing, although in mild districts seed is sometimes put in as early as January. Space the rows from nine to twelve inches apart, according to the character of the sort and the size of bulbs required. The drills must be drawn across the bed, at right angles to the alleys, for when drawn the other way it is difficult to keep the ground properly weeded. For a crop of Onions intended for storing, the seed should be only just covered with fine earth taken from the alleys and thrown over, after which the drills must be lightly trodden, the surface again touched over with the rake, and if the soil is dry and works nicely, the business may be finished by gently patting the bed all over with the back of the spade. If the ground is damp or heavy, this final touch may be omitted, as the Onion makes a weak grass that cannot easily push through earth that is caked over it. But speaking generally, an Onion bed newly sown should be quite smooth as if finished with a roller. To the beginner this will appear a protracted and complicated story, but the expert will attest that Onions require and will abundantly pay for special management.

As soon as possible after the crop is visible the ground between should be delicately chopped over with the hoe to check the weeds that will then be rising. Immediately the rows are defined a first thinning should be made with a small hoe, care being taken to leave a good plant on the ground. The next thinning will produce young Onions for saladings, and this kind of thinning may be continued by removing plants equally all over the bed to insure an even crop, the final distance for bulbing being about six inches. Keep the hoe at work, for if weeds are allowed to make way, the crop will be seriously injured. When Onions are doing well they lift themselves up and *sit* on the earth, needing light and air upon their bulbs to the very axis whence the roots diverge. If weeds spread amongst them the bulbs are robbed of air and light, and their keeping properties are impaired. But in the use of the hoe it is important not to loosen the ground or to draw any earth towards the bulbs. When all the thinning has been done, and the weeds are kept down, it will perhaps be observed that in places there are clusters of bulbs fighting for a place and rising out of the ground together as though enjoying the conflict. With almost any other kind of plant this crowding would bode mischief, but with Onions it is not so. Bulbs that grow in crowds and rise out of the ground will never be so large as those that have plenty of room, but they will be of excellent quality, and will keep better than any that have had ample space for high development. It is almost a pity to touch these accidental clusters, for the removal of a portion will perhaps loosen the ground, and so spoil the character of those that are left. Really fine Onions are rarely produced in loose ground, hence the necessity for care in the use of the hoe. Watering is not often needed, and we may go so far as to say that, in a general way, it is objectionable. But a long drought on light land may put the crop in jeopardy, unless watering is resorted to, in which case weak manure water will be beneficial. Still, watering must be discontinued in good time, or it will prevent the ripening of the bulbs, and if a sign is wanted the growth will afford it, for from the time the bulbs have attained to a reasonable size the water will do more harm than good.

The harvesting of the crop requires as much care as the growing of it. If all goes well, the bulbs will ripen naturally, and being drawn and dried on the ground for a few days with their roots looking southward, may be gathered up and topped and tailed or bunched as may be most convenient. But there may be a little hesitation of the plant in finishing growth, the result, perhaps, of cool moist weather, when dry hot weather would be better. In this case the growth may be checked by passing a rod (as the handle of a rake for example) over the bed to bend down the tops. After this the tops will turn yellow, and the necks will shrink, and advantage must be taken of fine weather to draw the Onions and lay them out to dry. A gravel path or a dry shed fully open to the sun will ripen them more completely than the bed on which they have been grown; but large breadths of Onions must be ripened where they grew, and experience teaches when they may be drawn with safety.

As to keeping Onions, any dry, cool, airy place will answer. But if a difficulty arises there is an easy way out of it, for Onions may be hung in bunches on an open wall under the shelter of the eaves of any building, and thus the outsides of barns and stables and cottages may be converted into Onion stores, leaving the inside free for things that are less able to take care of themselves. During severe frost they must be taken down and piled up anywhere in a safe place, but may be put on their hooks again when the weather softens, for a slight frost will not harm them in the least, and the wall will keep them comparatively warm and dry. When the best part of the crop has been bunched or roped, the remainder may be thrown into a heap in a cool dry shed, and a few mats put over them will prevent sprouting for at least three months. But damp will start them into growth, and the only way to save them then is to top and tail them again, and store as dry as possible in shallow baskets or boxes.

To grow large Onions the principles already explained must be carried into practice in a more intense degree. It will be necessary to devote extreme care to the preparation of the ground, and to give the plants more time to mature; much greater space must also be allowed than is usual for an ordinary crop. A good open position is imperative, and where the soil is sufficiently deep, trenching is desirable. Shallow soil ought to be thoroughly dug down to the last inch, and it will be an advantage to break up the subsoil by pickaxe and fork. Cover the subsoil with a thick layer of rotten manure before restoring the top soil. For light land farmyard manure is excellent, but stable manure is preferable for stiff cold soil. The usual time for trenching is October or November, leaving the surface rough for disintegration during winter. Nothing more need be done until the following March. Early in that month break the soil down to a fine tilth and make it quite firm by treading, or by rolling. Then broadcast over the plot a liberal dressing of ground lime and soot, using about three pounds of each per pole. Rake both in and leave the bed until the time arrives for planting out: this will depend on the weather.

Those who are accustomed to exhibit Onions at horticultural shows almost invariably sow very early in the year under glass and in due time transplant either from seed-pans or boxes. Of the two, properly prepared boxes are usually found most convenient. The dimensions are optional, but boxes about two feet long, one foot wide, and five inches deep answer admirably. Several holes are perforated in the bottom to insure efficient drainage. In every box place a thick layer of rotten manure and then fill with thoroughly rich soil firmly pressed down, leaving the surface quite smooth. One of the most successful growers sows seed in rather small boxes early in January, and about the middle of February the young Onions are pricked into boxes of the size we have named. Only the finest and most promising seedlings are used. When transferred, each Onion is allowed a space of three inches. The boxes are kept in a greenhouse, as near the glass as possible, in a temperature of about 50°. After sowing, very little water is given; but when transplanted, finish with a sprinkling from a fine rose. Every morning the plants will require spraying, but this must never be done at night or damping off may follow. All through their time in the greenhouse it is important to keep the boxes near the glass. Towards the end of March remove to cold frames, keeping the lights rather close for a few days, but gradually giving more air until the lights can be taken off for a short time daily.

In the south, about the middle of April is generally a suitable time for transplanting to open beds, but in the event of a cold east wind prevailing a brief delay is advisable and it is always an advantage to plant out on a dull day or in showery weather. Space the rows twelve to eighteen inches apart, and allow about fifteen inches between plants in the rows. In the actual work of transplanting take care to insert only the fibrous roots in the soil. To bury any portion of the stem results in thickened necks. Finish with a dusting of soot over the entire bed, including the Onions, and then well spray from a fine rose to settle the soil around the roots. Until the plants are established continue the spraying daily. After the middle of May renew the dusting of the bed with soot and repeat at fortnightly intervals. About the 20th of June feeding the Onions must commence. Peruvian guano and nitrate of soda are both excellent, but these powerful artificials need using with discretion, or the crop may be scorched instead of stimulated. It is often safer to employ them in liquid form than dry, and ten ounces of either, dissolved in ten gallons of water, will suffice for thirty square yards. Use the two articles alternately at intervals of ten days and cease at the end of July. If continued longer, some of the finest bulbs will split. The use of soot can, however, be regularly maintained. Should bulbs be required for autumn exhibition carefully lift them a week or ten days in advance of the show date. This has the effect of making the bulbs firm and reducing the size of the necks.

Supposing an attack of mildew to occur, a dusting of flowers of sulphur will prove effective if applied immediately the disease appears. Sulphide of potassium, one ounce to a gallon of water, is also a reliable remedy.

July and August Sowing.—During these months seed of the quick-growing types of Onion may be sown for producing an abundant supply of salading and small bulbs during the autumn and onwards. It is important to thin the plants early in order that those left standing in the rows may have every opportunity of developing rapidly.

Autumn-sown Onions, intended for use in the following summer, may also be sown in the same way as advised for spring sowing. The time of sowing is important, as the plants should be forward enough before winter to be useful, but not so forward as to be in danger of injury from severe frost. On well-drained ground all the sorts are hardy, and the finest types, which are so much prized as household and market Onions, may be sown in autumn as safely as any others. It may be well in most places to sow a small plot: in the latter part of July, and to make a large sowing of the best keeping sorts about the middle of August—say, for the far north the first of the month, and for the far south the very last day. Thin the plants in the rows and transplant the thinnings, if required, as soon as weather permits in February. In places where spring-sown Onions do not ripen in good time in consequence of cold wet weather, autumn sowing may prove advantageous, as the ripening will take place when the summer is at its best, and the crop may be taken off before the season breaks down.

Pickling Onions may be obtained by sowing any of the white or straw-coloured varieties that are grown for keeping, but the large sorts are quite unfit; the best are the Queen and Paris Silver-skin, as they are very white when pickled and are moderately mild in flavour. A piece of poor dry ground should be selected and made fine on the surface. Sow in the month of April thickly, but evenly, cover lightly, and roll or tread to give a firm seed-bed, and make a good finish. Be careful to keep down weeds, and do not thin the crop at all. If sown very shallow the bulbs will be round: if sown an inch deep they will be oval or pear-shaped.

The Potato or Underground Onion is not much grown in this country, in consequence of occasional losses of the crop in severe winters. In the South of England the rule as to growing it is to plant on the shortest day, and take up on the longest. It requires a rich, deep soil, and to be planted in rows twelve inches apart, the bulbs nine inches apart in the row. Some cultivators earth them up like Potatoes, but we prefer to let the bulbs rise into the light, even by the removal of the earth, so as to form a basin around each, taking care, of course, not to lay bare the roots in so doing. When the planted bulbs have put forth a good head of leaves, they form clusters of bulbs around them, and the best growth is made in full daylight, the bulbs sitting on and not in the soil.

The Onion Grub *(Phorbia cepetorum)* is often very troublesome to the crop, especially in its early stages, and its presence may be known by the grass becoming yellow and falling on the ground. It will then be found that the white portion, which should become the bulb, has been pierced to the centre by a fleshy, shining maggot, a quarter of an inch in length, this being the larva of an ashy-coloured, ill-looking, two-winged fly. Where this plague has acquired such a hold as to be a serious nuisance, care should be taken to clear out all the old store of Onions instantly upon a sufficiency of young Onions becoming available in spring, and to burn them without hesitation. If left to become garden waste in the usual way, these old Onions do much to perpetuate and augment the plague. A regular use of lime and soot will be found an effectual preventive. Other remedies are suggested in the article on Onion Fly, Page 420.

PARSLEY—*see* HERBS, *page* 68

PARSNIP

Pastinaca sativa

The Parsnip is one of the most profitable roots the earth produces. Probably its sweet flavour imposes a limit on its usefulness, but bad cooking doubtless has much to answer for, the people in our great towns being, in too many instances, quite ignorant of the proper mode of cooking this nourishing root. When cut in strips, slightly boiled and served up almost crisp, it is a poor article for human food; but when cooked whole in such a way as to appear on the table like a mass of marrow, it is at once a digestible dainty and a substantial food that the people might consume more largely than they do, to their advantage.

The Parsnip requires only one special condition for its welfare, and that is a piece of ground prepared for it by honest digging. Rich ground it does not need, but the crop will certainly be the finer from a deep fertile sandy loam than from a poor soil of any kind. But the one great point is to trench the ground in autumn and lay it up rough for the winter. Then at the very first opportunity in February or March it can be levelled down and the seed sown, and the task got out of hand before the rush of spring work comes on. A fine seed-bed should be prepared either in one large piece or in four-feet strips, as may best suit other arrangements. Sow in shallow drills eighteen inches apart, dropping the seeds from the hand in twos and threes at a distance of six inches apart; cover lightly, and touch over with the hoe or rake to make a neat finish. As soon as the plants are visible, ply the hoe to keep down weeds and thin the crop slightly to prevent crowding anywhere. The thinning should be carried on from time to time until the plants are a foot apart; or if the ground is strong and large roots are required, they may be allowed fifteen inches. Good-quality roots may be grown on the worst types of clay and on stony soils by boring holes and filling them in with fine earth, in the manner described for Beet and Carrot. The holes for Parsnip, however, should be rather larger and deeper, with more space allowed between. It may be well to lift some of the roots in November, a few spits of earth being removed first at one end or corner of the piece to facilitate removal without breaking the roots: these may be put aside for immediate use, but the general bulk of the crop should remain in the ground to be dug as wanted, because the Parsnip keeps better in the ground than out of it, and in the event of severe frost a coat of rough litter will suffice to prevent injury. Whatever remains over in the month of February should be lifted and trimmed up and stored in the coolest place that can be found, a coat of earth or sand being sufficient to protect the roots from the injurious action of the atmosphere.

GARDEN PEA

Pisum sativum

Thanks to the skill and enterprise of enthusiastic specialists, we have now the wrinkled as well as the round-seeded Peas for the earliest supply of this favourite vegetable. Not only can we commence the season with a dish possessing the true marrowfat flavour, but in the new maincrop varieties dwarf robust growth is combined with free-bearing qualities, while the size of both Peas and pods has been increased without in the smallest degree sacrificing flavour. On the contrary, there has been a distinct and welcome advance in all the special characteristics which have won for this vegetable its popular position, and so highly is the crop esteemed that it is usually regarded as a criterion by which the general management of a garden is judged.

As an article of food Peas are the most nutritious of all vegetables, rich in phosphates and alkalies, and the plant makes a heavy demand on the soil, constituting what is termed an exhausting crop. For this reason, and also because the time that elapses between sowing seed and gathering the produce is very brief, it is imperative that the land should be well prepared to enable the roots to ramify freely and rapidly collect the food required by the plant.

Treatment of Soil.—The soil for Peas must be rich, deep, and friable, and should contain a notable proportion of calcareous matter. Old gardens need to be refreshed with a dressing of lime occasionally, or of lime rubbish from destroyed buildings, to compensate for the consumption of calcareous matters by the various crops. For early Peas, a warm dry sandy soil is to be preferred; for late sorts, and especially for robust and productive varieties, a strong loam or a well-tilled clay answers admirably, and it is wise to select plots that were in the previous year occupied with Celery and other crops for which the land was freely manured and much knocked about. Heavy manuring is not needed for the earliest Peas, unless the soil is very poor, but for the late supplies it will always pay to trench the ground, and put a thick layer of rotten manure at the depth of the first spit, in which the roots can find abundant nutriment about the time when the pods are swelling. In all cases it is advisable not to enrich in any special manner the top crust for Peas. When the young plant finds the necessary supplies near at hand, the roots do not run freely but are actually in danger of being poisoned; but when the plant is fairly formed, and has entered upon the fruiting stage, the roots may ramify in rich soil to advantage. Hence the desirability of growing Peas in ground that was heavily manured and frequently stirred in the previous year, and of putting a coat of rotten manure between the two spits in trenching. As regards the last-named operation, it should be remarked that as Peas require a somewhat fine tilth, the top spit should be kept on the top where the second spit will prove lumpy, pasty, or otherwise unkind. In this case bastard trenching will be sufficient; but when the second spit may be brought up with safety, it should be done for the sake of a fresh soil and a deep friable bed. The use of wood ashes, well raked in immediately in advance of sowing, will prove highly beneficial to the crop, for the Pea is a potash-loving plant.

Method of Sowing.—It will always pay to sow in flat drills about six inches wide, but the V-shaped drill in which the seedlings are generally crowded injuriously is not satisfactory. Two inches apart each way is a useful distance for the seed, although more space may be given for the robust-growing maincrop and late varieties. It is wise policy, however, to sow liberally in case of losses through climatic conditions, birds or mice; and if necessary superfluous plants can always be withdrawn. The depth for the seed may vary from two to three inches: the minimum for heavy ground and the maximum for light land.

Early Crops (sown outdoors).—Early Peas are produced in many ways. The simplest consists in sowing one or more of the quick-growing round-seeded varieties in November, December, and January, on sloping sheltered borders expressly prepared for the purpose, and provided with reed hurdles to screen the plants from cutting winds. Where the assaults of mice are to be apprehended, it is an excellent plan to soak the seed in paraffin oil for twenty minutes, and then, having sown in drills only one inch deep, heap over the drill three inches of fine sand. If this cannot be done, sow in drills fully two inches deep, for shallow sowing will not promote earliness, but it is likely to promote weakness of the plant. It is not usual to grow any other crop with first-early Peas, but the rows must be far enough apart to prevent them from shading one another, and, if possible, let them run north and south, that they may have an equable enjoyment of sunshine. As soon as the plant is fairly out of the ground, dust carefully with soot, not enough to choke the tender leaves, but just sufficient to render them unpalatable to vermin. When they have made a growth of about three inches, put short brushwood to support and shelter them, deferring the taller sticks until they are required. Then fork the ground between, taking care not to go too near to the plant. Sticks must be provided in good time, lest the plant should be distressed, for not only do the sticks give needful support, but they afford much shelter, as is the case with the small brushwood supplied in the first instance.

On fairly warm soils the first opportunity should be taken to sow one of the early dwarf marrowfat varieties in the open ground. This may be in February or early March, but it will be useless to make the attempt until the ground is in a suitable condition. Sow in flat drills as already described, the distance from row to row depending upon future plans. If no intercropping is to be done, eighteen inches between the rows will generally suffice for dwarf-growing Peas, but many gardeners prefer to allow three feet and to take a crop of Spinach on the intervening space.

Early Crops (sown under glass.)—We now come to the modes of growing early Peas by the aid of glass. The surest and simplest method is to provide a sufficiency of grass turf cut from a short clean pasture or common. There is in this case a risk of wireworm and black bot; but if the turf is provided in good time and is laid up in the yard ready for use, it will be searched by the small birds and pretty well cleansed of the insect larvas that may have lurked in it when first removed. Lay the turves out in a frame, grass side downwards, and give them a soaking with water in which a very small quantity of salt has been dissolved. This will cause the remaining bots and slugs to wriggle out, and by means of a little patient labour they can be gathered and destroyed. In January or February sow the seed rather thickly in lines along the centre of each strip of turf, and cover with fine earth. By keeping the frame closed a more regular sprouting of the seed will be insured; but as soon as the plants rise, air must be given, and this part of the business needs to be regulated in accordance with the weather. All now depends on the cultivator, for, having a very large command of conditions, it may be said that he is removed somewhat from the sport of the elements, which wrecks many of our endeavours. There are now three points to be kept in mind. In the first place, a short stout slow-growing plant is wanted, for a tall lean fast-growing plant will at the end of the story refuse to furnish the dish of Peas aimed at. Give air and water judiciously, and protect from vermin and all other enemies. A little dry lime or soot may be dusted over the plants occasionally, but not sufficient to choke the leaves. All going well, plant out in the month of March or April, on ground prepared for the purpose, and laying the plant-bearing turves in strips, without any disturbance whatever of the roots. Then earth them up with fine stuff from between the rows, and put sticks to support and shelter them.

A more troublesome, but often a safer method, is to raise plants in pots, or in boxes about four and a half inches deep and pierced at the bottom to insure free drainage. Old potting soil will answer admirably, and the seeds should be put in one inch deep and two inches apart. Place the pots or boxes in any light cool structure as near the roof-glass as possible, but make no attempt to force either germination or the growth of the plants. When fair weather permits, transfer to the open in March or April. A good succession may be obtained by sowing a first-early dwarf variety and a second-early kind simultaneously.

Main crops require plenty of room, and that is really the chief point in growing them. Supposing the ground has been well prepared as already advised, the next matter of importance is the distance between the rows. The market gardener is usually under some kind of compulsion to sow Peas in solid pieces, just far enough apart for fair growth, and to leave them to sprawl instead of being staked, because of the cost of the proceeding. But the garden that supplies a household is not subject to the severe conditions of competition, and Peas may be said to go to the dinner table at retail and not at wholesale price. Moreover, high quality is of importance, and here the domestic as distinguished from the commercial gardener has an immense advantage, for well-grown 'Garden Peas' surpass in beauty and flavour the best market samples procurable. To produce these fine Peas there must be plenty of space allowed between the rows, and it will be found good practice to grow Peas and early Potatoes on the same plot, and to put short sticks to the Peas as soon as they are forward enough. By this management the first top-growth of the Potatoes may be saved from late May frosts, and the Peas will give double the crop of a crowded plantation. The general sowings of Peas are made from March to June, but as regards the precise time, seasons and climates must be considered. Nothing is gained by sowing maincrop Peas so early as to subject the plant to a conflict with frost. It should be understood that the finest sorts of Peas are somewhat tender in constitution, and the wrinkled sorts are more tender than the round. Hence, in any case, the wrinkled seeds should be sown rather more thickly than the round to allow for losses; but robust-habited Peas should never be sown so thickly as the early sorts, for every plant needs room to branch and spread, and gather sunshine by means of its leaves for the ultimate production of superb Green Peas.

Late Crops.—To obtain Peas late in the season sowings may be made in June and July, and preference should be given to quick-growing early varieties. Ground from which early crops of Cauliflower, Carrot, Cabbage, Potatoes, &c., have been removed is excellent for the purpose. In dry weather thoroughly saturate the trench with water before sowing, and keep the seedlings as cool as possible by screening them from the sun.

Staking.—This important operation must not be unduly deferred, as the plants are never wholly satisfactory when once the stems have become bent. Commence by carefully earthing up the rows as soon as the plants are about three inches high. In the case of early varieties, light bushy sticks of the required height, thinly placed on both sides of the row, will suffice. Maincrop and late Peas, however, should first be staked with bushy twigs about eighteen inches high, these to be supplemented with sticks at least one foot taller than the variety apparently needs, as most Peas exceed their recognised height in the event of a wet season. No attempt should be made to construct an impenetrable fence, for Peas need abundance of light and air. Neither should the stakes be arched at the top, but placed leaning outwards.

General Cultivation.—On the first appearance of the plant, a slight dusting of lime or soot will render the rising buds distasteful to slugs and sparrows, but this is more needful for the early than the later crops. When maincrop Peas have grown two or three inches, they are pretty safe against the small marauders. As the plant develops, frequently stir the ground between the rows to keep down weeds and check evaporation. The earthing up of the rows affords valuable protection to the roots of the plants, and a light mulch of thoroughly decayed manure will prove very helpful in a dry season. In the event of prolonged dry weather, however, measures must be taken to supply water in good time and in liberal quantity. The advantage of deep digging and manuring between the two spits will now be discovered, for Peas thus circumstanced will pass through the trial, even if not aided by water, although much better with it; whereas similar sorts, in poor shallow ground, will soon become hopelessly mildewed, and not even water will save them. In giving water, it will be well to open a shallow trench, distant about a foot from the rows on the shady side, and in this pour the water so as to fill the trench; by this method water and labour will be economised, and the plant will have the full benefit of the operation.

The enemies Of Peas are fewer in number than might be expected in the case of so nutritive a plant. Against the weevil, the moth, and the fly, we are comparatively powerless, and perhaps the safest course is occasionally to dust the plants with lime or soot, in which case the work must be carefully done, or the leaf growth will be checked, to the injury of the crop. Light dustings will suffice to render the plant unpalatable without interfering with its health, but a heavy careless hand will do more harm than all the insects by loading the leafage with obnoxious matter. The great enemy of the Pea crop is the sparrow, whose depredations begin with the appearance of the plant, and are renewed from the moment when the pods contain something worth having. Other small birds haunt the ground, but the sparrow is the leader of the gang. Ordinary frighteners used in the ordinary way are of little use; the best are lines, to which at intervals white feathers, or strips of white paper, or pieces of bright tin are attached. In the seedling stage the plants may be protected by wire guards, and even strands of black thread tied to short stakes will prove serviceable. We have found the surest way to guard the crop against feathered plunderers is to have work in hand on the plot, so as to keep up a constant bustle, and this shows the wisdom of putting the rows at such a distance as will allow the formation of Celery trenches between them. We want a crop to come off, and another to be put on while the Peas are in bearing; and early Potatoes, to be followed by Celery, may be suggested as a rotation suitable in many instances. Even then the birds will have a good time of it in the morning, unless the workmen are on the ground early. However, on this delicate point, the 'early bird' that carries a spade will have an advantage, because the sparrow is really a late riser, and does not begin business until other birds have had breakfast, and have finished at least one musical performance.

Early Peas under Glass.—So greatly esteemed are Peas at table that in many establishments the demand for them is not limited to supplies obtainable from the open ground. Sowings may be made from mid-November to mid-February, according to requirements and the extent of accommodation available, from which the crops may be expected to mature from mid-March onwards. Where a large glass-house, such as is used for Tomatoes, &c., is at command, early Peas may be grown without prejudice to other crops. Assuming that a good depth of soil exists, thoroughly trench and prepare it as for outdoor Peas. Select a tall-growing variety, of which there are a number that do well under glass. Sow in a triple row, placing the seeds about three inches apart each way, and in due course support the plants with stakes. A cool greenhouse or a frame will also carry through an early crop of Peas, but for these structures pots should be used and only dwarf-growing varieties sown. A ten-inch pot will accommodate about eight seeds, and these should be planted one and a half inches deep. When a few inches high insert a few bushy stakes to carry the plants. A compost consisting of two parts loam, one part leaf-soil or well-decayed manure, with a small quantity of wood ashes, will suit Peas admirably. At no time is a forcing temperature needed. From 50° to 55° at night, with a rise of about 10° by day will suffice, and free ventilation must be given whenever possible with safety. Apply water carefully, but never allow the roots to become dust-dry.

Peas for Exhibition.—On the exhibition table handsome well-grown Peas always elicit unstinted admiration, and the magnificent pods of the newer varieties are certainly worthy of the utmost praise bestowed upon them. In all cases where vegetables are grown for competition at Shows the amount of success achieved depends largely on the intensity of the cultivation adopted, and in this respect no other subject will respond more readily to liberal treatment than will the Garden Pea. Deep digging, generous manuring, and copious watering during dry weather, in the manner already described, are fundamental essentials. Another matter of no less importance is the selection of suitable varieties. It is now the general custom to start the early sorts in pots or boxes under glass (see page 104), and some growers treat mid-season Peas in the same manner. Of this system it may be said that it offers the fullest opportunity of giving attention to the young plants and allows of the strongest specimens being selected for transfer to open quarters. The number of sowings will, of course, depend on individual requirements. At the time of transplanting give each plant plenty of space for development, and it will be well to stake the rows immediately. Keep the plants under constant observation, especially while quite young, when they are liable to destruction by garden foes. The flowering should be limited to the fourth spike, and from the time the pods appear assistance must be given in the form of liquid manure or a mulching of well-rotted dung. Remove all lateral shoots and promote vigorous healthy growth at every stage. Some means should be adopted to prevent injury of any kind to the pods, which when gathered should be well filled, carrying a fine bloom free from blemish.

POTATO

Solarium tuberosum

The potato has been designated the 'King of the Kitchen Garden,' and perhaps 'the noble tuber' should be so regarded. Of its importance as an article of food it is impossible to speak too highly, and the dietetic value of the Potato appears to be always advancing. The known deficiency of flesh-forming constituents naturally associates this vegetable with meat of various kinds, poultry, game and fish, and in this proper association the root is probably capable of superseding all other vegetable foods, bread alone excepted. It is far from our intention to recommend abstention from Asparagus, Cauliflower, Peas, and Sea Kale, and to regard Potatoes as a sufficient substitute for these and other table delicacies; but it is well to remember that by virtue of its starchy compounds the Potato has a direct tendency to promote health and that freshness of complexion that generally prevails among well-fed people.

Forcing Potatoes.—The demand for new Potatoes exists long before the first of the outdoor crops grown in this country can be lifted. To meet such a demand is not a difficult matter where the necessary amount of glass is at command, and by adopting the method here given supplies may be maintained through the winter and onwards until the first-earlies from the open ground are available. It may be said at once that for culture in pots and boxes under glass a high temperature is neither requisite nor desirable. Sturdy healthy growth is essential to the formation of a crop of tubers, and if the plants be forced into an attenuated condition the labour will have been in vain. Another matter which needs to be specially mentioned is the choice of suitable varieties. Only dwarf-growing kinds, thoroughly adapted for forcing, should be considered. The date of planting will necessarily be regulated by the time at which the crop is required. But a few weeks in advance of planting, the sets should be sprouted by placing them on end in shallow boxes, packed with damp light soil and stood near the light in a slightly warm pit or house. When the sprouts are formed rub off all but the two strongest. Good turfy loam, a small quantity of manure from a spent Mushroom bed, and a little bone meal, will make an excellent compost for the pots or boxes. Two sets will suffice for a ten-inch or twelve-inch pot, or five tubers may be placed in a box measuring about four feet long by one foot wide. Perfect drainage must be insured.

Plant the sets with care, taking up as much soil as possible with the mass of fibrous roots which will have formed during the period of sprouting. The operation may best be accomplished by only half filling the pots or boxes at first, and when the sets are in position add a further two inches or so of soil. Water sparingly, especially at the outset. As root growth increases add more soil and give the plants an occasional application of tepid liquid manure. At all times avoid excessive heat, and if the crop can be finished off gradually in a cool house so much the better.

Where sufficient accommodation cannot be found for forcing Potatoes in pots or boxes, an excellent crop may be grown on a gentle hot-bed made up in the usual manner, and covered to a depth of at least nine inches with a compost of three parts light loamy soil to one part leaf-mould. After putting on the frame, keep the lights closed for a few days. But a great heat is not wanted, and undue forcing at any stage will lead to disaster. Partially exhausted hot-beds which have been used for other purposes will also be found to answer admirably. Prepare the sets in the manner already advised for pots and boxes, and plant them with the least possible disturbance to the fibrous roots, three inches deep, in rows fifteen inches apart, allowing twelve inches between the tubers in the row. Whenever the weather is fine afford the plants a little air. Increase the amount gradually as growth develops, but close the frames early in the afternoon and give them the protection of mats at night should the outside temperature be low. Water must be given in moderation. It should always be of the same temperature as the frame, and as soon as the haulm commences to turn yellow watering must be discontinued. Little earthing up is needed, but when the foliage is about nine inches high the addition of a small quantity of warm soil along the rows will be beneficial.

Early Potatoes outdoors are produced in various ways, and by very simple appliances. The Potato will not bear the slightest touch of frost. It is a sub-tropical plant, and will endure considerable heat if at the same time it can enjoy light, air, and sufficient moisture. In some respects it may be likened to the Lettuce, for if crowded or overheated, or subjected to sudden checks, it bolts—in other words, it produces plenty of top and no bottom, just as Lettuces similarly treated produce flowering stems and no hearts. We will here propose a very simple and practical procedure for obtaining a nice crop of Potatoes in the month of June. This system fairly mastered, endless modifications will be easily effected as circumstances and judgment may suggest.

Begin by selecting an early variety of the best quality. Some time towards the end of January the sets are packed closely in shallow boxes, one layer deep only, and these are placed in full daylight safe from frost, but are not subjected to heat in any way. Having started the sets into growth in full daylight, proceed with the preparation of the ground. This must be light, warm, dry and rather rich without being rank. If a length of wall is available, and perplexity arises concerning suitable soil for the early Potatoes, seize all the sandy loam that has been turned out of pots, and having mixed it with as much leaf-mould and quite rotten manure as can be spared, lay the mixture in a ridge at the foot of the wall. As walls do not anywhere run in such lengths as to provide for all the early Potatoes that are wanted, select a plot of ground lying warm and dry to the sun, and having spread over it a liberal allowance of decayed manure, and any light fertilising stuff, such as the red and black residue from the burning of hedge clippings, turf, and weeds, dig this in. The ground being ready, it is lined out in neat ridges two feet apart, running north and south. These ridges must be shallow, rising not more than six inches above the general level. On every fourth ridge sow early Peas that are not likely to grow more than two and a half to three feet in height. This being done in February, the land is ready for Potatoes in the first week of March. Plant on the fine stuff laid up next the wall in the first instance, and then on the ridges, where there is room for three rows of Potatoes between every two rows of Peas. In the process of planting, it will be advisable to rub off all the weak eyes and thin out those on the crown, two or three strong eyes being quite sufficient. This can easily be accomplished as the sets are laid into their places in a shallow drill opened on the top of the ridge. The sets may be put a foot apart, and have four inches of fine soil over them. Prick the ground over with a fork between the rows, leaving it quite rough, but regular and workmanlike. The Peas will soon be visible and require attention. Draw a little fine earth to them, and stake them carefully with small brushwood. If snails and slugs appear, give dustings of lime or soot, and as soon as possible supply stakes of sufficient height and strength to carry the crop. By the time the Potatoes begin to show their shaws the Peas will constitute an effectual shelter for them against east winds, and it will be found that the morning frosts that are often so injurious to Potatoes in the month of May will scarcely touch a crop that has the advantage of this kind of protection. But to that alone it is not wise to trust. One serious freezing that blackens the shaws will delay and diminish the Potato crop. Therefore, as the green tops appear, cover them lightly with fine earth from between the rows, and if necessary repeat this, always allowing the leaves to see daylight. When a sharp frost occurs, it will be advisable to cover the tops with a few inches of light dry litter in just the same way that a bed of Radishes is protected. There are many other methods of saving the rising shaws. A plank on edge on the east side of a row will suffice to tide through an ordinary white frost. Mats or reed hurdles laid on a few stout pegs will also answer admirably, but care must be taken that the plant is not pressed down, and the covering must be removed as soon as the danger is over.

Crops grown under walls will be ready first, and those in the beds will follow. Spaces between the trees of a fruit wall may be planted with Potatoes, without injury to the trees. Those grown on the south face of a good wall will be ready for table three weeks in advance of the earliest crops in the open quarters. But east and west walls may be made to contribute, and even north walls are useful, if planted a week later and a little deeper. In all cases the sets should be put close to the wall to enjoy the warmth, and dryness, and shelter it affords. When the crop is lifted, the soil specially laid up for it may be taken away, or scattered over the border. But the bulk will be so slight that it will not matter much what becomes of it. However, in a new place with a clay soil it may be prudent to remove it, and keep it ready as an aid in seed sowing, for there are times and places where a little fine stuff is worth a great deal to give a crop of some kind a proper start.

The main crop, as the source of supply for fully nine months out of twelve, deserves every attention. Potatoes are grown with advantage on so many diverse soils, and in such unlikely climates, that the plant appears, on a casual consideration, to be altogether indifferent to its surroundings. But it is none the less true that for the profitable cultivation of this crop certain conditions are absolutely essential. Among these an open situation and a well-drained soil are perhaps the most important. To this might be added favourable weather, because a bad season frustrates every hope and labour. Having an open situation and a well-drained soil, it is much to be preferred that the soil be of a deep, friable, loamy nature; in other words, a good medium soil, suitable for deep tillage, but neither a decided clay, chalk nor sand. A fertile sandy loam, lying well as regards sunshine and drainage, may generally be considered a first-rate Potato soil, and excellent crops have also been grown on thin soils overlying chalk and limestone. So again, fine crops are often taken from poor sandy soils, and from newly-broken bog and moss, as well as from clay lands that have had some amount of tillage to form a friable top crust. But when all is said the fact remains that the ideal soil for Potatoes is a deep mellow loam, and, failing this, preference should be given to calcareous and sandy soils rather than to clays or retentive soils of any kind.

Manures.—Much prejudice prevails against manuring land for Potatoes, and where the soil is good enough to yield a paying crop, it will be prudent to do without manure, and to dress generously for the next crop to restore the land to a reasonable state. Still it is the practice of many of the most successful growers for the early market to manure for this crop, and in some instances the manure is laid in the trenches at the time of planting. Generally speaking, land intended for Potatoes should be deeply dug, and, if needful, manured in the autumn. About twenty to thirty cartloads of half-rotten manure per acre may be dug or ploughed in to as great a depth as possible, consistent with the nature of the subsoil and the appliances at command. In breaking up pasture with the spade, bastard trenching will as a rule prove advantageous. The land is lined off in two-feet breadths, and the top spit of the first piece is removed to the last piece, which will often be close at hand by the rule of working a certain distance down and back again. The under spit will then be well broken up, the manure thrown in, and the top spit of the next piece will be turned in turf downwards, making a sandwich of the manure. If this is done in autumn, there will be a mellow top crust produced by the spring, and the best way to plant will be in trenches, unless the land is very light, in which case the dibber may be used.

As light lands are often profitably devoted to Potato culture, and more especially to the production of first-class early Potatoes for the markets, a few words on their management may be useful here. If on the light land there is a choice of aspects, by all means select the plots that slope to the south-west; the dangerous aspects are north and east. The ground should be ploughed up in autumn and left rough, but it is not economical to manure light lands in autumn. At the time of planting, the furrows should be cut with a plough fitted with a double mould-board, and the manure spread evenly along them previous to laying in the sets. A good dressing per acre will consist of fifteen loads of farmyard manure, and four cwt. of artificials, consisting of one and a half cwt. of guano, two cwt. of superphosphate of lime, and half a cwt. of muriate of potash. When the sets are laid, cover them by splitting the ridges with the plough. If planted early in March, the crop should come off in time for Turnips, for which the land will be in good heart, and the seed should be sown as quickly as possible after the clearing of the Potatoes.

Preparing the Sets.—Among the many subjects that open out before us at this point are the selection and preparation of the sets. Why are smallish tubers chosen in one case and planted whole? and why, in another case, are large tubers chosen and divided before planting, to make two or more sets of each? Because there is a principle on which sound practice rests, and it is this: the number of shoots starting from any one growing point must be limited, for if they become crowded the crop will be less than the land is capable of producing. Keeping this principle in view, we proceed to remark, in the first place, that carefully selected seed of moderate size may be planted as it comes from the store without any preparation whatever, and with a fair prospect of a profitable result. But certain varieties produce few tubers of seed size, and when large they must be divided in such a manner as to insure at least two eyes in each set. As a matter of fact, profitable crops are grown in the most simple way; the seed is neither sprouted nor disbudded, and with a well-made soil and a favourable season, the return is ample, and all claims are satisfied. Potato-growing entails much labour, therefore it is important to distinguish between tasks that are necessary and those that are optional.

But where the time and strength can be found for first-class cultivation, it should have the preference over the rough and ready methods that are satisfactory on a large scale. Exhibitions of Potatoes are for the most part sustained by persons who can find the time to do things with extra care, and they have their reward in their crops as well as in their prizes, for what may be styled Exhibition culture consists simply in growing the crop in the best possible way, and planting many sorts where in any other case a few would suffice. Here, then, on the best plan, we begin with sets most carefully selected, to insure true typical form and colour, and these are, some six weeks or so before planting time, put in shallow boxes or baskets, one layer deep, to sprout in full daylight, but quite safe from frost. In the first instance a number of sprouts appear, and a large proportion are rubbed off. The object of the cultivator is to secure two or three stout, short shoots of a green or purple colour; the long white threads that are often produced in the store being regarded as useless. When large sets are employed, they are allowed to make three or four stout shoots, and at the time of planting—not before—these sets are cut so as to leave to each large piece only one or two good sprouts or sprits. As for the smaller sets that are not to be divided, it is common practice to cut a small piece off each of these at the time of planting to facilitate the decay of the tuber when it has accomplished its work, for having nourished the first growth the sooner it disappears the better. Thus, with a little extra trouble, sound tubers have been prepared for planting, and the main reasons for taking this extra trouble are doubtless fully apparent. The best seed possible is wanted and the most suitable soil; these two items forming the first chapter. By sprouting the seed time is gained, which is equivalent to a lengthening of the season. By limiting the number of shoots an excess of foliage is prevented. Where the shoots are crowded the tubers will not be crowded, a few strong shaws with all their leaves exposed to the air and light being capable of producing better results than a large number contending for air and light that are insufficient for them all. And finally, by cutting the sets, whether to divide them, or simply to hasten their decay, we insure that they will not reappear with the young crop as useless, ugly things.

Distances for Planting.—The distance at which the sets are planted is of importance, for a crop too crowded will be of little value. But the ground must be properly filled. By wasting only a small space in each breadth, or in the spaces between the sets, the total crop will be many bushels short of the possible quantity. The guiding principle must be to allow to each plant ample room to spread, and absorb the air and sunshine, in accordance with the character of the sort and the condition of the soil. A considerable proportion of the losses from disease may be traced to overcrowding in the first instance; the tangled haulm being rendered weak through want of air, and then becoming loaded with water, and in contact with wet ground, the disease has made havoc where, had the management been founded on sound principles, there might have been a vigorous healthy growth. If a doubt arises, it is safer to allow too much rather than too little space, and in this respect the exhibition growers are very liberal. They often place the rows of strong-growing varieties four feet or more apart, and allow a space of three and a half feet for the more moderate growers. Even then, with good land, in a high state of preparation, the shaws sometimes meet across the rows, and enormous crops are lifted. For a very comprehensive rule, it may be said that the distance between the rows may vary from fifteen inches for the early sorts of dwarf growth, to forty inches for the vigorous-growing late sorts. Between these measurements, for varieties producing medium haulm, a distance of twenty-six to thirty-six inches may be allowed on good ground. The distance between the sets must in like manner be determined by the growth, and will range from nine inches for crops to be dug early, to sixteen or twenty inches for the robust kinds. The medium maincrop Potatoes will generally do well at twelve inches apart. Much, however, depends on the season, for when great space is allowed, and the season proves warm and showery, there will be more large tubers than the grower will care for; whereas, if planted somewhat closer, the crop would be smaller and more uniform in size. When planted, the tops of the tubers should be about four inches below the surface.

Time of Planting.—Under favourable conditions, it is possible to plant on a warm dry border as early as mid-February in very sheltered districts, but a supply of protecting material must be instantly available in the event of severe weather. As a rule, however, the opening of March is soon enough to plant early crops out of doors, always provided that the soil is light and the situation warm, but where these conditions do not exist it will be safer to wait until the middle of the month. Maincrops may be got in at the end of March and during April, according to the locality and the character of the soil. In any case, it is better to defer the operation for a week or so than to plant in heavy wet ground which quickly consolidates, making it impervious to air and unsuitable for root-penetration. Excellent crops may also be obtained by planting in July, preference being given to quick-growing early varieties. Old tubers only should be used and these must be carefully stored until required for planting.

Method of Planting.—On light soils, in a sufficiently dry condition, the dibber or planting stick may be used, but on heavy ground it is not satisfactory. A good method of planting for all classes of soil is to draw out a V-shaped drill of the requisite depth, place the sets into position and lightly return the earth. Another plan which is largely adopted is to insert the sets in the trenches as made during the operation of digging the ground in spring, a garden line being used to obtain the accurate alignment of the rows.

General Cultivation.—As soon as the shaws appear the ground should be hoed between the rows, and if there is any fear of frost the shaws should be lightly moulded over. As the growth advances the crop must be earthed up, care being exercised not to earth up too much, for, taking six inches as the best average depth, the crop will be diminished by an increase beyond this depth. One urgent reason for early work between the rows is that a prosperous crop will soon put a stop to it. The moment it becomes likely that the shaws will be bruised by traffic between the rows they must be left to finish their course in their own way, because the formation of tubers below will be in the ratio of the healthy growth above ground. The Potato may be said to be manufactured out of sunshine and alkaline salts. The green leaves constitute the machinery of the manufacture, for which the solar light from above, and the potash, phosphate of lime, phosphate of magnesia, and phosphoric acid from below are the raw materials.

Change of Ground and Seed.—In common with all other crops, the Potato needs as often as possible a fresh soil, and a renewal of seed from some distant source. The need for a change of soil is made apparent by an analysis of the root, which contains large proportions of potash, phosphorus, and sulphur, with smaller proportions of magnesia and lime, without which the plant cannot prosper. A succession of heavy crops of Potatoes on the same land may be said to take from the soil its available potash and phosphates, and this crop will not, like some others, take soda instead of potash when the last-named alkali runs short. Here then is a chemical reason for change of soil. Another reason is found in the history of the species of fungi that prey on the Potato when its growth is checked by heavy rains and a low temperature. These leave their spores in the soil, like wolves hiding in ambush, to destroy the next crop. They are powerless to attack any other crop; therefore a suitable rotation gives them time to die out and leave the land clean as regards the *Phytophthora* and other parasites that destroy Potato crops. The necessity for an occasional change of seed rests on old experience, and should scarcely need enforcing. One word may be said here by way of explanation, and it is this: the seed house that aims to put a good article in the market adopts measures which altogether differ from those followed by the majority of persons who have not been trained to the business. It is a common experience to find that those who save their own seed from year to year have as a result a constantly declining strain, so that every year the growth is weaker, less true, and less profitable. It is so all through, but is especially the case with Potatoes. We do not say that all who save their own seed act unwisely, for some are most expert in the business. But we do say that seed saving is not learned in a day, and many who think they save shillings when they save seeds, actually lose pounds by burdening themselves with a bad article. The art of 'roguing'—the elimination of plants which are untrue to type—is but one part of the seed-saving process. There is the proper storing, the selecting and sorting operations, to which eyes and hands must be trained, and there must be no scruple about the sacrifice of false, immature or diseased samples. The point we have in view is to advise the Potato grower to be sure of his seed, and when a doubt arises as to the purity and healthiness of the sample at command, it may be remembered that the seed merchant practises methods of purgation for insuring perfectly true stocks, while by growing in many different districts, and on diverse soils, he can furnish an admirable change of seed for any description of land.

The Potato Disease.—The culture of Potatoes cannot be dismissed without allusion to the destructive fungus which is never absent in dry seasons, and in wet summers does its deadly work on a vast scale. Scientific men have acquainted us with the history of the Potato fungus, and this may eventually result in as efficient a remedy as that which renewed the vineyards of France. Such a remedy for the Potato murrain has yet to be discovered. Meanwhile, we must continue to resist the foe with the plough, spade, draining tool, and above all with a wise selection of sorts. It is an acknowledged fact that many Potatoes that have been cultivated for a long time appear to have lost their vigour, and are liable to succumb to the disease; but several kinds that have been raised from seed in recent years possess a constitution which almost defies the virulent assaults of the *Phytophthora infestans.* Since the introduction of Sutton's Magnum Bonum Potato there has been a disposition to believe in 'Disease-proof Potatoes.' There is no such thing absolutely, and perhaps there never will be, any more than there is a disease-proof wheat, or dog, or horse, or man. But some varieties of Potatoes are known to be more susceptible to the ravages of disease than others, and it has been one of our aims to secure seedlings which combine the highest cropping and table qualities with the least tendency to succumb in seasons when conditions favour the spread of the fungus. Scientific men have not yet explained why the varieties differ in this respect, but practical men have discovered that initial vigour of growth is the main defence against the plague, and as the growing of a good Potato costs no more than the growing of a poor variety, the cultivator should bestow his care on the very best he can obtain. A little extra cost for seed in the first instance is as nothing to the multiplied chances of success a good variety carries with it. To sum up this subject, then, we say that disease may be avoided in the early crops by cultivating sorts which may be lifted before the plague generally appears; and on soils which will not produce an early crop, only such varieties should be grown for the main crops as have been proved to be most capable of standing uninjured until late in the season. Let there be a dry, warm bed, sufficient food, the fullest exposure to the life-giving powers of light, and conditions favourable to early ripening.

The Wart Disease (Black Scab) of Potatoes *(Synchytrium endobioticum*, Percival) is dealt with in the chapter on 'The Fungus Pests of certain Garden Plants.'

PUMPKIN—*see* GOURD, *page 63*

RADISH

Raphanus sativus

The Radish is often badly grown through being sown too thickly, or on lumpy ground, or in places not favourable to quick vegetation. Radishes grown slowly become tough, pungent and worthless. On the other hand, those which are grown quickly on rich, mellow ground are attractive in appearance, delicate in flavour, and as digestible as any salad in common use. It should be understood that earliness is of the very first importance, and that large Radishes are never wanted. To insure a quick growth and a handsome sample the ground must not only be good, but finely broken up.

Frame Culture.—For the earliest crops it is advisable to make a semi-hot-bed, by removing a portion of the surface soil, and laying down about two-feet depth of half-rotten stable manure, on which spread four inches of fine earth, and then cover with frames. Sow the seed thinly, and put on the lights. When the plants appear, give air at every opportunity to keep the growth dwarf, and cover with mats during frost, always taking care to uncover as often as possible to give light, for if the tops are drawn the roots will be of little account. Where the plants are crowded, thin them, allowing every plant just room enough to spread out its top without overlapping its neighbour. Sowings made in this way in December, January, and February will supply an abundance of beautiful Radishes in early spring, when they are greatly valued. To follow the outdoor crops frame culture will again be necessary in autumn.

Outdoor Culture.—The second crop (which in many gardens will be the first) may be sown on warm, dry borders in February. Within a few days after sowing, collect a quantity of dry litter, and lay it up in a shed ready for use. It happens often that we have warm, bright weather in February, and the Radishes start quickly and make good progress, and then may come a severe frost, when the litter must be spread as lightly as possible, three or four inches thick. These open-ground sowings will bear cold well, but they should not be allowed to get frozen, and therefore semi-hot-beds may be employed. If time and materials appear excessive for such a purpose, it should be remembered that this is a capital way of preparing for the next crop, whatever it may be, and is a particularly good method of preparing for Peas that are to be sown in the month of April, by which time the earliest sown Radishes will be off the ground. Successive sowings should be made from March to September in the coolest place that can be found for them, and the usual practice of four-feet beds will answer very well. In many gardens sufficient supplies of Radishes are obtained by sowing in the alleys between seed-beds, but care must be taken that this plan does not interfere with the proper work of hoeing, weeding, thinning, &c. When seed is sown on light soils a moderate firming with the back of the spade may be desirable, but generally speaking it is sufficient to cover the seed lightly, and so leave it. To thin the crop early is, however, of great importance, no matter how wasteful the process may seem, for wherever the plants are crowded they will make large useless tops, and small worthless roots, and prove altogether unprofitable. For the earliest sowings we have choice of many sorts, round, oval, and long; but the long Radishes are not well adapted for late sowing, whereas the round and oval sorts stand pretty well in hot weather, if on good ground in a cool situation, with the help of a slight amount of shade. As the year advances we return to the practice recommended for the earliest crops.

Winter Radishes.—These large-growing kinds are much prized by those who use them in winter in the preparation of salads. Seed may be sown in the open from June to August, in drills nine inches apart, and the plants thinned to six inches in the rows. The roots may be left in the ground and dug as required, or taken up and stored in sand. These Radishes may also be cooked in the same manner as Turnips and they make an excellent dish.

RHUBARB

Rheum hybridum

RHUBARB is so much valued that we need not recommend it. There are some remarkably fine sorts in cultivation, adapted for early work, main-crop, and late use.

Although an accommodating plant, Rhubarb requires for profitable production a rich deep soil, well worked, and heavily dressed with rotten manure, and a situation remote from trees, but in some degree sheltered. It will be observed that the markets are supplied from sheltered alluvial soils, that have been much cultivated, and kept in high condition by abundant manuring. On the other hand, the coarser kinds will make a free and early growth on a damp clay, if sheltered from the east winds that so often damage early spring vegetation. The shortest way to establish a plantation is to purchase selected roots of first-class named varieties, and plant them in one long row, three to four feet apart, or in a bed or compartment four feet apart each way. The smaller kinds will do very well at two and a half feet each way, but for large-growing sorts this would be injuriously close. Plant with the top bud two inches deep, tread in moderately firm, then lightly prick the ground over, and so leave it. Rhubarb may be planted at any time in spring or autumn but of the two the spring is preferable. In any case where a special cultivation is determined on, it will be found that bone manure has a wonderful effect on the growth of Rhubarb.

It is not sufficient to say that the plantation must be kept free from weeds, but the plant should be allowed to make one whole season's growth before a single stalk is pulled. And the pulling in the second season, and every season thereafter, should be moderate and careful, for every leaf removed weakens the plant, and it must be allowed-time to regain strength for the next season. Some people know not when to leave off pulling Rhubarb, but appear unwilling to cease until there is none to pull; and it is a pity this should happen, especially as after the delicate supplies of early spring are past, Rhubarb is a comparatively poor thing, and to ruin a plantation to get stalks for wine is great folly. For wine-making a special plantation should be made, from which not one stick should be taken for table use. The summer stalks will then be of a suitable character.

Rhubarb is easily forced in any place where there is a moderate warmth, and it is only needful to pack the roots in boxes with moss or any light soil, or even rough litter. The roots will push into any moist material and find sufficient food. If entirely exposed to the light, forced Rhubarb has a full colour; but the quality is better, and the colour quite sufficient, if it is forced in the dark; hence when put under the stage in a greenhouse, or any other place where there is a fair share of daylight, it is well to put an empty box or barrel over to promote a certain degree of blanching.

When raising Rhubarb from seed sow in spring in light soil, and the young plants should have frame culture until strong enough to plant out. If a great number are grown, they should all be kept in pots until the end of the season, and then the common-looking and unpromising plants should be destroyed, reserving the others for planting out in the following spring. A new type of Rhubarb which is readily raised from seed will remain in bearing continuously if put out on good ground and given protection during severe winter weather. Seed of this strain should be sown in March or April, in pots or boxes placed in a cold frame. Plant out the seedlings in May and these will generally yield sticks in the autumn. Seed may also be sown in the open ground in spring.

SALADS

Although the art of making Salads is to some extent understood in this country, it must be admitted that much has yet to be learned from the masters of Continental cookery, who utilise more plants than are commonly used on this side of the Channel, and who impart to their Salads an endless variety of flavourings. Here, however, we are only concerned with the plants that are, or should be, in requisition for the Salad-bowl at different seasons of the year. But it will not be irrelevant to allude to the fact, admitted by medical men of high reputation, that the appetite for fresh, crisp, uncooked vegetables is a really healthy craving, and that free indulgence in Salads is a means of supplying the human frame with important elements of plant-life. In the process of cooking, certain minerals, such as salts of potash, are abstracted from vegetables, while in Salads they are available, and contribute both to the enjoyment and the benefit of the consumer.

Our present object is to offer a reminder of the plants that must be grown in order to supply such a variety of Salads as will fairly meet the requirements of a generous table during the changing seasons of the year. The culture of all the following subjects will be found under their proper headings.

Beet.—For its distinct flavour and splendid colour Beet is highly valued as a component of Salads. As the roots are easily stored they are available for several months after the growing season has passed.

Celeriac is much used in French Salads, and some appreciation is now shown for it in this country. The roots or bulbs are trimmed, washed, and cooked in the same manner as Beet.

Celery.—This delicious Salad is in such general favour that no comment on its virtues is necessary.

Chervil.—The curled is far handsomer than the common variety, and is available for garnishing as well as for Salads.

Chicory.—The common Chicory *(Barbe de Capucin)* and the Brussels variety *(Witloof)* have attained to great popularity. Both are agreeable and wholesome, and a supply should be maintained from October to May.

Chives find acceptance at times when the stronger flavour of Onion is inadmissible.

Corn Salad.—The leaves should be gathered separately in the same manner as they are collected from Spinach.

Cress should be in continual readiness almost or entirely through the year.

Cucumber.—Everybody appreciates the value of this fruit, which is almost startling in its crisp coolness.

Dandelion.—The cultivated forms of this familiar plant are increasingly grown for use in the Salad-bowl.

Endive has a distinct flavour which is highly appreciated; and in winter the plant occupies the important position that Lettuce fills in summer and autumn.

Lettuce.—All the Cabbage varieties are in great demand for Salads, because they readily assimilate the dressing. But for delicious crispness the Cos varieties cannot fail to maintain their position of assured popularity.

Mustard needs only to be named. Like Cress, it is in continuous demand.

Nasturtium.—A few flowers may always be employed to garnish a Salad, for they are true Salad plants, and may be eaten with safety by those who choose to eat them.

Onion imparts life to every Salad that contains it; but for the sake of the modest people who do not fail to appreciate the advantage of its presence, although they scruple to avow their love, there must be discretion in determining the proportion.

Purslane.—The leaves and shoots are used for Salads, and the former should be gathered while quite young.

Radish finds a place on the tables of the opulent and of the humblest cottager.

Rampion.—The fleshy roots are employed in Salads in the natural state, and also when cooked.

Salsify is commonly known as 'Vegetable Oyster,' and is an excellent component of a Salad. The roots may also be allowed to put forth leaves in the dark to furnish blanched material.

Shallot.—A delicate substitute for Onion.

Sorrel possesses a piquant flavour that can be used by the skilful with most agreeable results.

Tomato has fought its way to popularity in this country, and now holds a commanding position.

Water Cress.—When the tender tops can be had they are seldom allowed to be absent from first-class Salads.

SALSIFY

Tragopogon porrifolius

Salsify may be sown from the end of March to May, but two sowings will in most cases be sufficient. Drill the seed in rows fifteen inches apart and one inch deep. Thin from time to time until the plants stand nine, ten, or in an extreme case twelve, inches apart. In ordinary soil nine inches will be sufficient. Hoe between frequently, but do not use a fork or spade anywhere near the crop, for the loosening of the ground will cause the roots to branch.

A deep sandy soil with a coat of manure put in the bottom of the trench will produce fine roots of Salsify. But there should be no recent manure within fifteen inches of the surface, or the roots will be forked and ugly. In a soil that produces handsome roots naturally the preparation may consist in a good digging only, but generally speaking the more liberal routine will give a better result.

In November dig a portion of the crop and store in sand, and lift further supplies as required. Some roots may be left to furnish Chards in spring. These are the flowering-shoots which rise green and tender, and must be cut when not more than five or six inches long. They are dressed and served in the same way as Asparagus.

Salsify is a root of high quality, the growing of which is generally considered a test of a gardener's skill. Perhaps the after-dressing and serving of Salsify may be a test of the skill of the cook, but upon that point we will not insist. It is a less troublesome root than Scorzonera, and superior to it in beauty and flavour—in fact, it is often dressed and served as 'Vegetable Oyster,' having somewhat the flavour of the favourite bivalve.

Salsify roots require to be prepared for use by scraping them, and then steeping in water containing a little lemon juice or vinegar. They are boiled until tender, and served with white sauce. To prepare them as the 'Vegetable Oyster' the roots are first boiled and allowed to get cold, then cut in slices and quickly fried in butter to a light golden brown, being dusted with salt and white pepper while cooking. Serve with crisped Parsley and sauce made with butter, flour, and the liquor from tinned or fresh oysters.

SAVOY—*see page 38*

SCORZONERA

Scorzonera hispanica

Scorzonera is not much grown in this country, but as it is prized on the Continent, it might be introduced to many English tables with advantage. The main point in the cultivation is to obtain large clean roots, for carelessly grown samples will be small, forked, and fibrous. Trench a piece of ground, and mix a good dressing of half-rotten manure with the bottom spit, taking care that there is none in the top spit. Make a nice seed-bed, and sow in the month of March in shallow drills fifteen inches apart, and as the plants advance thin them until they stand a foot apart in the drill. Keep the crop clean, and it will be fit for use in September. Lift as wanted in the same manner as Parsnips. Seed may also be sown in April and May.

To cook the roots they must first be scalded, then scraped and thrown into water in which there are a few drops of lemon juice. Let them remain half an hour; boil in salted water in the same way as Carrots until quite tender, and serve with white sauce. If left to get cold they can be sliced and fried in butter to make a good side dish.

SEA KALE

Crambe maritima

Many persons prefer Sea Kale to Asparagus, but the two differ so widely in flavour and general character that no comparison between them is possible. On two points, however, the advantage certainly rests with Sea Kale. It can be more easily grown, and, regarded solely as an article of food, it is the more profitable crop. This comparison has therefore a practical bearing. In forming a new garden, and in cases where it may not be possible to grow both these esculents satisfactorily, Sea Kale should have attention first, as a thing that will require but a small investment, and that will surely pay its way, with quick returns, to the general advantage of the household.

Outdoor Culture.—Sea Kale requires strong ground, fully exposed to the sun, and enriched with any good manure, that from the stable being undoubtedly the best. The most satisfactory way to begin is with well-grown roots, as they make a return at once with the least imaginable trouble. Let the ground be well dug two spits deep, and put a coat of manure between; or if it is a good substantial loam, plant without manure, and the results will be excellent. As the thriving plant covers a considerable space, and there must be a certain amount of traffic on the ground to manage it, there should be one row in the centre of a four-feet bed, with a broad alley on one side; or, better still, mark out a ten-feet space, with a three-feet alley on each side, and in this space plant three rows two and a half feet apart, and the roots one and a half to two feet apart. The planting may be done at any time after the leaves have fallen, late in autumn, and during winter and early spring. On warm, dry ground, winter planting answers perfectly, and enables the gardener to complete the task, for there is always enough to do in the spring months. But on damp ground and in exposed situations the best time to plant is the month of March. Put down the line, and open a trench one foot deep; plant the roots with their crowns two inches below the surface, filling in and treading firmly as each trench is planted. The precaution may be taken to pare off all the pointed prominent buds on each crown, as this will prevent the rise of flower-stems; but if this is neglected, the cultivator must take care to cut out all the flowering-shoots that appear, for the production of flowers will prove detrimental to the crop of Sea Kale in the following season. Our custom, when a plantation has been thus made, is to grow another crop with it the first season. The ground between the rows is marked out in narrow strips, and lightly forked over, and if a coat of rotten manure can be spared it is pricked in, and a neat seed-bed is made of every strip, eighteen to twenty-four inches wide. On this prepared bed sow Onions, Lettuces, and other light crops, and as the Sea Kale advances take care to remove whatever would interfere with their expansion, for the stolen crop should not stand in the way of that intended for permanent occupation. A crop of early Cauliflower, small Cabbage, or even Potatoes, may be taken, in which case there will be room for only one row alternately with each row of Kale, and perhaps one row also in the alleys.

The growth of the Kale should be promoted by all legitimate means, and in high summer it will take water, liquid manure, and mulchings of rich stuff, to almost any extent, with advantage. The irrigation that suits the Kale will probably also suit the stolen crop, but irrigation is not good for Onions or Potatoes; where these crops are grown care must be exercised to bestow the fluid on the Sea Kale only.

As the leaves decay in autumn they should be removed, and the ground kept thoroughly clean. When finally cleaned up, let it be forked over, but with care not to put the tool too near the plants; and if manure is plentiful, lay down a coat for a finish, or fork it in at the general clear up. There should now commence a systematic saving of clean leaves. Mere vegetable rubbish is not to be thought of. Proceed to cover the ground with leaves in heaps or ridges sufficient to make a coat finally of about one foot deep, or say nine inches at the very least. If there is any store of rough planking on the premises, let the planks be laid on the ridges of leaves on whichever side the prevailing wind may be. This will prevent the leaves being blown away, and the planks will be handy for the next stage in the business.

At the turn of the year put the planks on edge by driving posts down in any rough way that will hold them firmly for a brief season, and then spread the leaves equally. If there are not sufficient leaves to cover the bed for the requisite thickness, raise a good heap over each crown, and sprinkle a little earth to keep the heap together. But a better mode of procedure is to have a sufficiency of Sea Kale pots with movable covers, or in place of these large flower-pots, or old boxes. Put these over the crowns, and then heap the leaves over and around, and the preliminaries are completed. A very early growth will be the result, and the quality will be finer than that of forced Sea Kale. Uncover occasionally to see how the crop goes on, remembering that perfect darkness is needed to blanch it completely, and to produce a plump and delicate sample. Cut close over, taking a small portion of the woody part of the crown, and when all the growth of a crown is taken, remove the pot or box, but leave a thin coat of leaves on the cut crown to protect it, as at the time of cutting Sea Kale keen east winds are prevalent, and it is unfair to the plants to expose them suddenly. When the crop has been taken, remove the leaves and the planks, and dig in between the rows a thick coat of fat manure. The growth will be too strong now for a stolen crop, and will so continue for many years. After the crop has been secured, each crown will throw out a number of buds or shoots. These should all be removed except two or three of the strongest, which will form the crowns for cutting in the following year. At the same time take away any small blanched shoots that may have been left because they were too small or insignificant for table use. This proceeding will prevent the production of flower-stems, which is injurious to the plant, and there never need be any fear that the crop will be diminished, because plenty of buds around the crowns, that do not show themselves in the first instance, will come forward in due time.

Forcing.—It is so easy to force Sea Kale that the cultivator may safely be left to his own devices. But it will be well, perhaps, to say that perfect darkness is requisite, and the temperature should not exceed 60° at any time, this being the maximum figure. A rise above 60° will produce a thin or wiry sample. It is sufficient to begin with a temperature of 45°, and to rise no higher than 55°, to insure a really creditable growth. The market growers are not very particular as to temperature, but then they do not eat the crop, or know much of it after it has left their hands. With the gardener in a domestic establishment the case is different; and we venture to advise young men—to whom book advice is often valuable as entailing no obligations—that Sea Kale slowly forced may be nearly as good as that grown under pots in the open without any heat at all; better it cannot be. Any spare pits or odd places may be made use of for this crop, provided only that the heat is not too great. Pack the roots in mould or leaves, or even half-rotten manure, and shut them up to exclude light, and the crop will be ready in five or six weeks, unless forcing is commenced very early, in which case seven weeks at least must be allowed from the time of planting to that of the first cutting. Roots that have been lifted for forcing should be thrown away when the crop has been secured, but roots forced in the open ground suffer so little by the process that they may be forced for several years in succession ere it becomes necessary to renew the plantation, provided, of course, that the work is well done. The outdoor forcing is accomplished in the way described for growing the crop, with the aid of leaves only, but with certain differences. In the first place, care must be taken to let the plants feel the cold, but at the same time to prevent the ground becoming frozen. A touch of frost will render them more ready to grow when the cultivator brings his persuasions to bear by heaping hot manure over the pots, and covering the bed with a thick coat of the same. This is all that can be done, but it is sufficient. In cases where leaves and other suitable materials are not available, good Sea Kale may be grown by simply raising over each crown a heap of sand or sifted coal ashes, provided some clean material be interposed to keep the sand or ashes from actual contact with the plant. When this heap begins to crack at the top it will be worth while to examine it at the bottom, when there will be found a fine head of blanched Sea Kale, and the mound will have served its purpose.

To grow Sea Kale from seed is a simple matter enough, but there is a loss of a year as compared with growing it from roots. The ground should be rich and well worked, and the seed sown in March or April in drills one foot asunder if for planting out, or in patches about two and a half feet apart each way if to remain. It is believed by many that Sea Kale should stand where sown, and we admit that analogies are in favour of the proposal. But every year such fine produce is obtained from transplanted roots that we have not the courage to condemn a course of procedure which may not be theoretically correct. The fact is, the root is tough and enduring, and suffers but little by moderate exposure to the atmosphere if handled in a reasonable manner. But to return to the seeds: they sprout quickly, and, soon after, the plants make rapid progress. Let them have liberal culture, keep them scrupulously clean, and thin in good time. If quite convenient, give a light sprinkling of salt occasionally in the summer: they will enjoy it, and the leaves will not be injured in the least.

SHALLOT

Allium ascalonicum

The old-fashioned mode of culture is to plant on the shortest, and lift the crop on the longest, day; but that is only applicable to the milder parts of the country. As a rule, spring is the best time for planting, and it should be done as early as the ground can be got into working order—certainly not later than the middle of April. The soil should be in a friable condition, and it must be trodden firmly, after the manner usual for an Onion bed. Merely press the bulbs into the soil to keep them in position, and put them in rows one foot apart, and nine inches apart in the rows. They should not be earthed up, but, on the contrary, when approaching maturity the soil should be drawn away so as to expose the bulbs, for this facilitates the ripening process.

To store the roots for any length of time it will be necessary to have them well ripened, and this point demands consideration. If dry weather could be insured for harvesting the crop, it might be allowed to finish in the ground; but as this cannot be relied on, it is a wise precaution to lift the crop on some suitable opportunity before it is quite ready, and allow the ripening to be completed in a protected airy place.

SPINACH

Spinacia oleracea

Spinach plays an important part in the economy of the dinner table. There are unfortunate beings who cannot eat it, for they describe it as bitter, sooty, and nauseous. Probably an equal number of persons entertain a very high opinion as to its value. The rest of mankind proclaim it a wholesome, savoury, and acceptable vegetable. Spinach will grow anywhere and anyhow; but some little management is needed to keep up a constant supply of large, dark green leaves, that when properly cooked will be rich in flavour as the result of good cultivation. To produce first-class Spinach a well-tilled rich loam is needed, but a capital sample may be grown on clay that has been some time in cultivation.

Summer Spinach.—The early sowings of Round or Summer Spinach should be in a sheltered situation, but not directly shaded. Sow in drills twelve to fifteen inches apart, and one inch deep, beginning in January, although the first sowing may fail, and continue to sow about every fortnight until the middle of May. The earliest sowing should be on dry ground, but the later sowings will do well on damp soil with a little shade from the midday sun. It is important to thin the crop early, as it should not be in the least drawn. This is the only essential point in securing a fine growth, for if the plant cannot spread from the beginning it will never become luxuriant, and will soon run up to seed. Thin at first to six inches, and if large enough for use, send the thinnings into the house. Before the leaves overlap thin finally to twelve inches. Every plant will cover the space, and it will suffice to take the largest leaves, two or three only from each plant, and thus a basket may be filled in a few minutes with really fine Spinach.

As the heat of the summer increases, the crop will be inclined to bolt. The starved plant will bolt first; the plant in rich moist soil, with plenty of room to spread, will be more leisurely about it, and will give time for the production of a succession crop to take its place. The sowings from May to July should be small and numerous, and on rich moist land, to be aided, if needful, with water. In many gardens there is a sufficient variety of vegetables after the middle of June to render it unnecessary to keep up the supplies of Spinach, and it is best to dispense with it, if possible, during July and August.

Winter Spinach.—The sowing of Winter Spinach should commence in July, and be continued until the end of September, subject to the capabilities of the place. In gardens near towns, where the land is at all heavy, it is generally useless to sow after August, as the autumnal fogs are likely to destroy a plant that is only just out of the seed-leaf. But in favoured localities, with a warm soil and a soft air, seed may be sown up to the very end of the year with but little risk of loss. The winter crops are sometimes sown broadcast, but drilling is to be preferred, and the rows may be twelve to fifteen inches apart. Thin at first to three inches, and afterwards to six inches, and leave them at this distance, for Winter Spinach may be a little crowded with advantage, because the weather and the black bot will now and then remove a plant. Should ground vermin claim attention, the best way to proceed will be to scratch shallow furrows very near the plants, taking care not to injure them. This may be done with the hoe, but if time can be spared it will be better to do it with a short pointed stick, having at hand, as the work progresses, a vessel into which to throw the grubs as they come to light when the earth is disturbed. Where small birds are in sufficient numbers, they will observe the disturbance of the earth, and diligently search for the grubs at hours when the cultivator is no longer on the search himself.

The July sowings will be useful in the autumn and throughout the winter, as the weather may determine; the later sowings will be useful in spring. Plants may be drawn where they can be spared to make room for the remainder, but leaves only should be taken when the plant is large enough to supply them. When symptoms of bolting become visible in the spring, cut the plants over at the collar, and at once prepare the ground for another crop.

New Zealand Spinach *(Tetragonia expansa)*.—Gardeners are only too well acquainted with the difficulty of maintaining an unbroken supply of true Spinach during the burning summer months. But the weather which makes it almost impossible to produce a satisfactory crop of *Spinacia oleracea* brings New Zealand Spinach to perfection. The latter is prized by some persons because it lacks the peculiar bitterness of the former. The plant is rather tender, and therefore to obtain an early supply the seed must be raised in heat. It may be sown in pots or pans at the end of March or beginning of April. Transfer the seedlings to small pots immediately they are large enough, and gradually harden in preparation for removal to the open ground towards the end of May. They should be put into light soil in a sunny position, and be allowed three or four feet apart each way. It is not unusual to grow them on a heap of discarded potting soil, where they can ramble without restraint. The growth is rapid, and there must be no stint of water in dry weather. In five or six weeks the first lot of tender shoots will be ready for pinching off. Those who do not care to incur trouble under glass may sow in the open in the early part of May, and thin the plants to the distance named.

Perpetual Spinach, or Spinach Beet *(Beta Cicla)*.—A valuable plant for producing a regular supply of leaves which make an excellent Spinach at a period of the year when the ordinary Summer Spinach is past its prime. Although it is a true Beet, the roots are worthless, and there should be liberal treatment to insure an abundant growth of leaves. Seed may be sown from March to the end of July or beginning of August, in rows one foot apart. Thin the plants to a distance of six or eight inches in the rows. When the leaves are ready for gathering, they must be removed, whether wanted or not, to promote continuous growth.

Orache is frequently used as a substitute for Spinach where the ordinary variety fails. Seed should be sown during the spring months, and as the plant frequently attains a height of five feet allow a distance of at least three feet in each direction for development. Red Orache is useful for growing in ornamental borders, but it is not so suitable for culinary purposes as the white variety. The leaves only are eaten.

STACHYS TUBERIFERA

This vegetable is commonly known as the Chinese Artichoke, and from the peculiar form it is also called Spirals. A wide difference of opinion exists as to its value, but in its favour the fact may be stated that tubers are often exhibited in the finest collections of vegetables staged for competition.

The time for planting is early spring, in rows eighteen inches apart, allowing a distance of nine inches in the rows. The proper depth is four inches. The roots are quite hardy and the crop gives no trouble. After planting it is only necessary to keep the plot free from weeds.

The tubers do not mature until late in autumn, and as far as possible it is advisable to lift them when they are wanted. Should it be necessary for any reason to clear the ground, the Stachys must be covered with soil. When exposed to light and air they soon become discoloured and are then unfit for cooking. It is usual to boil them in the same manner as Potatoes, but the finish must be by steam alone. An agreeable variation consists in frying the boiled roots with butter until slightly brown, when the dish is considered by many connoisseurs to be very delicious and suitable for serving with poultry or joint.

STRAWBERRY

Fragaria

Probably the first thought will be that the Strawberry is a fruit, and that the consideration of its treatment is out of place in a series of articles on the culture of vegetables. The answer is that the plant forms an essential feature in every good Kitchen Garden, and the general routine of work has to be arranged with due regard to this crop, so that we need make no apology for alluding to it here.

When to Plant.—The Strawberry is the most certain of all our hardy fruits, and is much valued both for eating fresh as a summer luxury and as a preserve for winter use. Although it deserves the best of cultivation, its demands are few, for under the poorest system of management it is often extremely prolific, and not unseldom the most profitable crop in the garden. We have choice of seeds, divisions, and runners in making a plantation of Strawberries. The universal way is the best way, and it consists in planting rooted runners of named sorts in an open sunny spot in well-prepared ground any time during spring or autumn, when fresh and good runners are obtainable; but late planting is undesirable, for when the plants have not time to establish themselves before winter sets in many are lost. If, therefore, the planting cannot be accomplished at the latest by the beginning of October, it is better to defer the task until the spring. Plants put in at the latter time should have the flower-stems removed, and will then yield a heavy crop in the succeeding season.

Treatment of Soil.—The best soil for Strawberries is a rich, moist, sandy loam, but a heavy soil will answer perfectly if it is well prepared. The ground should be trenched and liberally enriched with rotten manure placed between the top and bottom spits, where the plants will reach it when they are most in need. In a new soil that is rather stiff it will be advisable, when the trenching has been completed, to put down the line and cut shallow trenches, which should be filled with any rather fine kindly stuff that may be at hand, such as old hot-bed soil, leaf-mould, or a mixture of material turned out of pots, with some good decayed manure. In this the young plants will root freely and quickly without becoming gross, for they should attain a certain degree of vigour; but an excessive leaf growth may result in losses during winter, and a small crop of fruit in the following year. Well-cultivated soils need no such special preparation, but in any case a good digging and a liberal manuring are absolutely necessary. And here it may be well to state that after the plants have obtained a firm hold on the soil it matters not how hard the ground becomes. The practice of some growers in running a plough lightly between the rows either for a mulch, or to give the plants the full benefit of rain, does not in the least degree upset this conclusion, for this only creates a loose and friable surface, and the operation is so managed that the soil near the roots remains undisturbed. It may be accepted as a secret of successful Strawberry culture that the bed should be firm and compact, and, in forcing, this principle is so far recognised that the soil is positively rammed into the pots.

Method of Planting.—If Strawberry plants come to hand somewhat dry, unpack them quickly, and spread them in small lots in a cool shady place, and sprinkle lightly with water to refresh them. A deluge of water is not needed, and in fact will do harm, but enough to moisten them will put them in a condition to begin growing as soon as they are properly located. In planting, a little extra care in the disposition of the roots in the soil will be well repaid, for plants merely thrust into the ground cannot develop that robust root growth on which the future of the crop largely depends. When preparing the positions it is an excellent plan to build in the centre of each excavation a mound of earth over which to spread the fibrous roots. Then return the soil and firmly tread down. As a finish give each plant a copious watering. On no account should the plant be deeply buried, but the crown should be left just clear of the surface level. The distances in planting will have to be determined by the relative vigour of the varieties and the nature of the ground. As a rule the rows should be two feet apart, and the plants eighteen inches in the rows, but some varieties require fully two and a half feet between the rows. It is good practice to leave a three-feet space between every two rows for necessary traffic. A modification of the plan consists in planting a foot apart each way; and immediately the first crop of fruit is off every alternate row is removed, and then every alternate plant in each row is also taken out. This places the remainder at two feet every way. The ground is then lightly forked and a heavy coat of manure put on.

The general management comprises keeping down weeds, supplying water abundantly in dry weather, especially when the berries are swelling, and removing runners as fast as they appear, for to allow them to get ahead is most injurious, and any serious neglect of this rule is likely to ruin the plantation. The Strawberry plant makes no proper return on a dry lumpy soil. Large plantations that cannot be watered must be aided in the height of the season by covering the ground with any light material which will prevent evaporation. As to obtaining runners, that is easy enough, but there is a good way and a bad way. To allow them to spread and root promiscuously is the bad way; it injures the plants, makes the bed disorderly, and does not produce good runners. At the time when runners begin to push, dig and manure the surrounding spaces, and allow a certain number of runners to come out from each side of the rows. As they approach maturity and are disposed to make roots, lay tiles or stones upon the runners near to the young plants to favour the process, but a neater way will be to peg them down. Or they may be fixed by short pegs in small pots, filled with light rich earth and plunged in the soil.

To keep the crop clean many plans are adopted, and the plant probably takes its name from the old custom of covering the ground with straw for the purpose. The cultivator must be left to his own devices, because of the difficulty in many places of obtaining suitable material. But we must warn the beginner in Strawberry culture against grass mowings as more or less objectionable. They sometimes answer perfectly, and at other times they encourage slugs and snails to spoil the crop, and if partially rotted by wet weather communicate to the fruit a bad flavour. There is a very simple means of feeding the crop and making a clean bed for the fruit. It consists in putting on a good coat of long, strong manure in February, and in doing this it is no great harm if the plants are in some degree covered. They will soon push up and show themselves, and by the time the fruit appears the straw will be washed clean, and the crop being thus aided will be a great one, weather permitting. As regards cutting off the leaves, we advise the removal of old large leaves as soon as the crop is gathered. But this should be done with a knife; to use a scythe amongst Strawberries is to ruin the plantation. The object of removing old leaves is to admit light and air to the young leaves, for on the free growth of these the formation of good crowns for the next year's use depends. By encouraging the young leaves to grow, root action is promoted, and the embryo buds are formed that will, in the next summer, develop into Strawberries.

Some gardeners recommend the removal of the Strawberry plantation every three years. It is a better plan to make a small plantation annually, and at the same time destroy an old plantation that has served its turn. But we are bound to say that Strawberry plantations, well made and well kept, will often last and prove profitable for six or even more years. But this will never be the case where there is a stint of manure or water, or where the runners are allowed to run in their own way to make a Strawberry mat and a jam of the wrong sort. The Strawberry fancier does not wish to keep a plantation any great length of time, and he must plant annually to taste the new sorts. This to many people is one of the chief delights of the garden, and it certainly has its attractions.

Forced Strawberries.—The high price realised on the market for the earliest supply of forced Strawberries is a sufficient proof that society is prepared to pay handsomely for this refreshing luxury. As the season advances and competition becomes keen the figure rapidly declines, but 'Strawberries at a guinea an ounce' has more than once appeared as a sensational head-line in the daily press.

The fruiting of Strawberries in pots is part of the annual routine of nearly all large establishments, but even with the most perfect appliances it must be admitted that to produce berries which win appreciation for their size, colour, and flavour demands both skill and patience, especially patience.

Strong well-rooted plants are essential to success, and no trouble should be spared to secure them from robust free-fruiting stocks. The earliest runners must either be layered on square pieces of mellow turf or over thumb pots filled with a good rich compost. When the runners are fairly rooted in the layers of turf or the thumb pots they should be transferred to pots of the fruiting size. No. 32 is generally used for the purpose. After the pots have been crocked some growers add a layer of half-inch bones, which aid the plants and insure free drainage. The most satisfactory soil is a rich fibrous loam, with the addition of one-fourth of well-rotted manure and a small proportion of sand, and the compost must be well firmed into the pots with the ramming stick.

The best place to keep the plants is an open airy situation, easily accessible, where the pots can stand on a bed of ashes. On the approach of frost they can be transferred to a cold frame, keeping them close to the glass, or they may be plunged in ashes in some sheltered position.

When the time arrives for forcing, it is usual to commence by plunging the pots in a bed of warm leaves or in a mild half-spent hot-bed. Immediately the plants show sign of blooming they must be shifted to warmer quarters. A shelf at the back of an early vinery or Peach-house, quite near the glass, is a suitable position. The temperature at starting should be 55° Fahr., rising gradually to 60° by the time the leafage is thoroughly developed.

The appearance of the flower trusses is a critical period. Liquid manure should then be given freely, and at the same time the plants must have abundance of light and a warm dry atmosphere. The blossoms need to be artificially fertilised with a camel's-hair pencil, choosing midday as the best time for this operation.

When the crop has set it must be thinned to about nine berries on each plant, and in due time the fruits should have the support of forked sticks. Care will be necessary to prevent injury to the stalks, or the flow of sap to the berries may be arrested. Syringe twice a day in dry weather; and on the first show of colour discontinue the manure-water and use pure soft water only. At this stage a night temperature of 65° must be maintained, giving all the air and light possible.

More failures in the pot culture of Strawberries are attributable to neglect in watering than to any other cause. The soil must never be allowed to become dry. Should the leaves once droop they seldom recover. At least twice a day the plants will need attention, and it is important that the water should be of the same temperature as the atmosphere. Always leave the cans full in readiness for the next visit.

Alpine Strawberries are very largely grown in France, probably more so than the large-fruited varieties which are popular in this country. The best method is to sow the seeds in January, in pans filled with a light rich compost and placed in a gentle heat. Prick out the plants on to a bed of light soil in a frame, or on a nearly exhausted hot-bed, whence they should be taken to the open ground. From these sowings fine fruits may usually be gathered in the following September. Seeds may also be sown outdoors in spring or in September in shallow drills, six inches apart, on a bed of light soil. Transplant in due course for fruiting in the succeeding Strawberry season. When a full crop has been gathered the plants should be destroyed, a succession being kept up by sowing annually. By slowly growing the plants from spring-sown seeds and potting in autumn, it is not a difficult matter to have Alpines in fruit under glass at Christmas.

SUNFLOWER

Helianthus annuus

Although the Sunflower is not utilised as food for man, the plant is frequently grown in the Kitchen Garden, partly as an ornament, and also for the production of seeds which are given to poultry.

As regards cultivation, sow in pans in April, and put on a gentle hot-bed, or shut up close in a sunny frame. The plants will soon appear. Give them light and air, and plant out when they are two or three inches high. But Sunflowers can be grown without any kind of artificial aid. A simple and effectual method is to make the spot intended for them very rich, and dibble the seed an inch deep on the first day of May.

TOMATO

Lycopersicum esculentum

The taste for Tomatoes often begins with a little antipathy, but it is soon acquired, and not infrequently develops into decided fondness for the fruit both cooked and in its natural condition. As a necessary article of food the call for it in this country is no longer limited to a select circle of epicures, for the value of its refreshing, appetising, and corrective properties is now widely recognised, and its advance in public favour has been accelerated by the improved quality, enhanced beauty, and increased variety effected by expert raisers.

The Tomato is a tender, but not a tropical plant, and it requires a moderately high temperature, free access of air, and above all a full flood of solar light to bring it to perfection. The necessary heat is easily managed in any garden equipped with ordinary forcing appliances; so also is a current of air in properly constructed buildings; but the deficiency of light during the darker months renders the task of producing fruit in midwinter less easy than at other seasons. By the introduction of varieties possessing increased powers of crop-setting, however, the difficulty of winter fruiting has been largely overcome, so that, with efficient management, it is now possible to send Tomatoes to table throughout the year.

Almost every imaginable glass structure can be employed for growing Tomatoes, from the small suburban greenhouse to the vast span-roof, hundreds of feet in length, devoted to their culture in the Channel Islands. And it is not essential that the crop should be grown alone. Potatoes, French Beans, Strawberries, and Vines may be forced in the same building, provided there be no obstruction to light and air, nor any interference with the conditions which experience has proved to be imperative for sustaining the plants in vigorous health. For winter and spring gathering there must be a service of hot-water pipes, but as the season advances it is easy to ripen fruit in cool houses, and later on plants outdoors will in favourable seasons yield an abundant return without artificial protection of any kind.

INDOOR CULTURE—Sowing and Transplanting.—Seed may be sown at almost any time of the year, but the most important months are January to March, August and September. In gardens favourably situated in the South of England and furnished with the most perfect appliances, seed is sown in all these months, and in others also; but in smaller gardens sowings are generally restricted to February and March. Whenever a start is made sow thinly and about half an inch deep, in pans or boxes, and do not allow the seedlings to remain in them for an unnecessary day. Immediately two or at most four leaves are formed either prick off into other pans or boxes, or transfer singly to thumb pots, and as a rule the pots will be found preferable. The soil for these pans or pots should be stored in the greenhouse a few days in advance of the transfer, so that the compost may acquire the proper temperature and save the plants from an untimely check. In small houses place the plants near the glass that they may remain short in the joint, but on cold nights they must be taken down to avoid injury from fluctuations of temperature. In large houses, where the light is well diffused, there is no need to incur this trouble, for the seedlings will do equally well on the ground level. In due time shift into six-inch pots, from which they can go straight to borders, or into a larger size if they are to be fruited in pots. About fourteen weeks will be required to prepare the plants for borders in the winter season, but a shorter period will suffice in spring and summer. Plants from an August or September sowing will not mature fruit in much less than six months, while a March sowing will yield a return in four months or less. A great deal depends on the character of the season, and more on skill and attention. Those who sow in January or February should sow again a fortnight later, and onwards until the end of April, according to requirements. For winter supplies a first sowing may be made in June, in a cold frame, and prepared for transfer to fruiting pots in September.

Treatment of Soil.—In the first instance there need be no anxiety about soil. Any fairly good sandy loam will answer for the seed-pans, and if too stiff it may be freely mixed with sharp sand or the sifted sweepings from roads and gravel walks. A fibrous loam, cut from a rich pasture, and laid up in a heap for twelve months, will, with an addition of wood ashes and grit, make an ideal soil for pots or borders. As the plants advance, leaf-mould or thoroughly decayed manure in moderate quantity should be supplied; but, instead of incorporating it with the loam in the usual way, it will be found advantageous to place the manure immediately above the crocks, and the roots will find it at the right time. But the quantity of manure must not be overdone, especially in the earlier stages of growth, because excessive luxuriance neither promotes fruitfulness nor conduces to early ripening. After the fruit has set, a mulch of decayed manure will aid the plants in finishing a heavy crop. Manure which is only partially fermented will not do at all. The ammonia it liberates exerts so deadly a power that the plants are quickly scorched.

In its demand for potash the Tomato closely resembles the Potato, and of the two the former is the more exacting. So quickly does this crop exhaust the soil, that in small houses it is usual to take out the earth to a depth of fifteen or eighteen inches every second or third year, and replace it with virgin loam. Others grow the Tomatoes alternately in the bed and in pots, but this is only a partial remedy. Constant dressings of farmyard or stable manure result in the formation of humus, which, as it becomes sour, has to be sweetened by the solvent influence of lime. The chief objection to the use of stable manure, however, even when well rotted, is that it induces a free growth of foliage instead of promoting an early development of fruit. The most enduring method is that which is based on chemical knowledge of the constituents of the soil, and the relation which the plant bears to it. One of the most successful growers for the London market almost entirely avoids the use of stable manure, and he is able, by applications of nitrate of potash, dissolved bones, and the occasional use of lime, to grow splendid crops in the same houses year after year.

All the conditions which answer for border work are applicable to pots, and a limited number of plants brought forward in succession will supply the requirements of a small household from early spring until near Christmas. The pot system is conducive to free setting and to early ripening, and for these reasons it is worth attention. The plants should be kept short in the joint by frequent shifts until the twelve-inch pot is reached, and this size will accommodate two cordons or one plant having two branches, each of which will require a separate stake for its support. Plunging the pots can be adopted to save labour in watering.

Temperatures.—No advantage is to be gained by attempting to force Tomatoes in a higher temperature than is consistent with healthy progress, although in winter there is great temptation in the direction of overheating. Full time for development in moderate heat will bring stout joints, and impart a vigorous constitution that materially aids the plants in resisting the insidious attacks of disease. The waning autumn and dull winter days are the most troublesome periods of management, and it is remarkable that of two days equal in duration and apparently in other conditions, the autumnal appears to be less favourable than the spring day. But if, on the one hand, a high temperature is injurious, a low temperature must be avoided; although for a time it may not appear to be harmful. A temperature of 60° or 65° suits the seed-pans, and after transfer to pots and the roots have become established, the thermometer should not register less than 55° during the night. It may rise 10° by means of fire heat in the daytime, and during bursts of sunshine another 10° or 15° will be quite safe, always assuming that the roots are not dry, and that the plants have free ventilation.

Watering.—The judicious administration of water forms an important feature in the culture of the Tomato. The plant is too succulent to endure drought with impunity, and it is mere folly to toy with the water-can. Saturate down to the roots, and then leave the plants alone until more water is wanted. No hard and fast rule can be stated as to frequency. It depends on the condition of the soil, the period of the year, and the age of the plants. Borders and soil for pots should be made sufficiently moist in advance, so that watering will not be necessary immediately after the plants are transferred. The prevalent opinion that excessive watering generates disease is not confirmed by our experience. Of course the watering should not be excessive for many reasons, but the diseases which are often attributed to over-watering are the result of atmospheric mismanagement.

General Treatment.—Authorities are not agreed as to whether branched plants or simple cordons yield the better results. In our judgment the single stem deserves preference, and it is now more extensively grown than any other form, although plants having two branches are almost equally popular. Certainly the cordon can be managed with extreme ease; it is admittedly the earliest producer, and there is a general consensus of opinion that the fruit it produces is unsurpassed in size and quality. The doubtful point is quantity, but even here the difference, if any, is too trifling to be worth the consideration of private growers. Cordons are formed by removing the laterals as fast as they appear, and when the fruit has set, or the requisite height is attained, the top is also pinched out.

The space allowed for each plant varies greatly, especially among growers for market. Under glass every branched Tomato should be allowed at least three feet each way. For cordons we advocate a distance between the rows of three feet, and a space of two feet in the row is not too much. The stems require support of some kind, and stakes are preferable to string; but of course the stems may be secured to wires whenever it is convenient to run the plants immediately under the glass.

Another point upon which authorities differ is the extent to which Tomatoes should be denuded of their foliage. Some growers condemn the procedure entirely; others reduce their plants to skeletons. Both extremes are objectionable, for when all the leaves are permitted to remain there is delay or partial failure in colouring the fruit, and the almost entire removal of foliage checks the root action injuriously. In practice it answers well to wait until the fruit has set, then by pinching out the leading point of each leaf, commencing at the bottom, ripening and colouring are promoted, and the health of the plant remains unimpaired.

In dull weather, and especially in short days, a difficulty is sometimes experienced in setting the fruit, particularly the first bunch. After fruit has begun to swell on one bunch, the remainder set with comparative ease. A rather higher temperature than usual combined with free movement of the atmosphere is generally sufficient to insure fertilisation. If assistance is necessary, however, water the plants early in the afternoon, and close the house rather before the usual time. The warm atmosphere will develop plenty of pollen, and a gentle shaking of the flower bunches with a slight touch from a hazel twig will liberate visible clouds, which will effectually set the fruit. Another method is to lift a flat label or paper knife against the flowers. The label becomes covered with pollen, and by gently touching each flower with a slight upward pressure a great number can be fertilised in a few minutes. A soft brush passed over the flowers daily has the same effect. Plants in the open ground need no such attention if they are in good health and the season is at all genial. When a bunch of flowers contains one that is fasciated or confused, the flower should be pinched out to prevent the formation of large and ugly fruit. The remainder of the bunch will be the finer for its absence.

OUTDOOR CULTURE.—For the open ground it is important to choose a variety that ripens early. The plants should be vigorous, and they must be carefully hardened before they are put out. Sow the seed in heat in February or March, and when large enough transfer the seedlings to single pots until wanted. Every effort should be made to avoid giving the plants a check, and if room is available they may be potted on to the six-inch size and allowed to form one truss of bloom before planting out, thus saving valuable time. The end of May is usually the right time for transfer to the open, but Tomatoes will not endure a keen east wind or nipping frost. During the prevalence of unfavourable weather it is advisable to wait a week or more rather than risk the destruction of the plants. When the temperature appears to be fairly reliable, put them into holes a foot deep and eighteen inches across, filled with light soil not too rich. For a few nights until the roots take hold slight protection should be at hand to assure safety; Sea Kale pots answer admirably, and are easily placed in position. In addition to beds all sorts of places are suitable for Tomatoes, such as under warm palings or walls, on sloping banks and in sheltered nooks, where they will thrive and yield valuable fruit. Stout stakes are required and should be promptly provided. Pinch out the lateral shoots, and as soon as the fruits commence to colour some of the largest leaves may be partially removed. Early in August nip out the tips of the leaders in order to encourage ripening. Thus in the open garden a supply of this delicacy may be insured for part of the year equal in quality to fruit which is grown under glass. (*See also page 181.*)

The diseases of the Tomato are dealt with in the chapter on The Fungus Pests of certain Garden Plants.

TURNIP

Brassica Rapa

The Turnip is not a difficult garden crop; indeed, the simplest management will produce an ample supply, and any fairly good ground will suffice for it. But whatever is worth doing is worth doing well, and a gardener may be pardoned for taking an especial pride in producing a sufficiency of handsome and tender Turnips. The great point is to insure a succession through a long season, or, say, the whole year round, for Turnips are always in request, and at certain periods of the year delicate young roots are greatly valued for the table.

The finest Turnips are grown in deep, sandy loam, kept in a high state of cultivation. Useful Turnips may be grown on any soil, but a handsome sample of the finest quality cannot be produced on heavy clay or thin limestone. In common with other fast-growing plants of the cruciferous order, Turnips must have lime in some form, and in many gardens it will occasionally be necessary to give a dressing of lime in addition to the ordinary manure. Superphosphate, bone, and old plaster or mortar from destroyed buildings, are all valuable in preparing the soil for this crop.

Times of Sowing.—An early crop of small bulbs may be grown by sowing in January on a very gentle hot-bed as prescribed for early Radishes, and it may be well to add, that in an emergency white Turnip Radishes may be made to take the place of Turnips, both to flavour soups and to appear as a dish in the usual way. Fast-growing Turnips may be sown on a sheltered warm border in February and March, to be carefully watched and protected when unkind weather prevails. In April and May sowings should be made consistently with the probable wants of the household, but the May sowings should comprise two or three sorts in the event of hot dry weather spoiling some of them.

The principal sowings for autumn and winter supplies are made in June and July, but seed may also be sown in August. Ground from which some crop, such as Peas, has just been cleared generally needs little preparation beyond breaking the surface with a hoe, followed by a good raking. Thin the plants early and let them stand finally at six to nine inches apart in the rows. For late crops seed is often sown broadcast, the roots being pulled as they mature.

General Culture.—It is advisable to sow Turnips in drills on a fine tilth, and it is an advantage to have a sufficiency of some stimulating manure near the surface to hurry the growth of the young plant, for the danger of fly belongs to the seed-leaf stage. Generally speaking, the Turnip fly does but little harm in gardens; but where it is much feared, the seed should be sown in prepared drills to encourage a quick growth. Draw the drills twelve to fifteen inches apart, three inches deep, and about the same width, and almost fill them with rotten manure, or with a mixture of earth and guano, or wood ashes; cover this with a little fine soil to prevent injury to the seed; then sow, and lightly conceal the seed with earth as a finish. If the ground is sufficiently moist, growth will commence almost immediately, and the plant will come up strong, and very quickly put forth rough leaves. In the general management more depends on timely and judicious thinning than upon any other point. If Turnips are not well thinned, so that each plant can spread its green head unimpeded by the leaves of a neighbour, a good growth cannot be expected; and thinning by the hoe should be commenced as soon as the rough leaves appear. The operation must be repeated until the plants are at a suitable distance, and then comes the process of singling, which should be done by hand. It will be found that in many cases two or three little plants stand together looking like one. There must be only one left at each station, and that should be the shortest. The distances may vary from four to ten inches, according to the vigour of the variety and the kind of Turnips required. An easy and profitable plan is to allow a certain number of bulbs to swell to supply young Turnips, and, by drawing these, leave room for the remainder of the crop to attain its proper size for storing.

The Turnip likes a light soil, but does not well endure the occasional dryness to which light soils are subject. This fact accounts for many failures of the crop in a hot dry season, for sunshine suits the Turnip, but it must have moisture or suffer deterioration in some way. If, therefore, the soil becomes dry, and there is no prospect of rain, the Turnips should have water, not simply to moisten the surface, but to go to the roots, for frequent watering is not good for the crop, as it tends to spoil the beauty of the bulbs, and promotes a rank leaf-growth which is not wanted. An occasional heavy watering in dry weather will also do much towards the repression of the many enemies that beset this useful root—the jumpers, the grubs, the weevils, and the rest of the vermin will be routed out of their snug hiding-places in the dusty soil when the watering takes place, and the death of many will follow. But so long as the soil is fairly moist at the depth the roots are ranging, there is no need for watering, and the time it would consume may be utilised for other work.

Lifting and Storing.—On the approach of winter a certain portion of the Turnip crop should be lifted and stored. In doing this the tops must be cut off, not too close, but just leaving a slight green neck, and the roots should be rather shortened than removed; at all events, to cut the roots off close is bad practice: when so treated the bulbs do not keep well. Any rough storage answers for Turnips, the object being to keep them plump by excluding the atmosphere, and at the same time render them safe against frost. The portion of the crop left in the ground may be lifted as wanted in the same way as Parsnips, but this should be done systematically, so that the ground which is cleared may be dug over and ridged up before winter. Those that remain will be in a piece, and will give a good crop of spring greens, after which they may be made use of as manure by putting them at the bottom of a trench.

Some of the foes that war against the Turnip crop are alluded to at greater length later on. Happily, the gardener has many friends that are insufficiently known to the farmer, not the least important being the starlings, song birds, and occasionally (but not often) the sparrows. Where the cultivation is good and small birds abound, the Turnip crop is pretty safe, and the general routine of culture sketched above will certainly promote, if it does not absolutely secure, its safety. The worst foes of the Turnip in the field are the fly and the caterpillar; but in the garden, and more especially the old garden, anbury is the most to be feared. When this happens the cultivator may rest satisfied that the soil is in fault, and this may be owing to a bad routine of cropping. Wherever anbury appears, whether on Cabbages or Turnips or any other cruciferous plant, there should be worked out a complete change in the order of cropping, taking care not to put any brassicaceous plants on the plots where the disease has occurred for two or three seasons, and allowing at least one whole year to pass without growing any of the cruciferous order upon them. In the meantime, for other crops the land should be well trenched and limed, and generously tilled. The result will be profitable crops of other kinds of vegetables and a refreshing of the soil that will enable it to carry brassicaceous plants again, with but little risk of the recurrence of anbury. Good cultivation is the only panacea known against the plagues that assail our crops. This does not surely secure them, for the elements are capricious and beyond our control; but where good cultivation prevails the failures are few, and even unfavourable seasons do not utterly obliterate the benefits of past labour.

Swede.—There are several advantages in growing Swedes as one of the garden crops. They are hardy in constitution and prolong the supply of a wholesome vegetable. In districts where Turnips are unsatisfactory,

Swedes prove successful, and are appreciated for their delicacy of flavour when grown from stocks which have been carefully selected for the purpose. The culture is in all respects the same as for Turnip. The date of sowing depends on the district. In the north it is safe to sow at the beginning of May, but in the midlands and southern counties of England the end of May or beginning of June is early enough.

VEGETABLE MARROW

Cucurbita Pepo ovifera

The Vegetable Marrow does not, in a general way, obtain the right kind of attention in gardens. It is very generally grown and is much valued as a summer vegetable. But too often the aim of the cultivator is to obtain large Marrows, that at the very best are coarse and troublesome to the cook and are always wanting in substance and flavour, instead of smallish Marrows, which are easily dressed, elegant on the table, and combine with a substantial and somewhat glutinous pulp a most delicious flavour. Two fears beset the average gardener: he is afraid to grow small sorts, and he is afraid to cut them when quite young. When he can overcome these fears he will appreciate the smaller Marrows that have of late years been secured by patient labour in cross-breeding, for while they are of the highest quality, they are also early and productive, far surpassing all the larger Marrows in quickness and usefulness. The market grower we do not pretend to advise, for he must grow what he can sell; and if the smaller Marrows are insufficiently appreciated in gardens, we cannot hope to see them on sale in shops.

The Vegetable Marrow will grow in any good soil, and although a tender plant, it is so accommodating that if the seed is sown on a piece of newly dug clay land in the latter part of May, or early in June, the plants will thrive and produce a heavy crop the same season. We put this as an extreme case, but we do not recommend such a careless mode of growing this valuable vegetable. The fact is, it pays better to grow it well than to grow it ill; and in a country where land and labour are costly, and the summer very uncertain, it is best to take such a thing in hand scientifically, and provide for it as many favourable conditions as possible. Three conditions are imperative: a moderate bottom heat from fermenting material; a kindly, loamy soil, quite mellow, in which the roots can run freely; and a sufficiency of water, for this is a thirsty plant. But the excessive use of manure is undesirable, as this only forces a rank growth of foliage at the expense of the fruit.

Frame culture is of some importance, because early Marrows are highly valued at good tables. For this business the neat-growing, small-fruited kinds should be chosen, as they yield a great crop in a small compass. The best place for an early crop of Marrows is a brick pit, with hot-water pipes for top heat, and a bed of fermenting materials for bottom heat. It is no difficult matter to obtain a supply in a house with Cucumbers, but it is better to grow the Marrows apart, as they require less heat and less moisture than Cucumbers. In making up the bed, it is well to employ leaves largely, say to the extent of one-half, the remainder being stable manure that has been twice turned. Such a bed will give a mild heat for a great length of time, and the plants can be put out upon it within three days of its being made up. When grown in a common frame, the arrangements are much the same as advised for the frame cultivation of the Cucumber, the chief points of difference being that Marrows should have less heat and more air. The temperature for Marrows under cover may range from 55° the minimum, to 80° the maximum; the safe medium being about 65° when the weather is cold and dull; running to 80° when strong sunshine prevails, and the plants are growing freely with plenty of air. As for the general management, a bed nine inches deep of good fibrous loam is required, with regular supplies of water of the same temperature as the pits, so that the bed is always reasonably moist, and every evening a slight syringing over the leaves and the walls before shutting up. The training out is a very simple matter. Let the vines run in their own way until they have made shoots eighteen inches long, then nip out the points. After this there must be no more stopping, but occasionally the laterals must be suppressed to prevent crowding. Give air freely at every opportunity, and be careful not to administer too much water, or the blunder will result in a deficiency of fruit.

To grow Marrows in the open air, the best course of procedure is to remove a portion of the top soil, to form a shallow trench four feet wide. Into this carry one foot to eighteen inches depth of half-rotten manure, or a mixture of equal parts of manure and leaves, and cover with the soil that was taken out. This will produce a very gentle hot-bed that will last until the natural ground heat is sufficient to keep the plants in vigorous health. The middle of May is quite early enough to make up the bed, and in the course of two or three days the plants may be put out. Cover with hand-lights or small frames, which on the following day should be tilted at bottom to admit a little air, and if strong sunshine occurs, a Rhubarb leaf may be laid over to subdue the glare upon the young plants. We will suppose these plants to have been raised in a Cucumber frame from seeds sown in April. If plants are not available, sow seeds in patches of two or three on the bed, and cover with inverted large flower-pots, and with a piece of tile to stop the hole. This plan hastens germination. Pots may also be used as protectors if glass frames are not at command, being taken off during the day and put on at night, the hole being left open to give a little air. During bad weather the pots should remain all day over the plants, but as soon as possible must be again taken off to keep the growth short, green, and vigorous. The plants should be put singly down the centre of the bed, three feet apart, and as a matter of course the seeds should be sown at the same distance, and each clump of two or three should be reduced to one when the plants are somewhat forward. It is advisable not to be in a hurry in thinning the plants, for the slugs will probably compel some modification of arrangements, so that sometimes it will be necessary to lift a clump, and divide the plants, to fill up gaps where the slugs have made a clearance. An occasional inspection in the after part of the day, and again in the early morning, will be the best course to keep down the slugs, as they may then be caught and disposed of; but a dusting of soot around each clump will do much to protect the plants against silent marauders. As for after-management, there is no occasion whatever for any stopping or training, but now and then a stout peg may be placed to keep some strong vine in order. The necessity for moisture must not be overlooked. If the ground becomes dry the plants will suffer, but with sufficient moisture they will continue growing and bearing until the frost destroys them. Cut the Marrows when quite young, for not only are they more useful on the table when small and tender, but the plants will bear five times as many as when a few are permitted to attain their full size. The explanation of the case is very simple. The production of the young fruits does not in any appreciable degree exhaust the plants; but when the fruits are allowed to develop, the plant is too severely taxed, and a succession is pretty well brought to a stop. The most delicately flavoured Marrows, as a rule, are the smallest; these when cooked should be served whole, or at most only cut into halves, and of course there is no occasion to remove the seeds.

A YEAR'S WORK IN THE VEGETABLE GARDEN

The following monthly notes are not intended to supersede the detailed instructions on the several kinds of Vegetables which appear in the preceding pages. The present object is to call attention to the work that must be done, and the work that must be prepared for, as the changes of the seasons require and the state of the weather may permit; yet some amount of detail is included. Merely to offer reminders would be to exclude the great mass of amateurs, and the less experienced of practical gardeners, from participation in the advantages of these monthly notes, and to restrict their use to a few practical men who are masters of every detail of the business of gardening. The routine under each month is generally in harmony-with that already recommended, but certain variations of practice are suggested which may prove of service in some districts and under particular circumstances.

A work on gardening demands of the reader the exercise of judgment. If blindly followed, it may prove as often wrong as right; for it is not in the power of the authors to influence the weather in favour of their directions, or to insure to those who may follow their guidance a single one amongst the many conditions requisite to success. Although the times named for certain operations are the best as an average, peculiarities of climate and of season will require some modifications, which each one must discover for himself; and after the seed of any vegetable has been sown it is not always needful to give subsequent reminders of successional sowings. These naturally follow in accordance with the requirements of each particular garden. With such allowances duly made, these notes will, it is hoped, prove thoroughly practical, and tend materially to aid the cultivator in obtaining from the vegetable garden an abundance of everything in its season, and of a quality of which he need not be ashamed.

JANUARY

Work in the garden during the opening month of the year is entirely dependent on the weather, and it is futile to enter on a vain conflict with Nature. When heavy rains prevail keep off the ground, but immediately it will bear traffic without poaching be prepared to take advantage of every favourable hour. Much may be done in January to make ready for the busy spring, and every moment usefully employed will relieve the pressure later on. Survey the stock of pea-sticks, haul out all the rubbish from the yard, and make a 'smother' of waste prunings and heaps of twitch and other stuff for which there is no decided use. If properly done, the result will be a black ash of the most fertilising nature, such as a mere fire will not produce. Should the soil be frost-bound wheel out manure and lay it in heaps ready to be spread and dug in where seed-beds are to be made. If the weather is open and dry, trench spare plots and make ready well-manured plots for sowing Peas and Beans. So far as may be convenient, all preparatory work should be pushed on with vigour, and every effort must be made to lay up as much land in the rough as possible; for the more it is frozen through the greater will be its fertility, and the more beautiful, as well as more abundant, the crops.

It is a matter of the most ordinary prudence to be prepared to resist the shock of a severe frost. When this event occurs, many suffer loss because they are not prepared for it. Good brick walls and substantial roofs are needed for the safe keeping of fruits and the more valuable kinds of roots; but when rough methods are resorted to, such as clamping and pitting, there should be a large body of stuff employed, for a prolonged frost will find its way through any thin covering, no matter what the material may be. As there is not much to do now out of doors, it is a good time to look over the notes which were made concerning various crops in the past season, and to attend to the seed list.

Seed sowing should be practised with exceeding caution; but great things may be done where there are warm, sheltered, dry borders, and suitable appliances for screening and forwarding early crops. Under these favourable conditions, we advise the sowing of small breadths of a few choice subjects towards the end of the month; and, this being done, every care should be taken to nurse the seedlings through the trying times that are before them. Such things as tender young Radishes, Onions, small Salads, Spinach, Cabbage, and Carrots never come in too early; the trouble often is that they are seen in the market while as yet they are invisible in the garden. Hedges of Hornbeam, Laurel, or Holly, to break the force of the wind, are valuable for sheltering early borders, and walls are great aids to earliness by the warmth they reflect and the dryness they promote.

The soil for these early crops should be light and rich, and the position extra well drained, to prevent the slightest accumulation of water during heavy rains. Supposing you have such a border, sow upon it, as early as weather will permit, any of the smaller sorts of Cabbage Lettuce, Onion, Long Scarlet Radish, Round Spinach, Cabbage, and Carrot. All these crops may be grown in frames with greater safety, and in many exposed places the warm border is almost an impossibility. Reed hurdles and loose dry litter should be always ready when early cropping is in hand; and old lights, and even old doors, and any and every kind of screen may be made use of at times to protect the early seed-beds from snow, severe frost, and the dry blast of an east wind.

Forcing is one of the fine arts in the English garden. It is an art easily acquired up to a certain point, but beyond that point full of difficulty. Every step in this business is a conflict with Nature, and in such a conflict the devices of man must occasionally fail. A golden rule is to be found in the proverb 'The more haste the less speed.' Whatever the source of heat, it should be moderate at first, and should be augmented slowly. The earlier the forced articles are required the more careful should be the preparation for them, and the more moderate the temperature in the first instance. There must be at command a constant as well as sufficient temperature: when a forced crop has made some progress a check will be fatal to success. The beginner should acquire experience with Rhubarb and Sea Kale, then with Asparagus and Mushrooms and Dwarf French Beans, and so on to 'higher heights' of this branch of practical gardening.

Artichokes, Globe, are not quite hardy, and must be protected with litter.

Asparagus beds to be heavily manured, if not already done, but the beds need not be dug. Be content to lay the manure on, and the rains will wash the stimulant down to the roots in due time. In gardens near the coast seaweed is the best of manure for Asparagus, and the use of salt can then be dispensed with.

Beans, Broad, may be sown in frames, and towards the end of the month in open quarters. For early crops select the Longpod varieties. Sow on ground deeply dug and well manured.

Cabbage may be planted out at any time when weather permits, provided you possess, or can obtain, the plants; and it is of the utmost importance to secure them from a reliable source, or varieties may be planted which will in a few weeks send up flower-stems instead of forming tender hearts. At every season of the year vacant plots should be kept going with a few breadths of Cabbage. With our variable climate they may be acceptable, even in the height of summer, if there has been a hard run upon other vegetables, or some important crop has failed outright.

Cauliflower may be sown on a gentle hot-bed, or in a pan in the greenhouse, or even in a frame, to make a start for planting out in March or April.

Cress, to be enjoyed, must be produced from a constant succession of small but frequent sowings. All the sorts are good, but different in flavour, and they should be used only while young and tender. Sow at intervals of a few days in pans, as in the case of Mustard, until it is possible to cultivate in the open air, and then give a shady position during summer on a mellow and rather moist soil.

Cucumbers are never ready too soon to meet the demand in early spring. They are grown in houses more or less adapted to their requirements, and also in frames over hot-beds. At this time of year, however, frames are somewhat troublesome to manage, and in trying weather they are a little hazardous, although later in the season there is no difficulty whatever with them. For the present, therefore, we shall confine our remarks to house culture. Almost any greenhouse may be made to answer, but the work can be carried on most successfully and with the greatest economy in houses which are expressly constructed for Cucumbers. For winter work a lean-to, facing south, possesses special advantages. But for general utility, if we had to erect a building on a well-drained soil, it should be dwarf, sunk three feet in the ground, with brick walls up to the eaves, and lighted only from the roof. Such a structure is less influenced by atmospheric changes than a building wholly above ground. The size, of course, is optional; and quite a small house will supply an ordinary family with Cucumbers. But a small house is not economical either in fuel or in labour. A building thirty feet long by twelve feet wide, six feet high at the sides, and eight and a half feet high at the ridge, will not only grow Cucumbers and Melons, but will also be of immense service for many other plants. A division across the middle by a wall rising four feet, surmounted with a glass screen fitted to the roof, and finished with a door partially of glass, will greatly augment its usefulness. There should be an alley down the centre four or five feet wide, bounded by walls reaching four feet above the floor. These walls should be nine inches thick for two feet six inches of their height, but for the upper parts the brickwork need only be four and a half inches thick. This arrangement will provide a ledge on the inner side of each wall, and the main walls should also have ledges corresponding in height, on which to lay slates to carry the soil. To insure drainage, allow a space of about an inch between the slates, and place tiles or an inverted turf over every opening to prevent the soil being washed away. The hot-water pipes will be in chambers immediately beneath the plants. Openings in the alley walls, fitted with sliding doors, will admit the heat direct into the house whenever it may be desirable. Ventilation should be provided for under the ridge at each end, as well as in the roof. In such a house it is easy to grow Cucumbers all the year round, except, perhaps, in the dead of winter, when the short, dark days render the task difficult, no matter how perfect the appliances at command. The division in the centre will be found valuable at all times, and especially when one set of plants is failing; for another set can be brought into bearing exactly when wanted. But whatever the structure may be, the mode of culture remains substantially the same in any case. Now, as to soil, a compost made of mellow turfy loam and leaf-mould in equal parts will be effective and sweet. In the absence of leaf-mould, use two parts of loam and one of thoroughly decayed manure with a few pieces of charcoal added. Sweetness is not absolutely necessary for success, but nevertheless we like to have it, so that a visit to the Cucumber-house may be a source of pleasure. This it cannot be if rank manure has been used. Raise the seed singly in small 60-pots, and sow enough, for however good the seed may be a proportion will almost certainly fail from some cause at this critical period. Give the plants one shift into the 48-size, to keep them going until they are ready for putting into the beds. Cucumbers grow with great rapidity, and should never know a check, least of all by starvation. Upon the slates make as many heaps of soil as are required, and in the centre of each heap put one plant. As the roots extend, add more soil until the heaps meet and finally become level with the top of the brickwork. This treatment will supply food as the roots develop, and help to maintain the plants in bearing for a long period. Stout wires running parallel with the length of the house, a foot below the glass, will carry the vines. Temperature should never fall below 60° at night; but as the season advances, if the thermometer registers 90° on sunny days, no harm will be done, provided the roots are not dry, and the air be kept properly moist by plying the syringe. On dull days one good sprinkling over the foliage will suffice, and it should be done in the morning. In warm sunny weather, however, two or three syringings will be beneficial; but the work must not be done so late as to risk the foliage being wet when night comes on. There will be occasions when it may be advisable to avoid touching the leaves with water, if there is no probability of their drying before nightfall. In such a case the moisture can be kept up by freely sprinkling the floor and walls. Cucumbers cannot thrive if they are dry at the roots, but although there should be no stint of water, it must be given with judgment; and it is of the utmost importance that the drainage should be effectual, for stagnant water is even more injurious than a dry soil. A few sticks placed in various parts of the bed, reaching down to the slates, will serve as indicators. Draw and inspect them occasionally, and a pretty correct idea of the condition of the soil will be obtained. The water should be of the same temperature as the house; if applied cold the plants will sustain a serious check. In the event of the bed falling somewhat below the proper temperature, the water may with advantage be a few degrees higher than usual.

Horse-radish should be planted early, to insure fine roots for next Christmas beef.

Leek.—Those who wish to produce stems of superb size and beautiful texture must sow in heat during this month or early in February, for a longer period of growth is requisite than for ordinary crops. When sufficient root growth has been made, transplant into larger pots, and in due course transfer these to a frame where the plants may be gradually hardened off for putting out into specially prepared trenches in April.

Lettuces will soon be in demand, and the early hearts will be particularly precious. Sow a few sorts in pans, in frames, or on gentle hot-beds, to be ready for planting out by-and-by.

Melon.—Although the Melon is a fruit, its culture naturally forms part of the routine of a vegetable garden. Up to a certain point it may be grown in the same house with Cucumbers; but after that point is reached, the two plants need widely different treatment. Cucumbers are cut when young, and must be grown in a warm and humid atmosphere from beginning to end. Melons need warmth, and at the commencement moisture also; but the fruit has to be ripened, and after it is set dry treatment becomes essential for the production of a rich flavour with plenty of aroma. In large gardens, three crops of Melons are usually grown in the same house in one season. A light soil is advisable at the beginning of the year, but later in the season a heavier compost may be employed. For the first sowing select an early variety, and at the beginning of this month put the seed in separate pots. Re-pot the plants once, and they will be ready for the beds by the first week of February. Melons from this sowing should be fit for table in May, which is quite as early as they can be produced with any sugar in them. Until the fruits begin to swell the treatment advised for Cucumbers will suit Melons also. Afterwards the watering will need careful management. It would be an advantage if the fruit could be finished off without a drop of water from the time they are about two inches in diameter, but the hot pipes render it almost impossible. Still, water must not be given more frequently than is actually necessary to keep the plants going, and when it is applied let there be a thorough soaking. At the same time ventilation will demand constant attention, and, provided the temperature can be maintained, it is scarcely possible to give air too freely. In the early stage of growth, and in mild weather, if the thermometer registers 65° at 9 P.M., the cultivator may sleep peacefully so far as Melons are concerned. As the season advances, the temperature may be increased to 70° by night, and 75° to 90° by day. With reference to stopping, it may be sufficient to say that it is a waste of energy to allow the plant to make a large quantity of vine, which has afterwards to be cut away. By judiciously pinching out the shoots, the plant can be equally spread over the allotted space. The flowers must be fertilised, and in this respect the treatment differs from that advised for Cucumbers. The practice has the advantage of allowing the fruits to be evenly distributed over the vine, and from four to six, according to the size of the variety, will be enough for each plant to ripen.

Mustard.—Those who care for salads need a supply of Mustard almost all through the year, and to secure a succession it will be necessary to sow at regular intervals. It is a good plan to keep a few boxes in use for the purpose in a plant-house or pit, sowing one or two at a time as required, and taking care not to sow wastefully. The seed may be sown out of doors all the summer, on a shady border, but nothing surpasses boxes or large pans under glass. Mustard and Cress should never be sown in the same row or in the same pan, but separately, because they do not grow at the same pace, and the former may be fit for use a week or so before the latter. Do not be content to use Rape, or any other substitute, but sow the genuine article.

Onion.—The modern practice of sowing Onion seed in boxes under glass is to be commended for several reasons. It insures a long season of growth and results in handsome bulbs far above the average in size. Transplanting affords the opportunity of selecting the strongest seedlings and of placing them at exact intervals in the bed. As a crowning advantage this system, to a large extent, prevents attack from the Onion Fly. Sow in boxes filled with rich soil and see that the plants have sufficient water, although very little is necessary until after transfer to other boxes.

Peas of the round-seeded class may be sown in open quarters, and the driest and warmest places must be selected. It is next to impossible to grow them too well; for if the haulm runs up higher than usual, the produce will be the finer. Remember, too, that if deep trenches are dug and a lot of manure is put in for Peas, the ground is so far prepared for Broccoli, Celery, and late Cauliflowers to follow; for the early-sown Peas will be off the ground in time for another paying crop. As everybody wants an early dish of Peas, sow one of the forward marrowfat varieties in pots, or on strips of turf laid grass-side downwards in boxes having movable bottoms that can be withdrawn by a dexterous hand when the transfer is made from frames to the open ground. Troughs for Peas can be made in very little time out of waste wood that may be found in the yard; or a few lengths of old zinc spouting blocked up at the ends will answer admirably. In the absence of such aids, flower-pots may be used. The seed should have the shelter of a frame or pit, but should have the least possible stimulus from artificial heat, except in cases where there is all the skill at command to promote very early production.

Potatoes are prized when they come in early, and may be forwarded on beds of leaves and exhausted hot-beds by covering with light rich soil, and employing old frames for protection, with litter handy in case of frost. For this early work select the earliest Kidneys and Rounds; the main-cropping varieties are not quick enough.

Radishes are more or less in demand for the greater part of the year. The early crops are, however, especially valued, and there need not be the least difficulty in producing a supply. A half-spent hot-bed, or, indeed, any position that affords shelter and warmth, will answer admirably for raising this crop until it may be trusted to a suitable position in the open.

Sea Kale may be covered with pots or a good depth of litter, or a combination of pots and litter. This should be done early, as at the first move of vegetation this delicious vegetable will come into use, and will generally be of finer quality than if forced. It happens, however, to be the easiest of all things to force, and so, wherever it is cared for, a plentiful supply may be maintained from Christmas (or earlier) until May. As the leaf-stems must be thoroughly blanched, covering is needful in all cases.

Spinach may be sown in open quarters. If the frost destroys the plant, sow again. Some risk must be encountered for an early dish of this highly-prized vegetable. Keep the autumn-sown Spinach clear of weeds, and in gathering (if it happens to be fit to supply a gathering), pick off the leaves separately with a little care.

Strawberries.—Seed of the Alpine varieties sown in pans this month, for transfer later to the open ground, usually produce fine fruits in September.

Tomato.—Of the immense value of the Tomato as an article of diet we need say nothing, but we may confidently affirm that its merits for decorative purposes have not as yet been fully recognised. Long racemes of brilliant glossy fruit are sometimes employed with striking effect in épergnes, and there is a natural fitness in using them for decorating the dinner table. All the Tomatoes can be grown and ripened under glass in almost any fashion which may suit the cultivator's convenience. Pits, frames, vineries, and Peach-houses will bring the fruit to perfection, either in pots or planted out. Magnificent crops are also grown in the manner usual with Cucumbers, but in a lower temperature; and those who have an early Cucumber house at liberty during the summer may turn it to good account for Tomatoes. The soil should be prepared and laid up in the autumn. It must not be too rich, or there will be much foliage and little fruit, and the flowering will also be late. A compost of leaf-mould and loam with an addition of sand suits Tomatoes admirably; but raw manure should be regarded as poison. Sow thinly in well-drained pots firmly filled with soil, and place in a temperature of 60° or 65°. When large enough to handle, transfer the seedlings to small pots, and, if necessary, shade them for a few days. Keep them near the glass until the roots are established, and allow them to suffer no check from first to last.

FEBRUARY

The work of this month is to be carried on as weather permits, but with greater activity and more confidence, for the sun is fast gaining power. Earnest digging, liberal manuring, and scrupulous cleansing are the tasks that stand forward as of pre-eminent importance. Many weeds, groundsel especially, will now be coming into flower, and if allowed to seed will make enormous work later on. It is well, however, to remember—what few people do remember, because the fact has not been pressed upon their attention—that weeds of all kinds, so long as they are not in flower, are really useful as manure when dug into the soil. Therefore a weedy patch is not of necessity going to ruin; but if the weeds are not stopped in time, they spread by their seeds and mar the order of the garden. Dig them in, and their decay will nourish the next crop. If early sowing is practised, and the earliest possible produce of everything is aimed at, there must be always at hand the means of protection, such as litter, spruce branches, mats, or other material, as circumstances require. The vigilant gardener is not surprised by the weather, but is always armed for an emergency. Read the notes for January before proceeding further; and in respect of what remains undone, spare the necessity of reminders here.

Frame Ground should be kept scrupulously clean and orderly. Many things will require watering now, but water must not be carelessly given, because damp is hurtful during frosty weather. Take care that the plants are not crowding and starving, or they will come to no good.

Artichoke, Globe.—Plants from a sowing made now in a frame, and transferred to the open at the end of April, will generally produce heads in the following August, September, and October.

Artichokes, Jerusalem, may be planted this month where it has been possible to prepare the ground. Use whole sets if convenient, or plant cut sets with about three eyes in each.

Beans, Broad, may be sown both for early and main crops now, and with but little risk of damage by spring frosts. The driest and warmest situation should be selected for the early sorts, and the strongest land for the late ones. If sowings were made in frames last month, take care to harden the plants cautiously preparatory to planting out; if caught by a sharp frost, every one will perish.

Beans, French.—To precede the outdoor crops make a sowing of Dwarf French Beans in frames, and of the Climbing French varieties in orchard-houses or other available spaces under glass.

Beet.—Sowings of the Globe variety may be made this month and in March, on a gentle hot-bed under frames, to provide roots in advance of the outdoor supplies.

Broccoli.—Sow on a warm sheltered border, and also in a frame. With such an important crop at this time of year, there should be at least two strings to the bow.

Brussels Sprouts.—For an early gathering of large buttons a sowing should be made now on the warm border. This vegetable requires a long period of growth to attain perfection, and those who sow late rarely obtain such fine buttons as the plant is capable of producing.

Cabbage may be sown in pans or boxes placed in a frame, to be planted out in due time for summer use, and from a quick-growing variety tender hearts may be cut almost as early as from autumn-sown plants. Where plantations stand rather thick, draw as fast as possible from amongst them every alternate plant, to allow the remainder ample space for hearting. It is well to remember that the small loose hearts of immature Cabbages make a more delicate dish than the most complete white hearts; but when grown for market, or to meet a large demand, there must be bulk and substance. Cabbages are in constant request to mend, and to make stolen crops, or take the place of anything that fails past recovery.

Capsicum and Chili should be sown now or in March on a hot-bed, and be potted on until the plants are fit to be placed in the greenhouse or conservatory.

Cauliflower.—Another sowing should be made under glass to supply a succession of plants.

Corn Salad thrives well in any soil not particularly heavy, the best being a sandy fertile loam. Sow in drills six inches apart; keep the hoe well at work, and when ready thin the plants out to six inches apart. They should be eaten young.

Couve Tronchuda produces two distinct dishes. The top forms a Cabbage of the most delicate flavour and colour, and furnishes the best possible dish of greens in autumn; and the midribs of the largest leaves may be cooked in the manner of Sea Kale, and will be found excellent. This delicious vegetable may be secured for use in summer and autumn and far on into the winter by successive sowings in February, March, and April; the first sowings to be assisted with heat. The plants should be put out as early as possible on rich soil at from two to three feet each way; they must have plenty of water in a dry summer. The season of Portugal Cabbage may be prolonged by taking up what plants are left before severe frost occurs, and heeling them into a bank of dry earth in a shed or outhouse.

Egg Plant.—The fruits of Egg Plants play a more important part in the cookery of the French and Italians than with us, and they make a delicious dish when properly cooked. Seed may be raised in heat, but when summer comes the plants thrive in rich soil at the foot of a wall facing south. The white and purple varieties are grown for ornament as well as for cooking. Sow now or in March in heat, and in June the plants should be ready for transferring to rich soil in a sheltered spot, allowing each one a space of two feet.

Garlic to be planted in rows, nine inches apart each way, and two inches deep in rich mellow soil.

Lettuce.—Sow again on a warm border and in frames. Plant out in mild weather any that are fit from frames and hot-beds, first making sure that they are well hardened.

Mustard.—It is easy work with a frame to have Mustard at any time; and many small sowings are better than large ones, which only result in waste to-day and want to-morrow.

Onion.—There is still time for sowing seed in boxes preparatory to planting out in April.

Parsley to be sown in the latter part of the month.

Parsnip should be sown as early as possible, on the deepest and best ground as regards texture; but it need not be on the richest, for if the roots can push down they will get what they want from the subsoil, and therefore it is of great importance to put this crop on ground that was dug twice in the autumn.

Pea.—Sow early sorts in quantity now, in accordance with probable requirements; but there will be a loss rather than a gain of time if they are sown on pasty ground or during bad weather. There are now many excellent sorts of moderate height, and these give the least trouble in their management; but a few of the taller varieties still remain in favour, because of their fine quality. However, there is time yet for sowing mid-season and late Peas; but the sooner some of the first-earlies are in, the better. It is customary to sow many rows in a plot rather close together, but it is better practice to put them so far apart as to admit of two or three rows of early Potatoes between every two rows of Peas. This insures abundance of light and air to the Peas, and the latter are of great value to protect the Potatoes from May frosts that often kill down the rising shaws. A warm, dry, fertile soil is needed for first-early Peas. Those already up and in a bad plight should be dug in and the rows sown again. It is worthy of note that if Peas are thoroughly pinched and starved by hard weather, they rarely prove a success; therefore, if they go wrong, sacrifice them without hesitation and begin again. Where early rows are doing well put sticks to them at once, as the sticks afford considerable protection, and the effect may be augmented by strewing on the windward side small hedge clippings and other light dry stuff.

Radishes, to be mild, tender, and handsome, must be grown rapidly. If checked, they become hot, tough, and worthless. Much may be done to forward a crop by means of dry litter and mats to protect the plants from frost, removing the protection in favourable weather to give the crop the fullest possible benefit of air and sunshine. Old worn-out frames that will scarcely hold together will pay their first cost over again, with the aid of a little skill, in growing Radishes.

Rhubarb should be taken up and divided, and planted again in rich moist soil, every separate piece to have only one good eye. Do not gather this season from the new plantation, but always have a piece one year old to supply the kitchen. This method will insure sticks to be proud of, not only for size, but for colour and flavour.

Savoys are valued by some when small, and by others they are prized for size as much as for their excellent flavour when well frosted. Large Savoys must have a long season of growth; therefore sow as soon as possible, either in a frame, or on a rich, mellow seedbed, and be ready to prick them out before they become crowded.

Sea Kale.—The plantations reserved for latest supplies should not be covered until they begin to push naturally, and then the coverings must be put on to blanch the growth effectually. Open-ground Sea Kale may be uncovered as soon as cut, but a little litter should be left to give protection and help the young shoots to rise, because after blanching the cutting is a severe tax on the plant, and it has to begin life afresh and prepare for the work of the next season.

Shallot.—When well grown the clumps are bigger than a man's fist, and each separate bulb thicker than a walnut. To grow them well they must have time; so plant early, on rich ground, in rows one foot apart and the bulbs about nine inches asunder. Press them into the earth deep enough to hold them firmly, but they are not to be quite buried.

Spinach.—Sow the Round-seeded plentifully; if overdone the extra crop can be dug in as manure, and in that way will pay.

Tomato.—In many gardens the first sowing is made this month, and when treated fairly, the plants come into bearing in about four months. Use good porous soil for the seed-pans. Sow very thinly in a temperature of 60° or 65°, and get the plants into thumb pots while they are quite small.

Turnip may be sown on warm borders, but it is too early for large breadths in open quarters.

MARCH

This is the great season for garden work, and the gardener must be up with the lark and go to bed with the robin, which is the latest of birds to bid farewell to a sunny day. The first care should be to make good all arrears, especially in the preparation of seed-beds, and the cleaning of plots that are in any way disorderly. Where early-sown crops have evidently failed, sow again without complaining; seed costs but little, and a good plant is the earnest of a good crop; a bad plant will probably never pay the rent of the ground it occupies. Keen east winds may cause immense damage, but a little protection provided in time will do wonders to ward off their effects, and the sunny days that are now so welcome, and that we are pretty sure to have, will afford opportunity for giving air to plants in frames, for clearing away litter, and for the regular routine work of the season.

Seed of almost every vegetable grown in the garden may be sown in the month of March. Make successional sowings of whatever it may be advisable to put under cover or on heat, and then proceed with open-ground sowings as weather and circumstances permit. The weather is the master of outdoor work, and it is sheer waste of time to fight against it. It is better to wait to the end of the month, or even far into the next, before sowing a seed than to sow on pasty ground. But it matters not how dry the ground may be, and if the wind blows keenly, that should only be an inducement to brisk action; for seeds well sown have everything in their favour if they are not too early for the district. Very important indeed it is now to secure a Hot-bed.—To make one is easy enough, but it is of no use to half make it; for half-acres in this department do not bear good corn. In the first place, secure a great bulk of manure, and if it is long and green, turn it two or three times, taking care that it is always moderately moist, but never actually wet. If the stuff is too dry, sprinkle with water at every turn, and let it steam away to take the rankest fire out of it. Then make it up where required in a square heap, allowing it to settle in its own way without treading or beating. Put on a foot-depth of light, rich soil after the frames are in their places, and wait a few days to sow the seed in case of a great heat rising. When the temperature is steady and comfortable, sow seeds in pots and pans, as needful, the quantity required of each separate crop, and stand them on bricks above the bed, and the heat will then be none too much for them. In the course of a few days finish the work by putting in a body of earth. Do not attempt to hurry the growth of anything overmuch, for undue haste will produce a weak plant; rather give air and light in plenty, but with care to prevent injurious check, and the plants will be short and healthy from the first.

Artichokes, Globe, to be cleared of protecting material as soon as weather permits, and fresh plantations made ready for suckers to be put in next month. A new plantation may also be formed by sowing seeds; in fact, a sowing ought to be made every year. Where early produce is required, the plants should be protected during winter to supply suckers in the spring; but, if late supplies suffice, the sowing of a few rows every year will reduce the labour, and render the production of Globe Artichokes a very simple affair.

Artichokes, Jerusalem, may be planted now advantageously. Strong, deep soil produces the best crop, and large roots are always preferred by the cook, because of the inevitable waste in preparing this vegetable. The Jerusalem Artichoke is certainly not properly appreciated, and one reason is that it is often carelessly grown in any out-of-the-way starving corner, whereas it needs a sunny, open spot, and a strong, deep soil, and plenty of room. To hide an ugly fence during summer no more useful plant is grown.

Asparagus.—Little attention is required as yet, except to remove every weed as soon as it can be seen. If the beds are dry, and there are no indications of coming rain, one good soaking of water or weak sewage will be very beneficial. Mark out and make beds for sowing seed next month.

Bean, Broad.—Plant out those raised in frames, and earth up those from early sowings that are forward enough. Sow for main crops and late supplies. In late districts a few of the earliest sorts may be sown to come in before the Windsor section.

Beet.—Sow a little seed for an early supply, in well-dug mellow soil. The crop will need protection in the event of frost.

Broccoli for autumn use to be sown early; and at the end of the month sow again in quantity for winter supplies. In mild weather, put out the plants from the earlier sowings made in frames as soon as they are fit and well hardened.

Brussels Sprouts.—Look after the bed sown last month, and sow again for the main crop. The best possible seed-bed is wanted and a rich well-tilled soil for the plants when put out.

Cabbage of two or three kinds should be sown now to supply plants for filling up as crops are taken off, and also to patch and mend where failures happen. Where the owner of a garden has opportunities of helping his poorer neighbours, he may confer a real benefit by supplying them with Cabbage and Winter Greens for planting in their garden plots. Cottagers too often begin with bad stocks—very much to their discouragement in gardening, and to the loss of wholesome food the garden should supply. The rankest manure may be employed in preparing ground for Cabbage, reserving the well-rotted manure for seed-beds and other purposes for which it will be required. A sowing of Red Cabbage now will insure heads for pickling in autumn.

Carrot.—Sow one of the quick-growing varieties at the first opportunity, but wait for signs of settled spring weather to sow the main crops of large sorts.

Cauliflower.—Plant out as weather permits from hand-lights and frames, choosing the best ground for this vegetable. In preparing a plot for Cauliflower, use plenty of manure; and if it is only half-rotten, it will be better than if it were old and mellow.

Celeriac.—So far as seed sowing is concerned, Celeriac may be treated in the same way as Celery.

Celery.—For the earliest supply, sow on the first of the month a pinch of seed of one or more of the smaller red or white sorts on a mild hot-bed, or in an early vinery. As soon as the plants are large enough to handle, prick them out three inches apart on a nice mellow bed of rich soil on a half-spent hot-bed; give them plenty of light, with free ventilation as weather allows, and constant supplies of water. About the middle of the month sow again and prick out as before; but if no hot-bed is available, a well-prepared bed in a frame in a sunny position will answer; or, if the season is somewhat advanced, a bed of rotten manure, two or three inches deep, on a piece of hard ground, will suffice, if the plants are kept regularly watered. From this bed they will lift with nice roots for planting out, scarcely feeling the removal at all.

Chives to be divided and re-planted on a spot which has not previously been occupied with the crop.

Cucumber.—The vines should now be in a flourishing condition, but it is necessary to look forward to the day when they will fall into the sere and yellow leaf. More seed sown singly in pots will provide a succession of plants. Re-pot them once or twice if desirable, and when large enough turn them out between the first lot. As the old plants fail, the new-comers will supply their places. Setting the bloom, as it is called, is not only useless, but is a mischievous procedure. It results in the enlargement of one end of the fruit, and ruins its appearance. If seed be the object, of course the process is justifiable; but for the table a 'bottle nose' cannot be regarded as an ornament. Besides, the ripening of seed in a single fruit will materially diminish the usefulness of the plant, and perhaps entirely end its career. Stopping the vine is a necessity, but it should not be done too soon. In the early stage of growth, it reduces the vigour of the plant and retards its fruiting; but when the fruit is visible, stopping aids its development and at the same time tends to regulate and equalise the growth.

Frame culture of Cucumbers is usually begun in March. There are men who can produce fruit from hot-beds all the year round, but it is a difficult task, and as a rule ought not to be expected. At this time of year, however, success is fairly within reach of ordinary skill. In quite the early part of the month put seed singly into pots which must be kept in a warm, moist place. The plants will then be ready for frames at the end of the month. The most important business is the preparation of the bed, and in this, as in all else, there is a right and a wrong way of doing the work. Accurately set out the space on which it is to be made. If there is plenty of manure, make the bed large enough to project eighteen inches beyond the lights all round. But if manure is scarce, cut the margin closer, and trust to a hot lining when the heat begins to flag. Commence with the outside of the bed, employing the long stuff in its construction; and keep this part of the work a little in advance of the centre until the full height is reached. A bed made in this way will not fall to pieces, and the heat will be durable in proportion to its size and thickness. Where fallen leaves are abundant, they should be used for the middle of the bed, and they will give a more lasting heat than short manure. When the bed has settled down to a steady temperature, add six or nine inches of mellow loam over the entire surface, upon which place the frames. To insure drainage, it is an excellent plan to lay common flake hurdles on the top of the heap before adding the soil. These do not in the least interfere with the free running of the roots. It is usual to have two plants under each light, but where the management is good, one is quite enough. The subsequent work consists of shading and sheltering, to prevent any serious check from trying weather, and in giving just water enough and no more. The fermenting material should sustain the temperature of the frame, even during frosty nights, and mats will screen off strong sunshine as well as cold winds. The plants will need stopping earlier than those grown in houses, and as there are no hot-pipes to dissipate the moisture, rather less water will be necessary, both in the soil and from the syringe. But the water employed should always be of the same temperature as the bed. This is easily managed by keeping a full can standing with the plants. In large frames, where there is a good body of manure and the loam is mellow and turfy, pieces of Mushroom spawn can be inserted all over the bed. The Mushrooms may appear while the bed is in full bearing; but if they do not they will come when the plants are cleared out, and pay well to keep the lights in use another month or so.

Garlic may still be planted, but no time is to be lost.

Herbs of many kinds may be sown or divided, and it will be necessary to look over the Herb quarter and see how things stand for the supplies that will be required. A little later, excess of work may prevent due attention to this department.

Horse-radish to be planted, if not done already.

Kohl Rabi, or Knol Kohl, to be sown in small quantity at the end of the month, and onwards to August, as required. If cooked while young, the bulbs are an excellent substitute for Turnips in a hot, dry season.

Leek.—Sow the main crop in very rich, well-prepared soil, and rather thickly, as the seedlings will have to be planted out. With a little management this sowing will yield a succession of Leeks.

Lettuce.—Plant out and sow again in quantity. All the kinds may be sown now, but make sure of enough of the Cos and smaller Cabbage varieties. In hot, dry soils, where Lettuces usually run to seed early, try some of the red-leaved kinds, for though less delicate than the green and white, they will be useful in the event of a scorching summer. Lettuces require a deep free soil with plenty of manure.

Melon.—Raise a few seeds singly in pots, in readiness for putting under frames on hot-beds next month. Re-pot the plants, and repeat the process if the beds are not ready, for Melons must not be starved, especially in the early stage of growth. Some growers make up the beds in March, and sow upon them when the heat becomes steady, but the practice is somewhat precarious. In a cold, late spring the heat may not last a sufficient time to carry the plants safely into warm weather. Hence it is more reliable to raise them now in a warm house, and make the bed at the beginning of April.

Onion.—The plants already raised in boxes to be removed to cold frames. If necessary, they should be pricked off into other boxes in order to avoid overcrowding. Keep the frames close at first, but give air with increasing freedom as the time approaches for transfer to the open ground. Sow the main crop in drills nine inches apart, and tread or beat the ground firm. This crop requires a rich soil in a thoroughly clean and mellow condition, and it makes a capital finish to the seed-bed to give it a good coat of charred rubbish or smother ash before sowing the seed.

Parsnip.—Sow main crop in shallow drills eighteen inches apart in good soil deeply dug. The seed should be lightly covered, and new seed is indispensable.

Pea.—Sow the finest sorts of the Marrowfat class. Take care to put them on the best seed-bed that can be made, and allow sufficient room between the taller sorts for a few rows of Cabbage, Broccoli, or Potatoes. A crowded quarter of Peas is never satisfactory; the rows smother each other, and the shaded parts of the haulm produce next to nothing.

Potato.—A small quantity for early use should be planted at the opening of the month when the ground is dry and the weather soft. If planted when frost or cold winds prevail, sets may become somewhat shrivelled before they are covered, and every care should be taken to prevent such a check to the initial vigour of the plant. The first-early sorts will necessarily have the chief attention now, and warm sheltered spots should be selected for them. Any fairly good soil will produce a passable crop of Potatoes; but to secure a first-class sample of any early sort, the ground should be made up with the aid of turfy soil and charrings of hedge clippings and other light, warm, nourishing material. Strong manures are not to be desired, but a mellow, kindly, fertile soil is really necessary, and it will always pay well to take extra pains in its preparation, because all the light rubbish that accumulates in yards and outhouses can be turned to account with only a moderate amount of labour, and the result of careful appropriation of such rubbish will be thoroughly satisfactory. Burn all the chips and sticks and other stubborn stuff, and lay the mixture in the trenches when planting, so that the roots may find it at their first start. As the Potato disease does not usually appear until late in summer, early planting is a safe precaution, for it insures early ripening of the crop. The planting of main crops may commence towards the end of March and be completed during April, according to the locality and the condition of the soil.

Radish.—From March to September make successive sowings in the coolest place that can be found for them.

Scorzonera to be treated much the same as Salsify. See note on the latter under April.

Sea Kale to be sown in well-prepared beds; or plantations may be made of the smaller roots of the thickness of a lead pencil, and about four inches in length. Plant them top end uppermost, and deep enough to be just covered.

Spinach.—Sow in plenty. The Perpetual or Spinach Beet should not be forgotten. This is one of the most useful vegetables known, as it endures heat and cold with impunity, and when common Spinach is running to seed the Perpetual variety remains green and succulent, and fit to supply the table all the summer long.

Spinach, New Zealand, is another excellent vegetable in high summer when the Round-seeded variety is worthless. The plant is rather tender, and for an early supply the seed must be sown in moderate heat, either in this month or in April. When large enough, get the seedlings into small pots, and gradually harden them before planting in the open about the end of May.

Strawberries.—Spring is undoubtedly preferable to autumn for planting, and results in a finer crop of fruit in the following year. Just as growth is commencing is the most favourable time, and this, of course, depends on the character of the season. Alpine Strawberries may be sown outdoors this month or in September for fruiting in the succeeding year.

Tomato.—In ordinary seasons and in the southern counties there is no difficulty in producing handsome Tomatoes in the open border; but to ripen the fruit with certainty it is imperative that an early variety be chosen. With the rise of latitude, however, the crop becomes increasingly precarious, until in the North it is impossible to finish Tomatoes without the aid of glass. For plants which are to ripen fruit in the open, a sowing should be made early in the month, in the manner advised under January. Plants which are ready should be transferred to small thumb pots. Put them in so that the first leaves touch the rim of the pot, and place them in a close frame or warm part of the greenhouse for a few days until the roots take hold. To save them from becoming leggy, give each plant ample space, and avoid a forcing temperature. A shelf in a greenhouse is a good position, and plants in a single row upon it will grow stout and short-jointed. Thrips and aphis are extremely partial to Tomatoes. Frequent sprinklings in bright weather will help to keep down the former, and will at the same time benefit the plants. Both pests can be destroyed by fumigating with tobacco, and when the remedy is to be applied water should be withheld on that day. A moderate amount of smoke in the evening and another application in the morning will be more destructive to the vermin, and less injurious to the plants, than one strong dose. The usual syringing must follow. Plants for the open ground must not be starved while in pots; they will need potting on until the 4-1/2-inch or 6-inch size is reached, and it is important that they should never be dry at the roots. Shading will only be necessary during fierce sunshine; in early morning and late in the afternoon they will be better without it.

Water Cress.—It is quite a mistake to suppose that a running stream is requisite for growing this plant, and it is equally a mistake to suppose that the proper flavour can be secured without the constant use of water.

Sow in a trench, water regularly and copiously, and mild and tender Water Cress will reward the labour.

Winter Greens of all kinds to be sown in plenty and in considerable variety; for in the event of a severe winter some kinds will prove hardier than others.

APRIL

Vegetation is now in full activity, the temperature increases rapidly, frosts are less frequent, and showers and sunshine alternate in their mutual endeavours to clothe the earth with verdure and flowers. The gardener is bound to be vigilant now to assist Nature in her endeavours to benefit him; he must promote the growth of his crops by all the means in his power; by plying the hoe to keep down weeds and open the soil to sunshine and showers; by thinning and regulating his plantations, that air and light may have free access to the plants left to attain maturity; by continuing to shelter as may be needed; and by administering water during dry weather, that vegetation may benefit to the utmost by the happy accession of increasing sunlight.

Artichoke, Globe.—Suckers to be put in the plantations prepared for them last month, in rows three to four feet apart each way.

Asparagus.—Rake off into the alleys the remnant of manure from the autumn dressing, and as soon as the weather is favourable give the beds a light application of salt. If new beds are required, there must be no time lost either to sow seed or get in plants. Our advice to those who require only one small plantation is to form it by planting strong roots; but those who intend to grow Asparagus largely may sow down a bed every year, until they have enough, and then leave well alone; for a bed properly made will last ten years at the very least, if taken care of. It has been clearly demonstrated that this much-esteemed vegetable may be grown to perfection in any garden with little more expense than attends other crops, provided only that a reasonable amount of skill is brought to bear upon the undertaking. A deep, rich, sandy loam suits it. Dig in a good body of manure, and provide a mellow seed-bed. This being done, care must be taken to sow thinly, and, in due time, to thin severely; for a crowded plant will never supply fat sticks. Beds may be made by planting roots instead of sowing seeds, but the roots must be fresh, or they will not prosper. The advantage of using plants is that 'grass' may be cut earlier than when produced from seed.

Bean, Broad.—Sowings may be made until the middle of this month, after which time they are not likely to pay, especially on hot soils. It is customary to top Beans when in flower, and the practice has its advantages. In case the black fly takes possession, topping is a necessity, for the insect can only subsist on the youngest leaves at the top of the plant, and the process pretty well clears them away.

Beans, Dwarf French, may be sown outdoors at the end of the month, but not in quantity, because of the risk of destruction by frost. Much may be done, however, to expedite the supply of this popular vegetable, and sowings in boxes placed in gentle heat or under the protection of a frame will furnish plants which may be gradually hardened off for transfer to the open in May. In proportion to the means at command, early sowings outdoors will live or die, as determined by the weather, although a very little protection is sufficient to carry the young plants through a bad time in the event of late frosts and storms. But sowings made at the end of the month will probably prosper.

Bean, Climbing French.—Sowings of the Climbing French Bean may be made this month as directed for the Dwarf French class: the earliest in gentle heat for transplanting, and later on in open quarters for succession crops.

Beet.—At quite the end of the month sow in drills, a foot or fifteen inches apart, on deep, well-dug ground, without manure. Large Beets are not desired for the kitchen; but rather small, deeply coloured, handsome roots are always valued, and these can only be grown in soil that has been stirred to a good depth, and is quite free of recent manuring.

Broccoli.—Make another sowing of several sorts, giving preference as yet to the early varieties. In particularly late districts, and, perhaps, pretty generally in the North, the late Broccoli should be sown now, but in the Midlands and the South there is time to spare for sowing. Be particular to have a good seed-bed, that the plants may grow well from the first; if the early growth be starved, the plants become the victims of club and other ruinous maladies.

Brussels Sprouts.—In many households late supplies of Brussels Sprouts are much valued, and as the crop is capable of enduring severe weather, a supplemental sowing should always be made during this month. Rich soil and plenty of room are essential.

Cabbage.—Sow the larger kinds for autumn use, and one or two rows of the smaller kinds for planting in odd places as early crops are cleared off. Cows, pigs, and poultry will always dispose of surplus Cabbage advantageously, so there can be no serious objection to keeping up a constant succession. Plant out from seed-beds as fast as the plants become strong enough, for stifling and starving tend to club, mildew, and blindness. Where Red Cabbage is in demand for use with game in autumn, seed should be sown now.

Cardoons to be sown on land heavily manured in rows three or four feet apart, the seeds in clumps of three each, eighteen inches apart. They are sometimes sown in trenches, but we do not approve of that system, for they do not require moisture to the extent of Celery, and the blanching can be effectually accomplished without it. Our advice is to plant on the level, unless the ground is particularly dry and hot, and then trenches will be of great service in promoting free growth. To insure their proper flavour, Cardoons must be large and fat.

Carrot.—Sow the main crops and put them on deeply dug ground without manure.

Cauliflowers to be planted out at every opportunity, warm, showery weather being most favourable. If cold weather should follow, a large proportion of the plants will be destroyed unless protected, and there is no cheaper protection than empty flowerpots, which may be left on all day, as well as all night, in extreme cases when a killing east wind is blowing. Sow now for late summer and autumn use, prick the plants out early to save buttoning, and they will make a quick return.

Celery.—Sow in a warm corner of the open ground on a bed consisting largely of rotten manure. It may happen in a good season that this outdoor sowing will prove the most successful, as it will have no check from first to last, and will be in just the right state for planting out when the ground is ready for it after Peas and other early crops. If Celery suffers a serious check at any time, it is apt to make hollow stems, and then the quality is poor, no matter to what size the sticks may attain. Prick out the plants from seed-pans on to a bed of rotten manure, resting on a hard bottom, in frames or in sheltered nooks, and look after them with extra care for a week or two. Good Celery cannot be grown by the haphazard gardener.

Endive.—Sow a small quantity in moderate heat for the first supply, in drills six inches apart, and when an inch high prick out on to a bed of rich light soil.

Herbs.—Chervil, Fennel, Hyssop, and other flavouring and medicinal Herbs, may be sown now better than at any other time, as they will start at once into full growth, and need little after-care other than thinning and weeding. Rich soil is not required, but the position must be dry and sunny.

Leek to be sown again if the former sowing is insufficient or has failed.

Lettuce to be sown for succession, the quick-growing, tender-hearted kinds being the best to sow now. Plant out from frames and seed-pans. A few forward plants may be tied, but as a rule tying is less desirable than most people suppose. Certainly, after tying, the hearts soon rot if not quickly eaten; and Lettuces as fine as can be desired may now be grown without tying, the close-hearting sorts being very much improved in that respect.

Melon.—Sow again for a second crop in houses, and grow the plants in pots until they reach a foot high. The early crop will then be ripe, and the house can be cleared and syringed for a fresh start. From this sowing fruit should be ready about the beginning of July. The frame culture advised for Cucumbers will be right for Melons, until the fruits attain the size of a small orange. Then a thorough soaking must be given, and under proper management no more water should be necessary. A dry atmosphere and free ventilation are essential to bring the fruit to perfection. Stopping must be commenced early by pinching out the leader, and only one eye should be allowed beyond the fruit which are to remain. Six will be enough for one plant to carry, and they should be nearly of a size, for if one obtains a strong lead, it will be impossible to ripen the others. The remainder should be gradually removed while young. The worst foe of the Melon is red spider, and it is difficult to apply a remedy without doing mischief. Water will destroy it, but this may have disastrous results on the fruit. The most certain preventive is stout well-grown plants. Weakly specimens appear to invite attack, and are incapable of struggling against it. Where plants are occasionally lost through decay at the collar, small pieces of charcoal laid in a circle round the stem have proved a simple and effectual antidote.

Onion.—The plants raised under glass in January or February should be ready for planting out on some favourable day about mid-April. If any mishap has befallen the sowings made in the open in March there must be no delay in resowing early in the present month, for Onions should have good hold of the ground before hot weather comes. Onions for pickling should be grown thickly on poor ground made firm. The plants are not to be thinned, but may be allowed to stand as thick as pebbles on the seashore. The starving system produces abundance of small handsome bulbs that ripen early, which are the very things wanted for pickling. The Queen and Paris Silver-skin are adapted for the purpose.

Parsley to be sown in quantity for summer and autumn supply; thin as soon as up, to give each plant plenty of room.

Peas to be sown again for succession.

Potato.—Take the earliest opportunity of completing the planting of main crops.

Salsify.—This delicious root, which is sometimes designated the 'Vegetable Oyster,' requires a piece of ground deeply trenched, with a thick layer of manure at the bottom of the trench, and not a particle of manure in the body of soil above it. The roots strike down into the manure, and attain a good size combined with fine quality. If carelessly grown, they become forked and fibrous, and are much wasted in the cooking, besides being of inferior flavour. Sow in rows fifteen inches apart, any time from the end of March to the beginning of May. Two sowings will generally suffice.

Spinach.—Sow the Long-standing variety, which does not run so soon as the ordinary kind. If a plantation of Spinach Beet has not been secured, sow at once, as there is ample time yet for a free growth and a valuable plant.

Turnip to be sown in quantity.

Vegetable Marrow.—An early sowing to be made in pots, in readiness for planting out immediately weather admits of it. Three plants in a pot are enough, and they must not be weakened by excessive heat.

Winter Greens.—A sowing of Borecole should be made, and if a supply is required in spring, it will be well to sow again in the first week of May.

MAY

High-Pressure times continue, for the heat increases daily, and the season of production is already shortened by two months. The most pressing business is to repair all losses, for even now, if affairs have gone wrong, it is possible to get up a stock of Winter Greens, and to sow all the sorts of seeds that should have been sown in March and April, with a reasonable chance of profitable results. It must not be expected, however, that the most brisk and skilful can overtake those who have been doing well from the first dawn of spring, and who have not omitted to sow a single seed at the proper time from the day when seed-sowing became requisite. The heat of the earth is now sufficient to start many seeds into growth that are customarily sown in heat a month or two earlier; and, therefore, those who cannot make hot-beds may grow many choice things if they will be content to have them a week or two later than their more fortunate neighbours. In sowing seeds of the more tender subjects, such as Capsicums, Marrows, and Cucumbers, it will be better to lose a few days, in order to make sure of the result desired, rather than to be in undue haste and have the seed destroyed by heavy rains, or the young plants nipped off by frost. Do not, therefore, sow any of these seeds in the open ground until the weather is somewhat settled and sunny, for if they meet with any serious check they will scarcely recover during the whole of the season.

Asparagus in seed-beds to be thinned as soon as possible, so that wherever two or three plants rise together, the number should be reduced to one. But there is time yet for seedlings to appear. The bearing beds are more attractive, for they show their toothsome tops. The cutting must be done in a systematic manner, and if practicable always by the same person. It is better to cut all the shoots as fast as they attain a proper size, and sort them for use according to quality, rather than to pick and choose the fat shoots and throw the whole plantation into disorder. Green-topped Asparagus is in favour in this country; but those who prefer it blanched have simply to earth it up sufficiently, and cut below the surface, taking care to avoid injuring the young shoots which have not pushed through. It is not for us to decide on any matter of individual taste, but we will give a word of practical advice that may be of value to many. It is not the custom to protect Asparagus in open beds, but it should be; for the keen frosts that often occur when the sticks are rising destroy a large number. This may be prevented by covering with any kind of light, dry litter, which will not in the least interfere with that full greening of the tops which English people generally prefer, because the light and air will reach the plant; but the edge of the frost will be blunted by the litter. If there is nothing at hand for this purpose, let a man go round with the sickle and cut a lot of long grass from the rough parts of the shrubbery, and put a light handful over every crown in the bed. The sticks will rise with the litter upon them like nightcaps, and will be plump and green and unhurt by frost.

Bean, Dwarf French.—The main crops should be got in this month, and successional sowings may be made until the early part of July. Dwarf Beans are but seldom allowed as much space as they require, and the rows therefore should be thinned early, for crowded plants never bear so well as those that enjoy light and air on all sides. In Continental cookery a good dish is made of the Beans shelled out when about half ripe. These being served in rich gravy, are at once savoury and wholesome. Almost all the varieties of the Dwarf and Climbing sections may be used in this way, and the Beans should be gathered when full grown, but not yet ripe. The self-coloured varieties are also grown for use as dry Haricots, in which case the pods should not be removed until perfectly ripe.

Bean, Climbing French.—Sow this month for the main crop, and onwards until June according to requirements. In a general way the treatment usual for Runners will answer well for outdoor crops of the Climbing French Bean.

Bean, Runner.—In the open ground sowings may be made as soon as conditions appear safe, but it is well to sow again at the end of the month or in June.

Beet.—The main crop should be sown in the early part of the month. Thin and weed the early sown, and if the ground has been suitably prepared, it will be needless to give water to this crop. As Beet is not wanted large, it is not advisable to sow any great breadth until the beginning of May, or it is liable to become coarse.

Broccoli to be sown for succession. Plant out from frames and forward seed-beds at every opportunity. About the middle of the month sow for cutting in May and June of next year.

Brussels Sprouts.—For the sake of a few fine buttons in the first dripping days of autumn, when Peas and Runners and Marrows are gone, put out as soon as possible some of the most forward plants, giving them a rich soil and sunny position.

Cabbage.—Plant out from seed-beds at every opportunity, choosing, if possible, the advent of showery weather. Sow the smaller sorts and Coleworts, especially in favoured districts where there is usually no check to vegetation until the turn of the year.

Capsicum can be sown out of doors about the middle of the month, and nice green pods for pickling may be secured in the autumn.

Carrot.—Thin the main crops early, and sow a few rows of Champion Horn or Intermediate, for use in a small state during late summer, when they make an elegant and delicate dish.

Cauliflowers must have water in dry weather; they are the most hungry and thirsty plants in the garden, but pay well for good living. Plant out from frames as fast as ready, for they do no good to stand crowded and starving.

Celery trenches must be prepared in time, though, strange to say, this task is generally deferred until the plants have really become weak through overcrowding. In a small garden it is never advisable to have Celery very forward, for the simple reason that trenches cannot be made for it until Peas come off and other early crops are over. To insure fine Celery the cultivator must be in advance of events rather than lag behind them. Plenty of manure must be used; it is scarcely possible, in fact, to employ too much, and liberality is not waste, because the ground will be in capital condition for the next crop. There are many modes of planting Celery, but the simplest is to make the trenches four feet apart and a foot and a half wide, and put the plants six to nine inches apart, according to the sorts. This work must be done neatly, with an artistic finish. In planting take off suckers, and if any of the leaves are blistered, pinch the blisters, and finish by dusting the plantation with soot. As Celery loves moisture, give water freely in dry weather.

Cucumbers of excellent quality may be grown on ridges or hills, should the season be favourable. Suppose the cultivator to have the means of obtaining plenty of manure, ridges, which are to run east and west, are preferable to hills. The soil should be thrown out three feet wide and two feet deep, and be laid up on the north side. Then put three feet of hot manure in the trench, and cover with the soil that was taken out, so as to form an easy slope to the south, and with a steep slope on the north side carefully finished to prevent its crumbling down before the season ends. The plants should be put out on the slope as soon as possible after the ridges are made ready, under the protection of hand-lights, until there is free growth and the weather has become quite summery. It is a good plan to grow one or two rows of Runner Beans a short distance from the ridge on the north side to give shelter, and in case of bad weather after the plants are in bearing, pea-sticks or dry litter laid about them lightly will help them through a critical time, but stable manure must not be used. In case manure is not abundant, make a few small hills in a sheltered, sunny spot, with whatever material is available in the way of turf, rotten manure, or leaf-mould, taking care that nothing injurious to vegetation is mixed with it. Put several inches of a mixture of good loam and rotten manure on the hills, and plant and protect as in the case of ridges. If plants are not at hand, sow seeds; there will still be a chance of Cucumbers during July, August, and September; for if they thrive at all, they are pretty brisk in their movements. Three observations remain to be made on this subject. In the first place, what are known as 'Ridge' Cucumbers only should be grown in the open air; the large sorts grown in houses are unfit. In the second place, the plants should only be pinched once, and there is no occasion for the niggling business which gardeners call 'setting the bloom.' Provide for their roots a good bed, and then let them grow as they please. In the third place, as encouragement, we feel bound to say that, as Cucumbers are grown to be eaten as well as to be looked at, those from ridges are less handsome than house Cucumbers, but are quite equal to them in flavour.

Dandelion somewhat resembles the Endive, and is one of the earliest and most wholesome additions to the salad-bowl. Sow now and again in June, in drills one foot asunder, and thin out the plants to one foot apart in the rows. These will be ready for use in the following winter and spring.

Gourd and Pumpkin.—An early show of fruit necessitates raising seeds under glass for planting on prepared beds, and the plants must be protected by means of lights or any other arrangement that can be improvised as a defence against late frosts. Of course the seeds can be sown upon the actual bed, but it is a loss of time. The rapidity with which the plants grow is a sufficient indication that generous feeding and copious supplies of water in dry weather are imperative.

Lettuce.—Sow for succession where the plants are to remain, and plant out the earlier sowings at every opportunity. To insure a quick growth, and prevent the plants from running to seed, extra care in giving water and shade will be necessary after transplanting. The larger Cabbage Lettuces will prove useful if sown now.

Maize and Sugar Corn may be grown in this country as an ornament to the garden, and also for the green cobs which are used as a vegetable. Sow early in the month on rich light soil, and in a hot season, especially when accompanied by moisture, there will be rapid growth. The cobs to be gathered for cooking when of full size, but while quite green.

Melon.—It is not too late to grow Melons in frames, provided a start can be made with strong plants.

Pea.—Sow Peas again if there is any prospect of a break in the supply. It is a good plan to prepare trenches as for Celery, but less deep, and sow Peas in them, as the trenches can be quickly filled with water in case of dry weather, and the vigorous growth will be proof against mildew.

Savoy sown now will produce small useful hearts for winter use. By many these small hearts will be preferred to large ones, as more delicate, and therefore a sowing of Tom Thumb may be advised.

Spinach, New Zealand, can be sown in the open ground in the early part of this month and should be thinned to about a yard apart. The growth somewhat resembles that of the Ice Plant. The tender young tops are pinched off for cooking, and they make an elegant Spinach, which is free from bitterness, and is therefore acceptable to many persons who object to the sooty flavour of ordinary Spinach.

Tomato.—By the third week in May the plants for the open border should be hardened. In a cold pit or frame they may be gradually exposed until the lights can be left off altogether, even at night. A thick layer of ashes at the bottom of the frame will insure drainage and keep off vermin. If the plants are allowed plenty of space, and are well managed, they will possess dark, healthy foliage, needing no support from sticks until they are in final quarters. Do not put them out before the end of the month or the beginning of June, and choose a quiet day for the work. If possible, give them a sunny spot under the shelter of a wall having a southern or western aspect. On a stiff soil it is advisable to plant on ridges, and not too deeply; for deep planting encourages strong growth, and strong growth defers the production of fruit. Tomatoes are sometimes grown in beds, and then it is necessary to give them abundant room. For branched plants three feet between the plants in the rows, and the rows four feet apart, will afford space for tying and watering. Each plant should have the support of a stout stake firmly fixed in the soil, and rising four feet above it; and once a week at least the tying should be attended to. As to stopping, the centre stem should be allowed to grow until the early flowers have set. It is from these early flowers that outdoor Tomatoes can be successfully ripened, and the removal of the main shoot delays their production. But after fifteen or twenty fruits are visible the top of the leading stem may be shortened to the length of the stake. The fruiting branches should also be kept short beyond the fruit, and large leaves must be shortened to allow free access of sunshine. Should the single-stem system be adopted, three feet between the rows and two feet between plants in the rows will suffice. On a light soil and in dry weather weak liquid manure may, with advantage, be alternated with pure water, but this practice must not be carried far enough to make the plants gross, or ripening will be delayed. Fruit intended for exhibition must be selected with judgment, and with this end in view four to six specimens of any large variety will be sufficient for one plant to bring to perfection.

Turnip to be sown for succession. It is well now to keep to the small white early sorts.

Vegetable Marrow.—In cottage gardens luxuriant vines may every year be seen trailing over the sides of heaps of decayed turf or manure. All forward vegetables are prized, and Marrows are no exception to the rule. An early supply from the open ground is most readily insured by raising strong plants in pots and putting them on rich warm beds as early as the season and district will permit. Late frosts must be guarded against by some kind of protection, and slugs must be deterred from eating up the plants.

JUNE

To some extent the crops will now take care of themselves, and we may consider the chief anxieties and activities of the season over. Our notes, therefore, will be more brief. We do not counsel the cultivator to 'rest and be thankful.' It is better for him to work, but he must be thankful all the same, if he would be happy in his healthy and entertaining employment. Watering and weeding are the principal labours of this month, and both must be pursued with diligence. But ordinary watering, where every drop has to be dipped and carried, is often injurious rather than beneficial, for the simple reason that it is only half done. In such cases it is advisable to withhold water as long as possible, and then to give it in abundance, watering only a small plot every day in order to saturate the ground, and taking a week or more to go over a piece which would be done in a day by mere surface dribblings.

Asparagus should be in full supply, and may be cut until the middle or end of the month. When cutting should cease depends on the district. In the South of England the 14th is about the proper time to make the last cut; north of the Trent, the 20th may be soon enough; and further north, cutting may be continued into July. The point to be borne in mind is that the plant must be allowed time to grow freely without any further check, in order to store up energy for making robust shoots next year. It is a good plan to insert stakes, such as are used for Peas, in Asparagus beds, to give support to the green growth against gales of wind; for when the stems are snapped by storms, as they often are, the roots lose their aid, and are weakened for their future work.

Beans, both Dwarf and Runner, may be sown about the middle of the month, to supply tender pods when those from the early sowings are past. A late crop of Runners will pay well almost anywhere, for they bear until the frost cuts them down, which may not happen until far into November.

Broccoli.—Take advantage of showers to continue planting out.

Cabbage.—Towards the end of the month sow a good breadth of small Cabbages and Coleworts. They will be immensely valuable to plant out as the summer crops are cleared away.

Capsicums may be planted out in a sunny sheltered spot.

Cauliflowers that are transferred now from seed-beds must have plentiful supplies of water, and be shaded during midday for a week. When the heads are visible it is customary to snap one of the inner leaves over them for protection.

Celery to be planted out without loss of time, in showery weather if possible; but if the weather is hot and dry, shade the plants and give water. The work must be well done, hence it is advisable to lift no more plants than can be quickly dealt with, for exposure tends to exhaustion, and Celery ought never to suffer a check in even the slightest degree. When planted, dust lightly with soot or wood-ashes. Pea-sticks laid across the trenches will give shade enough with very little trouble.

Chicory.—This wholesome esculent is used in a variety of ways, and is very much prized in some households. The blanched heads make an acceptable accompaniment to cheese, and are much appreciated for salading; they may also be stewed and served with melted butter in the same manner as Sea Kale. To grow large clean roots a deep rich soil is required. If manure must be added, use that which is well decayed, and bury it at least twelve inches, for near the surface it will produce fanged roots. Prepare the seed-bed as for Parsnips, sow in drills twelve inches apart, and thin the plants to nine inches in the rows. In October the roots will be ready for lifting, preparatory to being packed in dark quarters for blanching.

Cucumbers for Pickling may be sown on ridges.

Endive is not generally wanted while good Lettuces abound, but it takes the place of Lettuce in autumn and winter, when the more delicate vegetable is scarce. Sow in shallow drills six inches apart. Thin the plants, and transfer the thinnings to rich light soil. They must be liberally grown on well-manured land, with the aid of water in dry weather.

Lettuce to be sown and planted at every opportunity. A few rows of large Cos varieties should be sown in trenches prepared as for Celery, there to be thinned and allowed to stand. They will form fine hearts, and be valued at a time when Lettuces are scarce.

Melon.—For a final crop in houses sow as previously directed, and grow the plants on in pots, until the house can be cleared of the former set for their reception. The growth should be pushed forward to insure ripe fruit before the end of September. In the event of dull weather at the finish, there will be all the greater need of abundant but judicious ventilation, and of a warm dry atmosphere at night. Before they become heavy every fruit should have the support of nets or thin pieces of board suspended by wires from the corners.

Mushrooms may be prepared for now. The first step towards success is to accumulate a long heap of horse-droppings with the least possible amount of litter. Let this ferment moderately, and turn it two or three times, always making a long heap of it, which keeps down the fermentation. When the fire is somewhat taken out of it, make up the bed with a mixture of about four parts of the fermented manure and one part of turfy loam, well incorporated. Beat the stuff together with the flat of the spade as the work proceeds, fashioning the bed in the form of a ridge about three feet wide at the base, and of any length that may be convenient. Give the work a neat finish, or the Mushrooms will certainly not repay you. Put in rather large lumps of spawn when the bed is nicely warm, cover with a thin layer of fine soil, and protect with mats or clean straw. This is a quick and easy way of growing Mushrooms, and by commencing now the season is all before one. Nine times in ten, people begin preparations for Mushroom growing about a month too late, for the spawn runs during the hot weather, and the crop rises when the moderate autumnal temperature sets in.

Onions to be sown for salading. Forward beds of large sorts to be thinned in good time. The best Onions for keeping are those of moderate size, perfectly ripened; therefore the thinning should not be too severe.

Peas may still be sown, and as the season advances preference should be given to quick-growing early varieties.

Turnips may be sown in variety and in quantity after Midsummer Day. Sow on well-prepared ground, and put a sprinkle of artificial manure in the drills with the seed. By hastening the early growth of the plant the fly is kept in check.

JULY

For gardeners July is in one respect like January; everything depends on the weather. It may be hot, with frequent heavy rains, and vegetation in the most luxuriant growth; or the earth may be iron and the heavens brass, with scarcely a green blade to be seen. The light flying showers that usually occur in July do not render watering unnecessary; in fact, a heavy soaking of a crop after a moderate rainfall is a valuable aid to its growth, for it requires a long-continued heavy downpour to penetrate to the roots.

Summer-sown Vegetables for Autumn and Winter use.

As the month advances early crops will be finished and numerous plots of ground become vacant. In many gardens it is now the practice to sow in July and August seeds of quick-growing varieties of Vegetables and Salads to furnish supplies through the autumn and early winter months, and this system is strongly to be commended. These sowings not only increase the cropping capacity of the garden but they extend the use of many favourite Vegetables which from spring sowings customarily cease at the end of summer. Two things are essential to success. *Early-maturing varieties only should be sown and the plants must be thinned immediately they appear (thus avoiding transplanting), so that they receive no check in growth.* The following subjects are especially suited for the purpose: Dwarf French Beans (sow early in July), Beet, Cabbage, Carrot, Cauliflower (sow early in July), Italian Corn Salad, Cress, Endive, Kohl Rabi, Lettuce, Onion, Parsley, Peas, Radish, Spinach, and Turnip. Potatoes may also be planted in July, but only tubers of early varieties saved from the preceding year should be used.

Garden Rubbish is apt to accumulate in odd corners and become offensive. The stumps of Cabbages and Cauliflowers give off most obnoxious odours, and neighbours ought not to be annoyed by want of thought in one particular garden. The short and easy way with all soft decaying rubbish is to put it at the bottom of a trench when preparing land for planting. There it ceases to be a nuisance and becomes a valuable manure.

Beans.—A few Dwarf French Beans may still be sown to extend outdoor crops to the latest possible date. For autumn and winter supplies sowings of the Dwarf and Climbing classes may be made from mid-July to mid-September, the dwarfs in cold frames and the climbers on narrow borders in any house that can be spared for the purpose.

Broccoli to be planted out as before; many of the plants left over from former plantings will now be stout and strong, and make useful successions.

Cabbage.—The sowing of Cabbage seed at this period of the year entails consequences of such grave importance as to merit reconsideration. When the crop has passed the winter there is a danger that the plants may bolt, instead of forming hearts. In the great majority of such cases the loss is attributable to an unwise selection of sorts. For sowing in spring there is quite a long list of varieties, many of them possessing distinctive qualities which meet various requirements. It is otherwise now. The Cabbages that can be relied on to finish well in spring are comparatively few in number. But repeated experiments have demonstrated that loss and disappointment can be avoided by sowing only those varieties which show no tendency to bolt. Another, but minor, cause of Cabbages starting seed-stems is premature sowing. The exact date for any district must be determined by the latitude and the aspect of the place. In the North sowing will, of necessity, be earlier than in the Midlands or the South. Assuming, however, that suitable varieties are chosen, the whole difficulty can be disposed of, even on soils where Cabbages show an unusual tendency to send up seed-stems prematurely, by sowing in August instead of in July. The seed-bed should be nicely prepared, and any old plaster, or other rubbish containing lime, should be dug in. Sow thinly, for a thick sowing makes a weak plant, no matter how severely it may be thinned afterwards.

Cardoons to be thinned to one plant in each station, and that, of course, the strongest.

Carrot.—Frame culture of small sorts should commence, to produce a succession of young Carrots for table.

Celery to be planted out in showery weather. It is too late to sow now, except for soups, and for that purpose only a small sowing should be made, as it may not come to anything.

Chards.—Those who care for Chards must cut down a number of Globe Artichokes about six inches above ground, and, if necessary, keep the plants well watered to induce new growth, which will be ready for blanching in September.

Cucumbers on ridges generally do well without water, but they must not be allowed to suffer from drought. If watering must be resorted to, make sure first of soft water well warmed by exposure to the sun, and water liberally three or four evenings in succession, and then give no more for a week or so.

Endive to be sown for winter. It will be well to make two sowings, say on the first and last days of the month.

Garlic and Shallots to be taken up in suitable weather, and it may be necessary to complete the ripening under shelter.

Leeks to be planted out; and on dry soils, in trenches prepared as for Celery.

Parsley to be sown for winter use. It is a most important matter, even in the smallest garden, to have a constant supply.

Peas.—Only quick-growing early varieties should be sown now.

Potatoes.—Where there is a good crop of an early variety it should be lifted without waiting for the shaws to die down. The tender skins will suffer damage if the work is done roughly, but will soon harden, and the stock will ripen in the store as perfectly as in the ground. It needs some amount of courage to lift Potatoes while the tops are still green and vigorous, and it should not be done until the roots are fully grown and beginning to ripen. Quick-growing sorts may be planted to dig as new Potatoes later in the year.

Radish.—Sow the large-growing kinds for winter use.

Spinach.—Sow the Prickly-seeded to stand the winter, selecting for the seed-bed ground lying high and dry that has been at least twice dug over and has had no recent manure. The twice digging is to promote the destruction of the 'Spinach Moth' grub, which the robins and thrushes will devour when exposed by digging. These grubs make an end of many a good breadth of Winter Spinach every year, and are the more to be feared by the careless cultivator.

Turnips to be sown in quantity in the early part of the month; thin advancing crops, and keep the hoe in action amongst them.

Winter Greens of all kinds to be planted out freely in the best ground at command, after a good digging, and to be aided with water for a week or so should the weather be dry.

AUGUST

The importance of summer-sown Vegetables and Salads is dealt with under July, and seeds of most of the subjects there named may still be put in as ground becomes vacant. The supplies of the garden during the next winter and spring will in great part depend upon good management now, and the utmost must be made of the few weeks of growing weather that remain. One great difficulty in connection with sowing seed at this period of the year is the likelihood of the ground being too dry; yet it is most unwise to water seeds, and it is always better if they can be got up with the natural moisture of the soil alone. However, in an extreme case the ground should be well soaked before the seed is sown, and after sowing covered with hurdles, pea-sticks, or mats until the seeds begin to sprout.

Artichokes, Globe, to be cut down as soon as the heads are used.

Broccoli to be planted out. As the Sprouting Broccoli, which belongs to the class of 'Winter Greens,' does not pay well in spring unless it grows freely now, plant it far enough apart; if crowded where already planted to stand the winter, take out every alternate plant and make another plantation.

Cabbage.—In many small gardens the August sowing of Cabbages is made to suffice for the whole year, and in the largest establishments greater breadths are sown now than at any other period. But whether the garden be small or large, it is not wise to rely exclusively on the sowing of any one kind. At least two varieties should be chosen, and as a precaution each variety may be sown at two dates, with an interval of about a fortnight between. The wisdom of this arrangement will be evident in nine seasons out of ten. It allows for contingencies, prolongs the season of supply, and offers two distinct dishes of a single vegetable—the mature hearts, and the partially developed plants, which differ, when served, both in appearance and in flavour. Where the demand is extensive, or great diversity is required, three or four kinds should be sown, including Red Cabbage to produce fine heads for pickling next year.

Cardoon.—Commence blanching if the plants are ready.

Cauliflower.—Seed sown now will produce finer heads in spring and early summer than are generally obtained from a January or February sowing. The time to sow must be determined by the climate of the district. In cold, late localities, the first week is none too early; from the 15th to the 25th is a good time for all the Midland districts; and the end of the month, or the first week of September, is early enough in the South. In Devon and Cornwall the sowing is later still. But whatever date may suit the district, the seed should be sown with care, in order that a healthy growth may be promoted from the first. Winter the plants in frames or by other convenient means, but it is important to keep them hardy by giving air at every favourable opportunity.

Celery to be carefully earthed up as required. It takes five weeks or more to blanch Celery well, and as the earthing up checks growth, the operation should not be commenced a day too soon. Take care that the earth does not get into the hearts.

Corn Salad should be sown during this month and September to produce plants fit for use in early spring. In the summer months the whole plant is edible, but in winter or spring the outer leaves only should be used.

Cucumber.—For a supply of Cucumbers during the winter months the general principles of management are identical with those given under January and March, with one important exception. At the commencement of the year a continued increase of light and warmth may be relied on. Now there will be a constant diminution of these vital forces. Hence the progress of the plants will gradually abate as the year wanes, and due allowance must be made for the fact. So much depends on the character of the autumn and winter that it will be unwise to risk all on a single sowing. Seed put in on two or three occasions between the end of August and the end of October will provide plants in various stages of growth to meet the exigencies of the season. The production of Cucumbers will then depend on care and management. In very dull cold weather it may be dangerous to syringe the foliage, but the necessary moisture can be secured by sprinkling the floor and walls.

Endive.—Make a final sowing, and plant out all that are large enough, selecting, if possible, a dry, sloping bank for the purpose.

Lettuce to be sown to stand the winter, choosing the hardiest varieties. In cold districts the middle of the month is a good time to sow; in favoured places the end of the month is preferable.

Onion.—For many years the Tripoli section enjoyed pre-eminence for sowing at this season, the opinion prevailing that other kinds were unsuitable. But it is found that several varieties which may with propriety be described as English Onions are as hardy as the Tripolis, and therefore as well adapted for sowing at this season. Thus, instead of sorts that must be used quickly, we may command for summer sowing the best of the keepers, and the result will be heavier crops and earlier ripening, with plentiful supplies of 'thinnings' for salads all through the autumn and winter. Two sowings—one at the beginning, the other at the end of the month—may be adopted with advantage. The storage of Onions is often faulty, and consequently losses occur through mildew and premature growth. If any are as yet unripe, spread them out in the sun in a dry place, where they can be covered quickly in case of rain. In wet, cold seasons, it is sometimes necessary to finish the store Onions by putting them in a nearly cold oven for some hours before they are stored away.

Pea.—Crops coming forward for late bearing should have attention, more especially to make them safe against storms by a sufficiency of support, and in case of drought to give abundance of water.

Strawberry Plants may be put in should the weather prove favourable; but next month will answer. In burning weather it is well worth while to bed the plants closely in a moist shady place until rain comes, and then plant out.

Tomatoes to be gathered as soon as ripe. If bad weather interferes with the finishing of the crop, cut the full-grown fruit with a length of stem attached, and hang them up in a sunny greenhouse, or some other warm spot in full daylight. Seed sown now or in September will produce plants that should afford fine fruit in March, and it will need care and judgment to carry them safely through the winter.

Turnip may be sown in the early part of the month. The best sorts now are White Gem, or Snowball. All the Year Round will please those who like a yellow Turnip.

SEPTEMBER

Weeds will be troublesome to the overworked and the idle gardener, while the best-kept land will be full of seeds blown upon it from the sluggard's garden, and the first shower will bring them up in terrific force. All that we have to say about them is that they must be kept down, for they not only choke the rising crops in seed-beds and spoil the look of everything, but they very much tend to keep the ground damp and cold, when, if they were away, it would get dry and warm, to the benefit of all the proper crops upon it. Neglect will make the task of eradication simply terrible, and, in the meantime, every crop on the ground will suffer. The two great months for weeds are May and September; but often the September weeds triumph, because the mischief they do is not then so obvious to the casual eye. As there are now many used-up crops that may be cleared away, large quantities of Cabbage, Endive, Lettuce, and even thinnings of Spinach may be planted out to stand the winter.

Cabbage.—We advocate crowding the land now with Cabbage plants, for growth will be slow and the demands of the kitchen constant. Crowding, however, is not quite the same thing as overcrowding, and it is only a waste of labour, land and crop to put the plants so close together that they have not space for full development. The usual rule in planting out the larger sorts of Cabbage at this time of the year is to allow a distance every way of two feet between the plants. The crowding principle may be carried so far as to put miniature Cabbages between them, but only on the clear understanding that the small stuff is all to be cleared off before spring growth commences, and the large Cabbages will then have proper space for development.

Cauliflower.—Sow again in a frame or in a pan in the greenhouse.

Celery.—Continue to earth up, selecting a dry time for the task.

Chards take quite six weeks to blanch by means of straw, covered with earth.

Cucumbers for the winter need careful management and suitable appliances. See the remarks on this subject under August.

Endive to be planted out as directed last month. Plant a few on the border of an orchard-house, or in a ground vinery, or in old frames for which some lights, however crazy, can be found.

Lettuces should be coming in from the garden now in good condition, but the supply will necessarily be running short. Sowings of two or three sorts should be made partly in frames and partly on a dry open plot from which a crop has been taken. The ground should be well dug but not manured. Sow thinly, so that there will not be much need for thinning, and confine the selection to sorts known to be hardy. The August sowings will soon be forward enough for putting out, and it will be advisable to get the work done as early as possible, to insure the plants being well established before winter.

Parsley.—The latest sowing will require thinning, but for the present this must not be too strictly carried out; between this and spring there will be many opportunities. Thin the plot by drawing out complete plants as Parsley is demanded for the kitchen. If no late sowing was made, or, having been made, has failed, cut down to the ground the strongest plants, that a new growth may be secured quickly. A few plants potted at the end of the month, or lifted and placed in frames, may prove exceedingly valuable in winter.

Potatoes that are ready should be taken up with reasonable care. It is not wise to wait for the dying down of the shaws, because, when the tubers are fully grown, they ripen as well in the store, out of harm's way, as in the ground, where they are exposed to influences that are simply destructive.

Spinach.—In favourable seasons and forward localities Winter Spinach sown in the first half of this month will make a good plant before winter. Thin the plants that are already up to six inches apart.

OCTOBER

Weeds and falling leaves are the plagues of the season. It may seem that they do no harm, but assuredly they are directly injurious to every crop upon the ground, for they encourage damp and dirt by preventing a free circulation of air amongst the crops, and the access of sunshine to the land. Keep all clean and tidy, even to the removal of the lower leaves of Cabbages, where they lie half decayed upon the ground.

The heavy rains of this month interfere in a material degree with outdoor work, and are often a great impediment to the orderly management that should prevail. The accumulation of rubbish anywhere, even if out of sight, is to be deplored as an evil altogether. The injury to vegetation is as great as that inflicted on our own health when dirt poisons the air and damp hastens the general dissolution. It is therefore above all things necessary to keep the garden clean from end to end. All decaying refuse that can be put into trenches should be got out of sight as soon as possible, to rot harmlessly instead of infecting the air, and leaves should be often swept up into heaps, in which form they cease to be injurious, although, when spread upon the ground and trodden under foot, they are breeders of mischief. If in want of work, ply the hoe amongst all kinds of crops, taking care not to break or bruise healthy leaves, or to disturb the roots of any plant. Dig vacant plots, and lay the land up in ridges in the roughest manner possible. Heavy land may be manured now with advantage, but it is not desirable to manure light land until spring.

Cabbages to be planted out as advised last month.

Cardoon.—Blanching must be continued.

Carrots.—Lift the roots and store in sand.

Cauliflowers to be prepared for the winter.

Celeriac.—Part of the crop should be lifted and stored in sand; the plants left in the ground to be protected by earthing over.

Celery must be earthed up, and protecting material got ready to assure its safety during frost.

Chicory.—Raise about a dozen plants at a time as required, cut or wrench off the foliage, and pack the roots, crown upwards, in boxes with moist leaf-mould or soil. They must be stored in absolute darkness in some cellar or Mushroom-house which is safe from frost, but a forcing temperature is detrimental to the flavour. Gathering may commence about three weeks after storing. The yield is abundant, and is of especial value for salading through the autumn and winter months.

Endive to be blanched for use as it acquires full size, but not before, as the blanching makes an end of growth.

Lettuce.—Continue to plant as before advised, and make a final sowing in frames not later than the middle of the month.

Parsnips may be dug all the winter as wanted. Although a slight frost will not injure them when left in the ground, protection by rough litter is needful in very severe weather. It often happens that they grow freely soon after the turn of the year, and then become worthless.

Potatoes to be taken up and stored with all possible speed.

Rhubarb for forcing should be taken up and laid aside in a dry, cool place, exposed to the weather. This gives the roots a check, and constitutes a kind of winter, which in some degree prepares them for the forcing pit.

Roots, such as Beet, Salsify, and Turnip, to be taken up as soon as possible, and stored for the winter.

Winter Greens may still be transplanted, and it is often better to use up the remainder of the seed-beds than to let the plants stand. In the event of a severe winter, these late-planted Greens may not be of much value; but in a mild growing winter they will make some progress, and may prove very useful in the spring.

NOVEMBER

The remarks already made on the necessity for tidiness and the quick disposal of all decaying refuse apply as forcibly to this month as to October. The leaves are falling, the atmosphere is moist, and there should be the utmost care taken not to make things worse by scatterings of vegetable rubbish. Now we are in the 'dull days before Christmas' the affairs of the garden may be reviewed in detail, and this is the best period for such a review. Sorts that have done well or ill, wants that have been felt, mistakes that have been made, are fresh in one's memory, and in ordering seeds, roots, plants, &c., for next season's work, experience and observation can be recorded with a view to future benefit. Consistently with the revision of plans by the fireside, revise the work out of doors. Begin to prepare for next year's crops by trenching, manuring, planting, and collecting stuff to burn in a 'smother.' Land dug now for spring seeds and roots, and kept quite rough, will only require to be levelled down and raked over when spring comes to be ready for seed, and will produce better crops than if prepared in a hurry. Protecting material for all the needs of the season must be in readiness, in view of the fact that a few nights of hard frost may destroy Lettuces, Endives, Celery, and Cauliflowers worth many pounds, which a few shillings'-worth of labour and litter would have saved. Earthwork can generally be pushed on, and it is good practice to get all road-mending and the breaking up of new ground completed before the year runs out, because of the hindrance that may result from frost, and the inevitable pressure of other work at the turn of the spring. The weather is an important matter; but often the month of November is favourable to outdoor work, and labour can then be found more readily than at most other seasons.

Artichokes, Globe, must be protected ere frost attacks them. Cut off the stems and large leaves to within a foot of the ground; then heap up along each side of the rows a lot of dry litter consisting of straw, pea haulm, or leaves, taking care in so doing to leave free access to light and air. The hearts must not be covered, or decay will follow.

Artichokes, Jerusalem, may be dug as wanted, but some should be lifted and stored in sand for use during frosts.

Asparagus beds not yet cleaned must have prompt attention. Cut down the brown grass and rake off all the weeds and rubbish, and finish by putting on a dressing of seaweed, or half-rotten stable manure.

Bean, Broad.—It is customary on dry warm soils to sow Beans at the end of October or during November for a first crop, and the practice is to be commended. On cold damp soils, and on clay lands everywhere, it is a waste of seed and labour to sow now, but every district has its peculiar capabilities, and each cultivator must judge for himself. In any case, Beans sown during this month should be put on well-drained land in a sheltered spot.

Broccoli.—In inclement districts lay the plants with their heads facing the north.

Carrot to be sown in frames, and successive sowings made every three or four weeks until February.

Cauliflowers will be turning in, and possibly those coming forward will be all the better off for being covered with a leaf to protect the heads from frost. If the barometer rises steadily and the wind goes round to north or north-east, draw all the best Cauliflowers, and put them in a shed or any out-of-the-way place safe for use.

Celery.—Hard frost coming after heavy rain may prove destructive to Celery; and it is well, if there is a crop worth saving, to cut a trench round the plantation to favour escape of surplus water. If taken up and packed away in a dry shed, the sticks will keep fresh for some time.

Horse-radish to be taken up and stored ready for use, and new plantations made as weather permits and ground can be spared.

Pea.—The sowing of Peas outdoors now is not recommended for general practice, but only for those who are so favourably circumstanced as to have a fair prospect of success. If it is determined to sow, select for the purpose a dry, light, well-drained sunny border, and make it safe from mice, slugs, and sparrows. The quick-growing round-seeded varieties must be chosen for the purpose, and it will be advisable to sow two or three sorts rather than one only. Peas to be grown entirely under glass may be started now.

Sea Kale to be lifted for forcing. This delicious vegetable may, indeed, be forced for the table in this month; but it is not advisable to be in such haste, for a fine sample cannot be secured so early. Sea Kale is the easiest thing in the world to force; the only point of importance is to have strong roots to begin with. Any place such as Mushroom-houses, cellars, pits, or old sheds, where it is possible to maintain a temperature of 45° to 55°, may be utilised for the purpose. Put the plants thickly into pots or boxes, or plant them in a bed, and it is essential to exclude light to insure blanching. By these simple means a regular supply may be obtained until the permanent beds in the open ground come into use.

DECEMBER

The best advice that can be given for this month is to be prepared for either heavy rain or sharp frost, so that extreme variations of temperature may inflict the least possible injury in the garden. Let the work be ordered with reference to the weather, that there may be no 'poaching' on wet ground, or absurd conflict with frost. Accept every opportunity of wheeling out manure; and as long as the ground can be dug without waste of labour, proceed to open trenches, make drains, and mend walks, because this is the period for improving, and the place must be very perfect which affords no work for winter weather. Dispose of all rubbish by the simple process of putting it in trenches when digging plots for early seeds. In sheds and outhouses many tasks may be found, such as making large substantial tallies for the garden; the little paltry things commonly used being simply delusive, for they are generally missing when wanted, from their liability to be trodden into the ground or kicked anywhere by a heedless foot. Make ready pea-sticks, stakes of sizes, and at odd times gather up all the dry stuff that is adapted for a grand 'smother.' A careful forecasting of the next year's cropping will show that even now many arrangements may be made to increase the chances of success.

Warm Border to be prepared for early work by digging and manuring. All the refuse turf and leaf-mould from the potting-shed and the soil knocked out of pots may be usefully disposed of by adding it to this border, which cannot be too light or too rich, and a good dressing of manure will give it strength to perform its duties.

Beans, Broad, to be earthed up for protection and support.

Celery to be earthed up for the last time. In case of severe weather, have protecting material at hand in the shape of dry litter or mats. Pea-sticks make a capital foundation on which to throw long litter, mats, &c., for quickly covering Celery, the protection being as quickly removed when the frost is over, and costing next to nothing.

Endive will be valued now, and must be blanched as required. Place a few in frames and other protected spots. In the unused corners of sheds and outhouses they may be safer than out of doors.

Parsley.—In all cold districts it is wise to secure a bed of Parsley, in a frame or pit, or if a few plants were potted in September, they may be wintered in any place where they can have light and air freely. It is so important to have Parsley at command as wanted, that it may be worth while to put a frame over a few rows as they stand in the open quarter, rather than risk the loss of all in the event of severe weather.

Radish.—Sow one of the long sorts for a first supply in some warm spot, to secure quick growth.

Underground Onions to be planted in rows one foot apart. They should not be earthed up, for the young bulbs form round the stems in full daylight.

THE ROTATION OF CROPS IN THE VEGETABLE GARDEN

This is a subject worthy the attention of those who aim at the largest possible production and the highest possible quality of every kind of kitchen-garden crop, for it concerns the natural relations of the plant and the soil as to their several chemical constituents. The principle may be illustrated by considering the demands of two of the most common kitchen-garden crops. If we submit a Cabbage to the destructive agency of fire, and analyse the ashes that remain, we shall find in them, in round numbers, eight per cent. of sulphuric acid, sixteen per cent. of phosphoric acid, four per cent. of soda, forty-eight per cent. of potash, and fifteen per cent. of lime. It is evident that we cannot expect to grow a Cabbage on a soil which is destitute of these ingredients, to say nothing of others. The obnoxious odour of sulphur emitted by decaying Cabbages might indicate, to anyone accustomed to reflect on ordinary occurrences, that sulphur is an important constituent of Cabbage. If we submit a Potato tuber to a similar process, the result will be to find in the ashes fifty-nine per cent. of potash, two per cent. of soda, six per cent. of sulphuric acid, nineteen per cent. of phosphoric acid, and two per cent. of lime. The lesson for the cultivator is, that to prepare a soil for Cabbage it is of the utmost importance to employ a manure containing sulphates, phosphates, and potash salts in considerable quantity; as for the lime, that can be supplied separately, but the Cabbage must have it. On the other hand, to prepare a soil for Potatoes it is necessary to employ a manure strongly charged with salts of potash and phosphates, but it need not be highly charged with soda or lime, for we find but a small proportion of these ingredients in the Potato. There are soils so naturally rich in all that crops require, that they may be tilled for years without the aid of manures, and will not cease to yield an abundant return. But such soils are exceptional, and those that need constant manuring are the rule. One point more, ere we proceed to apply to practice these elementary considerations. In almost every soil, whether strong clay, mellow loam, poor sand, or even chalk, there are comminglings of all the minerals required by plants, and, indeed, if there were not, we should see no herbage on the downs, and no Ivies climbing, as they do, to the topmost heights of limestone rocks. But usually a considerable proportion of those mineral constituents on which plants feed are locked up in the staple, and are only dissolved out slowly as the rain, the dew, the ever-moving air, and the sunshine operate upon them and make them available. As the rock slowly yields up its phosphates, alkalies and silica to the wild vegetation that runs riot upon it, so the cultivated field (which is but rock in a state of decay) yields up its phosphates, alkalies and silica for the service of plants the more quickly because it is the practice of the cultivator to stir the soil and continually expose fresh surfaces to the transforming power of the atmosphere. It has been said that the air we breathe is a powerful manure. So it is, but not in the sense that is applicable to stable manure or guano. The air may and does afford to plants much of their food, but it can only help them to the minerals they require by dissolving these out of pebbles, flints, nodules of chalk, sandstone, and other substances in the soil which contain them in what may be termed a locked-up condition. Every fresh exposure of the soil to the air, and especially to frost and snow, is as the opening of a new mine of fertilisers for the service of those plants on which man depends for his subsistence.

The application to practice of these considerations is an extremely simple matter in the first instance, but it may become very complicated if followed far enough. Here we can only touch the surface of the subject, yet we hope to do so usefully. Suppose, then, that we grow Cabbage, or Cauliflower, or Broccoli, on the same plot of ground, one crop following the other for a long series of years, and never refresh the soil with manure, it must be evident that we shall, some day or other, find the crop fail through the exhaustion of the soil of its available sulphur, phosphates, lime, or potash. But if this soil were allowed to lie fallow for some time, it would again produce a crop of Cabbage, owing to the liberation of mineral matters which, when the crops were failing, were not released fast enough, but which, during the rest allowed to the soil, accumulated sufficiently to sustain a crop. Obviously this mode of procedure is unprofitable and tends of necessity to exhaustion, although it must be confessed that utter exhaustion of any soil is a thing at present almost unknown. But, instead of following a practice which impoverishes, let us enrich the soil with manure, and change the crops on the same plot, so that when one crop has largely taxed it for one class of minerals, a different crop is grown which will tax it for another class of minerals. Take for a moment's consideration one of the necessary constituents of a fertile soil, common salt (chloride of sodium). In the ash of a Cabbage there is about six per cent. of this mineral, in the Turnip about ten per cent., in the Potato two to three per cent., in the Beet eighteen to twenty per cent. On the other hand the Beet contains very little sulphur, but both Turnip and Beet agree in being strongly charged with potash and soda. It follows that if we crop a piece of ground with Cabbage, and wish to avoid the failure that may occur if we continue to crop with Cabbage, we may expect to do well by giving the ground a dressing of common salt and potash salts, and then crop it with Beet.

The whole subject is not exhausted by this mode of viewing it, for all the facts are not yet fully understood by the ablest of our chemists and physiologists, and crops differ in their methods of seeking nourishment.

We might find two distinct plants nearly agreeing in chemical constitution, and yet one might fail where the other would succeed. Suppose, for instance, we have grown Cabbage and other surface-rooting crops until the soil begins to fail, even then we might obtain from it a good crop of Parsnips or Carrots, for the simple reason that these send their roots down to a stratum that the Cabbage never reached; and it is most instructive to bear in mind that although the Parsnip will grow on poor land, and pay on land that has been badly tilled for years, yet the ashes of the Parsnip contain thirty-six per cent. of potash, eleven per cent. of lime, eighteen per cent. of phosphoric acid, six per cent. of sulphuric acid, three per cent. of phosphate of iron, and five per cent. of common salt. How does the Parsnip obtain its mineral food in a soil which for other crops appears to be exhausted? Simply by pushing down for it into a mine that has hitherto been but little worked, though Cabbage might fail on the same plot because the superficial stratum has been overtaxed.

Having attempted a general, we now proceed to a particular application. In the first place, good land, well tilled and abundantly manured, cannot be soon exhausted; but even in this case a rotation of crops is advisable. It is less easy to say why than to insist that in practice we find it to be so. The question then arises—What is a rotation of crops? It is the ordering of a succession in such a manner that the crops will tax the soil for mineral aliments in a different manner. A good rotation will include both chemical and mechanical differences, and place tap-roots in a course between surface roots, as, for example, Carrot, Parsnip, and Beet, after Cabbage, Cauliflower, and Broccoli; and light, quick surface crops, such as Spinach, to serve as substitutes for fallows. The cropping of the kitchen garden should be, as far as possible, so ordered that plants of the same natural families never immediately succeed one another; and, above all things, it is important to shift from place to place, year after year, the Cabbages and the Potatoes, because these are the most exhaustive crops we grow. In a ton of Potatoes there are about twelve pounds of potash, four pounds of sulphuric acid, four pounds of phosphoric acid, and one pound of magnesia. We may replace these substances by abundant manuring, and we are bound to say that the best rotation will not obviate the necessity for manuring; but even then it is well to crop the plot with Peas, Spinach, Lettuce, and other plants that occupy it for a comparatively brief space of time, and necessitate much digging and stirring; for these mechanical agencies combine with the manure in preparing the plot to grow Potatoes again much better than if the land were kept to this crop only from year to year. If we could mark out a plot of ground into four parts, we should devote one plot to permanent crops—such as Asparagus, Sea Kale, and Rhubarb—and on the other three keep the crops revolving in some such order as this: No. 1, Potatoes, Celery, Leek, Carrot, Parsnip, Beet, &c. No. 2, Peas, Beans, Onions, Summer Spinach, &c., followed by Turnips for winter use, Cabbage for spring use, and Winter Spinach. No. 3, Brassicas, including Broccoli, Brussels Sprouts, Kale, &c. In the following year the original No. 1 would be cropped as No. 2, and No. 2 as No. 3. In the third season corresponding changes would be made, constituting a three-course system. The cultivator must use discretion in cropping vacant ground. As an example it will be obvious that land cleared of Early Potatoes will be very suitable for planting Strawberries. Another point is worth attention: Peas sown on the lines where Celery has been grown will thrive without any preparation beyond levelling the ground and drawing the necessary drills. This is a West of England custom, and it answers exceedingly well.

THE CHEMISTRY OF GARDEN CROPS

A Consideration of the chemistry of the crops that engage attention in this country will afford an explanation of one great difference between farming and gardening. And this difference should be kept in mind by all classes of cultivators as the basis of operations in tillage, cropping, and the order and character of rotations. The first thing to discover in the cropping of a farm is the kind of vegetation for which the land is best adapted to insure, in a run of seasons, fairly profitable results. If the soil is unfit for cereals, then it is sheer folly to sow any more corn than may be needful for convenience, as, for example, to supply straw for thatching and litter, and oats for horses, to save cost of carriage, &c. On large farms that are far removed from markets it is often necessary to risk a few crops that the land is ill fitted for, in order to satisfy the requirements of the homestead, and to save the outlay of money and the inconvenience of hauling from distant markets. But everywhere the cropping must be adapted to the soil and the climate as nearly as possible, both to simplify operations and enlarge to the utmost the chances of success. In the cropping of a garden this plain procedure cannot be followed. We are compelled certainly to consider what the soil and climate will especially favour amongst garden crops, but, notwithstanding this, the gardener must grow whatever the household requires. He may have to grow Peas on a hot shallow sand; and Potatoes and Carrots on a cold clay; and Asparagus on a shallow bed of pebbles and potsherds. To the gardener the chemistry of crops is a matter of great importance, because he cannot restrict his operations to such crops as the land is particularly adapted for, but must endeavour to make the land capable of carrying more or less of all the vegetables and fruits that find a place in the catalogue of domestic wants. That he must fail at certain points is inevitable; nevertheless his aim will be, and must be, of a somewhat universal kind, and a clear idea of the relations of plants to the soil in which they grow will be of constant and incalculable value to him.

We are bound to say at the outset that a complete essay on the chemistry of vegetation is not our purpose. We are anxious to convey some useful information, and to kindle sufficient interest to induce those who have hitherto given but slight attention to this question to inquire further, with a view to get far beyond the point at which we shall have to quit the subject.

Plants consist of two classes of constituents—the Inorganic, which may be called the foundation; and the Organic, which may be considered the superstructure. With the former of these we are principally concerned here. A plant must derive from the soil certain proportions of silica, lime, sulphur, phosphates, alkalies, and other mineral constituents, or it cannot exist at all; but, given these, the manufacture of fibre, starch, gum, sugar, and other organic products depends on the action of light, heat, atmospheric air, and moisture, for the organic products have to be created by chemical (or vital) action within the structure, or, as we sometimes say, the tissues of the plant itself. To a very great extent the agencies that conduce to the elaboration of organic products are beyond our control (though not entirely so), whereas we can directly, and to a considerable degree, provide the plant with the minerals it more particularly requires; first, by choosing the ground for it, and next by tilling and manuring in a suitable manner. A clay soil, in which, in addition to the predominating alumina, there is a fair proportion of lime, may be regarded as the most fertile for all purposes; but we have few such in Britain, our clays being mostly of an obdurate texture, retentive of moisture, and requiring much cultivation, and containing, moreover, salts of iron in proportions and forms almost poisonous to plants. But there are profound resources in most clays, so that if it is difficult to tame them, it is also difficult to exhaust them. Hence a clay that has been well cultivated through several generations will generally produce a fair return for whatever crop may be put upon it. Limestone soils are usually very porous and deficient of clay, and therefore have no sustaining power. Many of our great tracts of mountain limestone are mere sheep-walks, and would be comparatively worthless except for the lime that may be obtained by burning. On the other hand, chalk, which is a more recent form of carbonate of lime, is often highly productive, more especially where, through long cultivation, it has been much broken up, and has become loamy through accumulation of humus. Between the oldest limestone and the latest chalk there are many intermediate kinds of calcareous soils, and they are mostly good, owing to their richness in phosphates, the products of the marine organisms of which these rocks in great part, and in some cases wholly, consist. For the growth of cereals these calcareous soils need a certain proportion of silica, and where they have this we see some of the finest crops of Wheat, Trifolium, Peas and Beans in these islands. If we could mix some of our obdurate clays with our barren limestones, the two comparatively worthless staples would probably prove remarkably fertile. Although this is impossible, a consideration of the chemistry of the imaginary mixture may be useful, more especially to the gardener, who can in a small way accomplish many things that are impracticable on a great scale. Sandy soils are characterised by excess of silica, and deficiency of alumina, phosphates and potash. Here the mechanical texture is as serious a matter as it is in the case of clay. The sand is too loose as the clay is too pasty, and it may be that we have to prevent the estate from being blown away. It is especially worthy of observation, however, that sandy soils are the most readily amenable of any to the operation of tillage. If we cannot take much out of them, we can put any amount into them, and it is always necessary to calculate where the process of enrichment is to stop. It is not less worthy of observation that sandy soils can be rendered capable of producing almost every kind of crop, save cereals and pulse, and even these can be secured where there is some basis of peat or loam or clay with the sand. The parks and gardens of Paris, Versailles, and Haarlem are on deep sands that drift before the wind when left exposed for any length of time with no crop upon them; and not only do we see the finest of Potatoes and the most nutritious of herbage produced on these soils, but good Cauliflowers, Peas, Beans, Onions, fruits, and big trees of sound timber.

Garden soils usually consist of loam of some kind, the consequence of long cultivation. Natural loams are the result of the decay and admixture of various earths, and they are mostly of a mellow texture, easily worked and highly productive. They are, as a rule, the best of all soils, and their goodness is in part due to the fact that they contain a little of everything, with no great predominance of any one particular earth. Cultivation also produces loam. On a clay land we find a top crust of clayey loam, and on a lime or chalk land a top crust of calcareous loam. Where cultivation has been long pursued the staple is broken and manures are put on, and the roots of plants assist in disintegration and decomposition. Thus there is accumulation of humus and a decomposition of the rock proceeding together, and a loam of some sort is the result. Hence the necessity of caution in respect of deep trenching, for if we bury the top soil and put in its place a crude material that has not before seen daylight, we may lose ten years in profitable cropping, because we must now begin to tame a savage soil that we have been at great pains to bring up, to cover a stratum of a good material prepared for us by the combined operations of Nature and Art during, perhaps, several centuries. But deep and good garden soils may be safely trenched and freely knocked about, because not only does the process favour the deep rooting of the plants, but it favours also that disintegration which is one of the causes of fertility. Every pebble is capable of imparting to the soil a solution—infinitesimal, perhaps, but not the less real—of silica, or lime, or potash, or phosphates, or perhaps of all these; but it must be exposed to light and air and moisture to enable it to part with a portion of its substance, and thus it is that mechanical tillage is of the first importance in all agricultural and horticultural operations.

The principal inorganic or mineral constituents of plants are potash, soda, lime, iron, phosphorus, sulphur, chlorine, and silica. Clays and loams are generally rich in potash, sulphur, and phosphates, but deficient in soluble silica and lime. Limestone and chalk are usually rich in lime and phosphates, but deficient in humus, silica, sulphur, and alkalies. Sandy soils are rich in silica, but are generally poor in respect of phosphates and alkalies. Therefore, on a clay or loam, farmyard manure is invaluable, because it contains ingredients that all crops appreciate, and also because it is helpful in breaking up the texture of the soil. The occasional application of lime also is important for its almost magical effect on garden soil that has been liberally manured and heavily cropped for a long term of years. Calcareous soils are greatly benefited by a free application to them of manure from the stable and cow-byre; but as a rule it would be like carrying coals to Newcastle to dress these soils with lime. Clay may be put on with advantage; and nothing benefits a hot chalky soil more than a good dose of mud from ponds and ditches, which supplies at once humus, alumina, and silicates, and gives 'staple' to the soil, while preventing it also from 'burning.' In the manuring of sandy soils great care is requisite, because of their absorbing power. In the bulb-growing districts of Holland, manure from cowsheds is worth an enormous price for digging into loose sand for a crop of Potatoes, to be followed by bulbs. Sandy soils are generally deficient in phosphates and alkalies; hence it will on such soils be frequently found that kainit (a crude form of potash) and superphosphate of lime will conjointly produce the best results, more especially in raising Potatoes, Onions, and Carrots, which are particularly well adapted for sandy soils. Probably one of the best fertilisers is genuine farmyard manure from stall-fed cattle, for it contains phosphates, alkalies, and silicates in available forms. For similar reasons Peruvian Guano is often useful on such soils. Artificial manure should be selected with a view to correct the deficiencies of the soil, and to satisfy the requirements of the crops to be grown on it.

While we have thus dealt principally with the Inorganic or mineral constituents of plants, and the way in which the deficiencies of the soil in respect of any of them may be supplied by artificial applications, we must not ignore the other class of constituents, the Organic. These are supplied almost entirely from the atmosphere itself, though, to a limited extent, the presence in the soil of humus or vegetable matter contributes also. Yet this latter, as seen in the case of land heavily dressed with farmyard or stable manure, vegetable refuse, &c, exercises important functions in other directions. Not only are mineral constituents, in forms available for assimilation, supplied, but soils so treated derive peculiar advantages as regards their mechanical state and improved physical conditions, chiefly in respect of retention of moisture, warmth, &c. Thus, sandy soils, which are very apt, through poverty in humus, to lose their moisture readily and to 'burn,' are rendered more retentive of moisture and fertilising constituents by the use of farmyard manure, &c., and have more 'staple' or substance given to them, while heavy, tenacious clays are opened out, lightened, and rendered more amenable to the influences of drainage, aeration, &c., and so become less cold and inactive.

For the present purpose the principal garden crops may be grouped in two classes, in accordance with their main characteristics and the predominance of certain of their mineral elements. The figures given on the following page show the average percentage proportions of the several minerals in the ashes of the different plants.

In Class I. Phosphates and Potash predominate. This class consists of the less succulent plants, and includes the following: The Pea: containing, in 100 parts of the ashes, phosphates, thirty-six; potash, forty. Bean: phosphates, thirty; potash, forty-four. Potato (tubers only): phosphates, nineteen; potash, fifty-nine; soda, two; lime, two; sulphuric acid, six. Parsnip: phosphates, eighteen; potash, thirty-six; lime, eleven; salt, five.

Carrot: phosphates, twelve; potash, thirty-six; soda, thirteen; sulphuric acid, six. Jerusalem Artichoke: phosphates, sixteen; potash, sixty-five.

In Class II. Sulphur, Lime and Soda Salts are predominant. This class consists of the more succulent plants, and includes the following: Cabbage: containing, in 100 parts of the ashes, phosphates, sixteen; potash, forty-eight; soda, four; lime, fifteen; sulphuric acid, eight. Turnip: phosphates, thirteen; potash, thirty-nine; soda, five; lime, ten; sulphuric acid, fourteen. Beet: phosphates, fourteen; potash, forty-nine; soda, nineteen; lime, six; sulphuric acid, five.

As a matter of course, Lentils and other kinds of pulse agree more or less with Peas and Beans in the predominance of phosphates and potash. So, again, all the Brassicas, whether Kales, Cauliflower, or whatever else, agree nearly with the Cabbage in the prominent presence of lime and sulphur; ingredients which fully account for the offensive odour of these vegetables when in a state of decay. Fruits as a rule are highly charged with alkalies, and are rarely deficient in phosphates; moreover, stone-fruits require lime, for they have to make bone as well as flesh when they produce a crop. As regards the alkalies, plants appear capable of substituting soda for potash under some circumstances, but it would not be prudent for the cultivator to assume that the cheaper alkali might take the place of the more costly one as a mineral agent, for Nature is stern and constant in her ways, and it can hardly be supposed that a plant in which potash normally predominates can attain to perfection in a soil deficient in potash, however well supplied it may be with soda. The cheaper alkali in combination as salt (chloride of sodium) may, however, be usually employed in aid of quick-growing green crops; and more or less with tap-roots and Brassicas. Salt, too, is very useful in a dry season by reason of its power of attracting and retaining moisture. As regards Potatoes, it is worthy of observation that they contain but a trace of silica, and yet they generally thrive on sand, and in many instances crops grown on sand are free from disease and of high quality, although the weight may not be great. The mechanical texture of the soil has much to do with this; and when that is aided by a supply of potash and phosphates, whether from farmyard manure or artificials, sandy soils become highly productive of Potatoes of the very finest quality. On the other hand, Potatoes also grow well on limestone and chalk, and yet there is but little lime in them. Here, again, mechanical texture explains the case in part, and it is further explained by the sufficiency of potash and phosphates, as also of magnesia, which enters in a special manner into the mineral constitution of this root.

Thus far we have not even mentioned nitrogen, or its common form of salts of ammonia; nor have we mentioned carbon, or its very familiar form of carbonic acid. These are important elements of plant growth; and they account for the efficacy of manures derived directly from the animal kingdom, as, for example, the droppings of animals, including guano, which consisted originally of the droppings of sea-birds. Some of the nitrogen in these substances, however, is of an evanescent character, and rapidly flies away in the form of carbonate of ammonia; hence, a heap of farmyard manure, left for several years, loses much of its value as manure, and guano should be kept in bulk as long as possible, and protected from the atmosphere, or its ammonia will largely disappear. One difficulty experienced by chemists and others in preparing artificial manures is that of 'fixing' the needful ammonia, so that it may be kept from being dissipated in the atmosphere, and at the same time be always in a state in which it can be appropriated by the plant. In all good manures, however, there is a certain proportion of it in combination, and in many instances the percentage of nitrogen is made the test of the value of a manure.

The importance of humus—the black earthy substance resulting from the decay of vegetation—in a soil is that it contains in an assimilable form many of the ingredients essential to plant life. Humus when it decomposes gives off carbonic acid, which breaks up the mineral substances in the soil and renders them available as plant food. When vegetable refuse is burned, the nitrogen—one of the costliest constituents—is dissipated and lost. But by burying the refuse the soil gets back a proportion of the organic nitrogen it surrendered and something over in the way of soluble phosphatic and potassic salts; and as this organic nitrogen assumes ultimately the form of nitric acid, it can be assimilated by the growing plant, to the great benefit of whatever crop may occupy the ground.

The practical conclusion is, that in the treatment of the soil a skilful gardener will endeavour to promote its fertility by affording the natural influences of rain, frost and sun full opportunity of liberating the constituents that are locked up in the staple; by restoring in the form of refuse as much as possible of what the soil has parted with in vegetation; and by the addition of such fertilising agents as are adapted to rectify the natural deficiencies of the soil. Thus, instead of following a process of exhaustion, the resources of the garden may be annually augmented.

ARTIFICIAL MANURES AND THEIR APPLICATION TO GARDEN CROPS

Plants, like animals, require food for their sustenance and development, and when this is administered in insufficient quantities, or unsuitable foods are supplied, they remain small, starved, and unhealthy.

The chemical elements composing the natural food of ordinary crops are ten in number, viz.—carbon, hydrogen, oxygen, nitrogen, sulphur, phosphorus, potassium, calcium, magnesium, and iron. These are obtained from the soil and air, and unless all of them are available plants will not grow. The absence of even one of them is as disastrous as the want of all, and a deficiency of one cannot be made up by an excess of another; for example, if the soil is deficient in potassium the crop suffers and cannot be improved by adding iron or magnesium. All the food-elements are found in adequate quantities in practically all soils and the surrounding air, except three—nitrogen, potassium, and phosphorus. These are often present in reduced amount, or in a state unsuited to plants; in such cases the deficiency must be made up before remunerative healthy crops can be grown, and it is with this express object that manures are added to the soil.

One of the best known substances employed in this way is farmyard manure, which is indirectly derived from plants and contains all the elements needed for the growth of crops. It is, however, of very variable composition and rarely, or never, contains these elements in the most suitable proportions, and its value can always be greatly improved by supplementing its action with one or other of the so-called artificial manures or fertilisers. Although it is strongly advisable to add farmyard manure or vegetable composts to the soil of all gardens now and again, in order to keep the texture of the soil in a satisfactory condition, excellent crops can be grown by the use of artificial fertilisers alone. To obtain the best results from these some experience is of course necessary, but the following details regarding the nature and application of the commoner and more useful kinds should prove a serviceable guide in the majority of cases.

Artificial manures may be divided into three classes:—

1. The Nitrogenous class, of which nitrate of soda and sulphate of ammonia are examples.
2. The Phosphatic class, such as superphosphate, basic slag, and steamed bone flour.
3. The Potash class, including kainit and sulphate of potash. The several examples of each class contain only one of the three important plant food-elements, and as a single element can only be of use when the others are present in the soil, it is generally advisable to apply one from each class, either separately or mixed, in order to insure that the crop is supplied with nitrogen, phosphates, and potash.

Nitrogenous manures specially stimulate the growth of the foliage, stems, and roots of plants, and are therefore of the greatest benefit to Carrots, Parsnips, Turnips, Beet, Celery, Asparagus, Rhubarb, all the Cabbage tribe, and leafy crops generally.

Nitrate of soda supplies the single plant food-element, nitrogen, and the soda for all practical purposes may be disregarded. It dissolves very easily in water and is taken up immediately by growing plants, its effect being plainly seen a few days after application. As this artificial readily drains away from uncropped land it should only be administered to growing plants. It is best applied in spring and summer and in small quantities; for example, at the rate of one pound per square rod, repeated at intervals of two or three weeks, rather than in a single large dose. Nitrate of soda must not be mixed with superphosphate, but it may be added to basic slag and the potash manures.

Sulphate of ammonia is another nitrogenous fertiliser, similar in its effects to nitrate of soda, but slower in action since its nitrogen must undergo a change into nitrate before it is available for plants. It is held by the soil, and can therefore be applied earlier in spring than nitrate of soda without fear of loss. The continued use of this manure, however, is liable to make the soil sour, and consequently it should only be employed on ground containing lime, or to which lime has been added. Never mix sulphate of ammonia with basic slag or with lime, but it may be mixed with superphosphate and the potash manures.

Phosphatic manures have the opposite effect to the nitrogenous fertilisers, checking rampant growth and encouraging the early formation of flowers, fruit, and seeds. They are comparatively inexpensive and should be liberally applied to all soils for all crops. *Superphosphate* is an acid manure and best suited for use on soils containing lime. *Basic slag* is a better material for ground deficient in lime, or where 'club-root' is prevalent. It is less soluble and therefore slower in action than superphosphate. Both these fertilisers should be dug into the soil some time before the crop is planted or seed sown—superphosphate at the rate of two to three pounds per square rod; basic slag in larger amount, five to six pounds per square rod. Superphosphate may also be employed as a top-dressing and worked into the surface around growing plants with the hoe. *Steamed bone meal* or *flour* is another useful phosphatic fertiliser, valuable on the lighter classes of soil.

Potash manures are of benefit to plants in all stages of growth. They are particularly valuable to Potatoes, leguminous crops, Carrots, Parsnips, Turnips, and Beet. Like the phosphatic manures they should be worked into the soil before seeds are sown or plants are put out. *Kainit* is best applied in autumn, for it contains a considerable amount of common salt and magnesium compounds which are sometimes deleterious and best washed away in the drainage water during winter. It should be dug in at the rate of about three pounds per square rod. *Sulphate of potash* is three or four times as rich in potash as kainit, and is correspondingly more expensive; apply in spring and summer, a little in advance of sowing or planting, at the rate of about one pound per square rod.

Lime.—A word or two must be said about lime, which is a natural constituent of all soils. In many instances there is sufficient for the needs of most plants, but where lime is deficient in quantity it must be added before healthy crops can be raised. Old gardens to which dung has been freely applied annually require a liberal dressing of lime every few years, or the ground becomes sour and incapable of growing good crops of any kind. To insure the proper action of whatever manures are used and to secure healthy crops, an application of slaked quicklime, at the rate of fourteen to twenty pounds per square rod, is strongly recommended. As a remedy against 'clubbing' or 'finger-and-toe' disease of the Cabbage tribe of plants it is indispensable; it also neutralises the baneful acidity of the land, and opens up stiff soils, making them more easily tilled, more readily penetrated by the air, and warmer by the better drainage of water through them.

The following suggestions for the manuring of the different crops mentioned will be found effective. It is, however, not intended that they should be slavishly followed, for useful substitutions may be made in the formulæ given, if the nature of the various fertilisers is understood and an intelligent grasp is obtained of the principles of manuring enunciated in this and the preceding chapter.

In place of nitrate of soda, a similar quantity of sulphate of ammonia may be used.

Instead of superphosphate, the following may be advantageously employed: phosphatic guano, or mixtures of basic slag and superphosphate, or bone meal and superphosphate; or basic slag may be applied alone on land deficient in lime.

Four pounds of kainit may also take the place of one pound of sulphate of potash in the suggested mixtures mentioned below.

Where dung is recommended, twenty to twenty-five loads per acre is meant; larger quantities are frequently applied, but these are uneconomical and much less efficient than more moderate amounts supplemented with artificial fertilisers.

All the manures should be worked into the soil before sowing or planting out, except the nitrate of soda, which is best applied separately to the growing plants, preferably in small doses at intervals of two to four weeks.

In all cases the quantities of artificials named are intended for use on one square rod or pole of ground.

PEAS AND BEANS.—These leguminous plants are able to obtain all the nitrogen they need from the air. They should, however, be amply supplied with potash and phosphates, a good dressing being:—

1

2-3/4 to 3-1/2 lb. superphosphate

3/4 lb. sulphate of potash

DWARF BEANS are sometimes benefited by the addition of 1/2-lb. to 1 lb. of nitrate of soda.

ASPARAGUS.

A dressing of dung

2 lb. nitrate of soda

3-1/2 to 4 lb. superphosphate

3 lb kainit

The kainit contains a considerable amount of salt, which is of value to this crop.

BEET.—For a fine crop a moderate amount of well-decayed dung applied in autumn is almost essential, as well as 3 to 4 lb. of superphosphate per square rod in spring. On land previously dressed with dung for a former crop, the following may be used, especially on the lighter class of soils:—

1-1/2 lb. nitrate of soda when the plants are well
up, and a similar amount a fortnight
after singling
4 to 5 lb. superphosphate
4 lb. kainit

BROCCOLI AND CAULIFLOWER.

With dung.
2 to 3 lb. nitrate of soda
2 to 3 lb. superphosphate
3/4 lb. sulphate of potash

Without dung.
4 to 5 lb. nitrate of soda
4 to 5 lb. superphosphate
3/4 lb. sulphate of potash

CABBAGE, KALE, AND BRUSSELS SPROUTS.—These Brassicas require considerable quantities of nitrogen and phosphates. For spring Cabbage planted in autumn, land well dunged for the previous crop gives good results with the addition of the artificials mentioned below: for the autumn crop, dung should be applied before planting out in the early part of the year.

With dung.
2 to 3 lb. nitrate of soda
4 to 5 lb. superphosphate
3/4 lb. sulphate of potash

Without dung.
4 lb. nitrate of soda
5 to 6 lb. superphosphat
3/4lb. sulphate of potash

CARROT AND PARSNIP.—A good dressing of dung applied to the previous crop is a valuable preparation where Carrots and Parsnips are to be grown. In addition, one of the following mixtures should be used:—

(1)

3/4 lb. nitrate of soda
3 to 4 lb. superphosphate
3/4 lb. sulphate of potash

(2)

3/4 lb. nitrate of soda
2 lb. superphosphate
1 to 2 lb. basic slag
3 lb. kainit

CELERY requires the use of dung more than almost any other crop, and it is little affected by artificial manures, except phosphates, which may be given in the form of superphosphate at the rate of 2-1/2 to 3-1/2 lb per square rod.

LETTUCE.

With dung.
3 to 4 lb. superphosphate
1/2 to 1 lb. nitrate of soda

Without dung.
3 to 4 lb. superphosphate
1 to 1-1/2 lb. nitrate of soda
1 lb. sulphate of potash

ONIONS never succeed without an ample supply of potash. This crop should therefore have farmyard dung, or the special potash fertilisers in adequate quantity.

With dung.
3/4 lb. nitrate of soda
4 to 5 lb. superphosphate
3/4 lb. sulphate of potash

Without dung.
1-1/2 to 2-1/2 lb. nitrate of soda
5 lb. superphosphate
1 lb. sulphate of potash

LEEKS require the same fertilisers as Onions, but will need little or no nitrate if good dung is used.

POTATO.—For good yield, high quality, and freedom from disease, Potatoes are dependent upon a good supply of potash. They do best when supplied with a moderate amount of farmyard manure, supplemented by suitable artificials, but can be grown on some soils with artificials alone.

With dung.
3/4 lb. sulphate of ammonia
3 lb. superphosphate
3/4 lb. sulphate of potash

Without dung.
1-1/2 lb. sulphate of ammonia
3-1/2 lb. superphosphate
1 to 1-1/2 lb. sulphate of potash

Instead of superphosphate, a mixture of this fertiliser with an equal amount of bone meal or basic slag may be used, and either 4 lb. of kainit and 1 lb. of muriate of potash instead of 1 lb. of sulphate of potash.

RHUBARB.—An annual dressing of dung is beneficial, together with 6 lb. of basic slag, 1 lb. of sulphate of potash, and 4 lb. of nitrate of soda, half the nitrate being applied when growth commences and the remainder a fortnight later.

SPINACH.

With dung.
3 to 4 lb. superphosphate
2 to 3 lb. nitrate of soda

Without dung
4 to 5 lb. superphosphate
1 lb. sulphate of potash
3 to 4 lb. nitrate of soda

TOMATOES need large supplies of potash and phosphates to induce stocky growth and abundance of flowers and fruit. Nitrogenous manures should be withheld until the flowering stage, for they stimulate the production of rank succulent stems and leaves which are specially liable to attacks of fungus pests. After the fruit is set the application of small doses of nitrate of soda, or sulphate of ammonia, as advised below, greatly assists the swelling of the crop. The following mixtures worked into the soil will be found beneficial for Tomatoes:—

5 to 6 lb. superphosphate 7 to 8 lb. basic slag
1 lb. sulphate of potash *or* 1 lb. sulphate of potash

Nitrate of soda, or sulphate of ammonia, at the rate of 1-1/2 to 2 lb. per square rod, may be given with advantage as soon as the fruit is set.

TURNIP AND SWEDE.—For the development of fine roots a liberal supply of phosphates is essential.

With dung.
1 lb. nitrate of soda
3 to 4 lb. superphosphate
3/4 lb. sulphate of potash

Without dung
2 lb. nitrate of soda
4 to 5 lb. superphosphate
1 lb. sulphate of potash

THE CULTURE OF FLOWERS FROM SEEDS

Whether the modern demand for flowers has created the supply, or the supply has found an appreciative public, we need not stay to discuss. The fact remains that the last four or five decades have witnessed a phenomenal extension in the use of flowers by all classes of the community, for the decoration of the house no less than for beautifying the garden. Primarily, this advance of refinement in the popular taste is traceable to the skill and enthusiastic devotion of the florists who have supported in all their integrity the true canons of floral perfection, and whose labours will continue to be imperative for maintaining the standards of quality. By their severe rules of criticism the florists further the ends of floriculture subjectively, and by the actual results of their labours they render objective aid, their finest flowers serving not only as types, but as the actual stud for perpetuating each race. Hence the decline of floriculture would imply the deterioration of flowers, and the prosperity of floriculture involves progress not only in those subjects which lie within the florists' domain, but of many others to which they have not devoted special attention. Yet the acknowledgment must be made that, brilliant as their triumphs have been, the methods they practised have in some instances entailed very severe penalties. Continuous propagation for many generations, under artificial conditions, so debilitated the constitution of Hollyhocks, Verbenas, and some other subjects, that the plants became victims of diseases which at one time threatened their existence. To save them from annihilation it was necessary to desert the worn path of propagation, and raise plants possessing the initial vigour of seedlings. In stamina these seedlings proved eminently satisfactory, although in other respects they were at first sadly disappointing. It then became clear that before show flowers could be obtained from seedlings judgment and skill must be devoted to the art of saving seed. This was necessarily a work of time, demanding great patience and rare scientific knowledge. The task was undertaken with enthusiasm in many directions, and the results have more than justified this labour of love. Formerly, the universal mode of perpetuating named Hollyhocks was by the troublesome process of cuttings, or by grafting buds on roots of seedlings in houses heated to tropical temperature. In many places it was the custom to lift the old plants, pot them, and keep them through the winter in pits. All this was found requisite to insure fine flowers. While the burden of the work was thus rendered heavy, the constitution of the plant became enfeebled, and at one time the fear was entertained that its extinction was at hand. But the new system has preserved the Hollyhock, and at the same time afforded a striking example of the principle that seed saved scientifically is found to reproduce the varieties it was taken from. Seedling Hollyhocks now give double flowers of the finest quality; and the seedling plants are less liable to disease. So with the Verbena. From suitable seed plants can be raised that will produce the most resplendent flowers, and instead of propagating a stock to keep over winter, to be stricken with mildew and cost no end of care, only to become diseased at last, a pinch of seed is sown in January or February, and soon there is a stock of healthy plants possessing the vigour peculiar to seedlings. These, being bedded out at a proper time, flower far more freely than plants from cuttings, and produce trusses twice the size.

To illustrate the change of method still further we may instance the Cineraria. Formerly this was a troublesome plant to grow, because it was considered necessary to propagate named varieties by divisions and suckers. The restricted system was reflected in limited cultivation. Few were willing to venture on a task known to be hedged about with difficulties. By degrees it was discovered that the finest Cinerarias might be secured by simply sowing seed, and giving the plants the usual cultivation of tender annuals. This has brought the Cineraria within the reach of thousands who would not attempt to grow it under the old system, and the consequent gain to society is immense.

What has been done with the Cineraria has its parallel in quite a number of the most elegant decorative flowers. Brilliant results have been achieved with Begonias, Calceolarias, Cyclamens, Gloxinias, Primulas, and Schizanthus. It has also ceased to be needful to keep such large stocks of bedding and other plants through the winter, for Ageratums, Lobelias, and Pansies have proved amenable to the new treatment, and very much of the accustomed labour in striking and potting cuttings, as well as the expense of glass, fuel, and the frequent purchase of high-priced plants, have been rendered unnecessary. Even among the flowers which are properly designated annuals, new and delightful variations have been obtained from original types. Of these we have examples in Aster, Godetia, Larkspur, Mignonette, Phlox Drummondii, Poppy, Stock, Sweet Pea, and many others. In some instances the increase in the size of the flowers is remarkable, and in others the development of new tints will surprise those who are not familiar with the labours of modern hybridisers.

Thus a revolution has been accomplished in the economy and complexion of the English Flower Garden, a revolution which has reduced and simplified the gardener's labours, augmented the number and enhanced the beauty of many flowers, effected a marked saving in the cost of garden pleasures, and brought the culture of a large number of the most attractive subjects within the means of those who had neither the facilities nor the knowledge requisite for pursuing the florist's methods. There appear to be no limits to further progress. All that we can do is to experiment and gather knowledge, and those who love gardening may assist in extending the area of this new and cheap system of producing some of the most elegant garden flowers in one season from seed alone.

The time and the method of sowing flower seeds must in each case be regulated by considerations as to their nature. Seeds of tender plants are usually sown in pots or pans and placed on a moderate hot-bed or in a propagating house early in spring, and in this case the plants have greenhouse cultivation until the time arrives for hardening them off preparatory to final planting. But seeds of many hardy flowers may be treated in the same way, when a long season of growth is necessary for their development. Thus Phloxes, Verbenas, and Hollyhocks, plants that differ immensely in habit and constitution, may all be sown in February, and put side by side in the same warm pit or vinery, or even in the warmest corner of any greenhouse, and the very same treatment will suit them equally well. The soil should be principally loam and sand, with a little old thoroughly well-rotted manure from a hot-bed or compost heap; and light, air, and moisture must be regulated with a view to insure a free and vigorous growth from the first, with the least possible amount of artificial heat. In some cases, however, the sowing should be deferred to March or April, and the result will be far more satisfactory than the growth made under the stimulus of artificial heat earlier in the season. But in every case the plants must have sufficient time; for although the rapid system has been developed, the constitution of the plants remains unchanged, and those which have heretofore been classed as biennials and perennials need a long season when treated as annuals.

A considerable proportion of the finest flowers may be raised from seed by the aid of a frame and a little careful management. We will take as an example a very restricted garden. Here is a small frame and some packets of seed, and the month of February or March has arrived. The pans and pots are made ready with sweet sandy compost, and the seeds are sown and labelled, and the pans and pots are packed together in the frame on a bed of clean coal ashes, or some slates, or tiles, or bricks laid on the soil, to promote warmth and cleanliness and to prevent the intrusion of worms among the seeds. By simple management almost as quick a growth of seeds can be insured in this frame as with the aid of a hot-bed, and the secret consists in careful storage of the heat of the sun. Lay over the seed-pans sheets of glass to prevent evaporation, and let the sun shine full upon them. Be careful as to moisture: they must never be wet, never dry, and the water must not be slopped about carelessly. It is a good rule to immerse the pots or pans in a vessel containing soft water, slightly tepid. When the seedlings begin to appear, give a little air and lay sheets of paper tenderly over them during the hour or two at midday when the sun may be shining brightly. But keep them from the first as 'hard' as possible with plenty of light and air, always taking care that they are neither roasted, nor blown away by the cruel east wind, nor nipped at night by a killing frost. A few old mats or light loppings of trees laid over the frame from sundown to sunrise will be sufficient protection at those trying times; and when spring frosts are making havoc with the tender sprouting leaf and bloom in every part of the garden those little things will be safe under their glass cover, and slight experience will show that a common frame may become a miniature hot-house in the hands of one who has learned to make failure the stepping-stone to success. We must not omit to mention that the owner of such a garden, or, indeed, of any garden, will be prudent to take advantage of the first fine weather to sow in the open ground whatever flower or vegetable seeds should be sown at that season. The frame garden can be reserved, if needful, for wet weather, because it is of the utmost importance to sow a good breadth of seeds in the open ground as early as possible in the month of March.

Turning from this small example to the great garden, it will be obvious that to those who always have heavy work on hand the advantages of this transference, of labour from the old system to the new are immense. Both to employers and gardeners the advantages are of importance; the propagation of bedders by cuttings, and of florists' flowers by suckers and divisions and layers and pipings, will not, of course, be completely abolished; but for all ordinary purposes the ends in view may be accomplished more simply, more expeditiously, and more cheaply than heretofore. The pits hitherto appropriated to bedders, and the like, may to a great extent be liberated, and there will be no difficulty in finding for them more profitable occupants. While Mushrooms and early Potatoes and winter salads are in request, it will be a gain to many a garden to have reduced the summer display of flowers to a simple system of seed-sowing, at an expense that may be described as merely nominal.

Before dealing specifically with certain flowers, it may be advisable to say a few words generally concerning the culture of Annuals—Hardy, Half-hardy, and Tender—and also on hardy Biennials and Perennials.

Annuals.—Although the most popular kinds of annuals are largely employed in the embellishment of flower gardens, they are adapted for many uses to which they may with advantage be more frequently applied. A few misconceptions prevail as to the relative merits of this class of plants. By some they are regarded as 'weedy' and 'short-lived.' Their very cheapness, and the relatively small amount of skill required in their cultivation, tend in some degree to detract from their value in public estimation. We will not be so rash as to say that a more extended use of annuals would render unnecessary the cultivation of what are especially known as 'bedding plants'; but there is something to be said on behalf of annuals that may be worth the consideration of all who are interested in the development of freshness, variety, and richness of colour in the flower garden. In the first place, these plants come into flower within a comparatively short period of time from the sowing of the seed, and it is a matter of considerable importance that a large proportion of the best continue beautiful until the very close of the season. Sometimes in the autumn Geraniums become literally washed out, while Tom Thumb Nasturtiums may be ablaze with colour, and continue so when the Geraniums are housed for the winter. A large number of showy and long-lasting annuals are adapted for employment in bedding, and by a little management those that do not last the season out may be replaced by others for succession; thus affording the advantage of increased variety, and making no demand for glass and fuel to keep them through the winter as do the ordinary bedders. We have had great and glorious sheets of Candytufts, snow-white, rich crimson, and bright carmine; and when they began to wane they were removed, and the ground planted with Asters, and very soon there was another display, so fresh and bright and various that no greenhouse bedders could surpass them. Great hungry banks, that would have swallowed many pounds' worth of greenhouse plants to cover them, have been made delightfully gay at a very trifling cost by sowing upon them Tropæolums, Tom Thumb Nasturtiums, *Bartonia aurea*, the dwarf varieties of *Lupinus*, Virginian Stock, *Collinsia bicolor*, Convolvuluses, Candytufts, Eschscholtzias, Poppies, and Clarkias; and damp, half-shady borders have been delicately tessellated by means of Forget-me-nots, Venus' Looking-glass, Pansies, the Rosy Oxalis, Nemophilas, Godetias, Silenes, Coreopsis, and Scabious.

For the more important positions in the flower garden we have choice of many really sumptuous subjects, such as Stocks, Asters, Balsams, Drummond's Phlox, Lobelias, the lovely new varieties of Antirrhinums, Dianthus, Portulacas, Zinnias, tall Stock-flowered Larkspurs, Nemesias, and many other flowers equally beautiful and lasting. We do not hope by these brief remarks to change the prevailing fashion—indeed, we have no particular wish that way—but we feel bound to observe that it is sufficient for the beauty of the garden that the greenhouse bedders should be confined to the parterre proper. It is waste of space and opportunity to place them in the borders everywhere, as is too commonly done. In sunny borders, annual and perennial herbaceous plants are far more appropriate.

Some time since, while walking over a large garden, we left the rich colouring of the geometric beds to discover what should make the wondrous glow of crimson on a border far away; and to our surprise it proved to be a clump of the Indian Pink, which had been sown as an annual with other annuals, and was there shining in the midst of a constellation of the loveliest flowers of all forms and hues, the result simply of sowing a few packets of seed. No one can despise the Wallflower in the spring, and the heavenly-blue flowers of *Nemophila insignis* in early summer will tempt many a one to walk in the garden who would care little for sheets of scarlet and yellow that in full sunshine make the eyes ache to look upon them. It must be remembered, too, that among annuals are found many most richly-scented flowers; others, like the everlastings and the grasses, are valuable to dry for winter use for employment in bouquets, and garlands in Christmas decorations; and the Sweet Peas, and *Tropæolum canariense,* and climbing Convolvulus may be employed to cover arbours and trellises with the best effect possible, and may even be allowed to hang in festoons about the sunny parts of rockeries, or trail over the ground to make genuine bedding effects. Another important matter must have mention here, and we commend it to the consideration of gardeners who are severely taxed to secure extensive displays of flowers during the summer season. It is that a number of plants of highly ornamental character, usually treated as perennials, are really more effective, besides occasioning less labour to produce them, when cultivated as annuals. The Dianthus and its several splendid varieties do better as annuals than biennials. For all the ordinary purposes of display, Lobelias may be as well grown from seed as from cuttings, and in every garden will be found proof of the small amount of care they require; for we find stray, self-sown plants in pots of Geraniums and other places, and these, if left alone, become perfect bushes, and are a mass of flowers all the summer. Many annuals commonly reputed to be tender and usually raised in heat do very well indeed on a more rough and ready method. In proof of this, sow *Perilla nankinensis* in the first week of May where it is required, and in the month of July you will probably be convinced that Perilla does not always need careful nursing in heated houses through the spring. Even the really tender Castor-oil Plant will thrive if sown in the open ground the first week in May. Having no check, as plants put out from pots must have, the growth will be regular and sturdy, and attain magnificent dimensions.

Perhaps the most effective way of growing annuals is to arrange them in harmonious blendings or contrasts of colour. The wide choice of varieties available admits of an almost endless number of combinations, and the following tables, classified according to colour, will no doubt afford some serviceable suggestions, although these by no means exhaust the list. The height is indicated in feet and Climbers as 'Cl.'

WHITE, AND CREAM SHADES.

TALL.

Chrysanthemum	coronarium,	Princess May	3
Chrysanthemum	coronarium,	Double white	3
Cornflower,	White		3
Helichrysum,	Silver Globe		3
Larkspur,	Stock-flowered,	White	3
Lavatera	alba	splendens	3
Poppy,	Giant Double,	White	3
Poppy	Giant Single,	White	3
Scabious,	Snowball		3
Chrysanthemum	carinatum	album	2-1/2
Chrysanthemum	Dunnetti,	Double white	2-1/2
Nasturtium,	Tall,	Pearl	Cl.

MEDIUM.

Clarkia	elegans,	Snowball	2
Lupinus	Hartwegii,	White	2
Malope,	White		2
Poppy,	White Swan		2
Shirley,	Double White		2
Calendula	pluvialis		1-1/2
Chrysanthemum	inodorum	plenissimum	
Clarkia,	Double White		1-1/2
Gilia	nivalis		1-1/2
Gypsophila	elegans		1-1/2
Hawkweed,	White		1-1/2
Hawkweed	Silver		1-1/2
Jacobea,	Double,	White	1-1/2
Sweet Sultan,	Giant	White	1-1/2
Chrysanthemum	coronarium,		
Dwarf	double	white	1-1/4

DWARF.

Acroclinium,	Single White	
Candytuft,	Improved	White Spiral
Clarkia,	Dwarf white	
Clarkia	Double	dwarf white
Convolvulus	minor,	White
Eschscholtzia	crocea	alba
Godetia,	Duchess	of Albany
Layia	elegans	alba
Linaria,	Snow-white	
Nasturtium,	Dwarf,	Pearl
Platystemon	californicus	
Viscaria,	Pure White	
Alyssum,	Sweet	
Chrysanthemum	inodorum	plenis-simum,
Bridal Robe		
Collinsia	candidissima	
Godetia,	Dwarf White	
Swan River	Daisy,	White
Swan River	Daisy	Star White
Venus'	Looking-glass,	White
Venus'	Navel-wort	
Virginian Stock,	White	
Candytuft,	Little Prince	
Nemophila	insignis	alba
Alyssum	minimum	
Silene,	Dwarf White	

YELLOW AND ORANGE SHADES

TALL.

Sunflower,	Giant Yellow		10
Sunflower	Primrose	Perfection	6
Sunflower	Miniature		4
Sunflower	Stella		4
Sunflower	Primrose Stella		4
Chrysanthemum	coronarium,		
Chrysanthemum	Double yellow		3
Chrysanthemum,	Golden Queen		3
Coreopsis	tinctoria		3
Helichrysum,	Golden Globe		3
Sunflower,	Dwarf Double		3
Sunflower	Single Dwarf		3
Chrysanthemum	Dunnettii,		
Chrysanthemum	Double Golden		2-1/2
Marigold,	African		2-1/2
Nasturtium,	Ivy-leaved	Golden Gem	Cl.
Nasturtium,	Tall,	Yellow	Cl.

MEDIUM.

Hibiscus	africanus	major	2
Bartonia	aurea		1-1/2
Chrysanthemum,	Star	varieties	1-1/2
Coreopsis	Drummondii		1-1/2
Coreopsis	coronata		1-1/2
Erysimum,	Orange Gem		1-1/2
Hawkweed,	Yellow		1-1/2
Leptosyne	Stillmani		1-1/2
Lupinus	Menziesii		1-1/2
Sweet Sultan,	Yellow		1-1/2

DWARF.

Calendula,	Orange King		1
Calendula	Lemon Queen		1
Cheiranthus	Allionii		1
Chrysanthemum	coronarium,		
Chrysanthemum	coronarium,	Dwarf double yellow	1
Dimorphotheca	aurantiaca		1
Eschscholtzia	californica		1
Escholtzia,	crocea		1
Escholtzia,	crocea	fl. pl.	1
Eschscholtzia,	Mikado		3/4
Layia	elegans		1
Lupinus,	Dwarf yellow		1
Nasturtium,	Dwarf,	Cloth of Gold	1
Nasturtium,	Dwarf,	Yellow	1
Tagetes	signata	pumila	1
Tagetes	Mandarin		3/4
Linaria,	Golden Gem		3/4
Marigold,	Miniature	orange	3/4
Marigold	Miniature	orange	3/4
Eschscholtzia,	Miniature Primrose		1/2
Limnanthes	Douglasii		1/2
Sanvitalia	procumbens,	Single	1/2
Sanvitalia	Double		1/2
Leptosiphon	aureus		1/4

BLUE, MAUVE, AND PURPLE SHADES.

TALL.

Cornflower,	Blue		3
Larkspur,	Stock-flowered,	Blue	3
Larkspur,	Stock-flowered,	Pale Mauve	3
Lupinus,	Tall dark blue		3
Poppy,	Giant Double, Mauve		3
Scabious,	Mauve		3

MEDIUM.

Godetia,	Double Mauve		2
Lupinus	Hartwegii,	Azure Blue	2
Poppy,	Mauve Queen		2
Sweet Sultan,	Purple		2
Xeranthemum	superbissimum		2
Xeranthemum	imperiale		2
Anchusa,	Annual Blue		1-1/2
Gilia	capitata		1-1/2
Gilia	tricolour		1-1/2
Jacobea,	Double, Purple		1-1/2
Nigella,	Miss Jekyll		1-1/2
Phacelia	tanacetifolia		1-1/2
Salvia,	Blue Beard		1-1/2
Sweet Sultan,	Giant Delicate Mauve		1-1/2
Sweet Sultan,	Giant Mauve		1-1/2

DWARF.

Asperula	azurea	setosa	1
Candytuft,	Lilac		1
Convolvulus	minor,	Dark blue	1
Convolvulus		Sky-blue	1
Cornflower,	King of	Blue Bottles	1
Eutoca	viscida		1
Linaria,	Mauve		1
Lupinus,	Dwarf	rich blue	1
Mathiola	bicornis		1
Phacelia	congesta		1
Viscaria,	Bright Blue		1
Whitlavia	gloxinioides		1
Cornflower,	Victoria,	Dwarf blue	3/4
Leptosiphon	androsaceus		3/4
Nigella,	Double dwarf		3/4
Phacelia	campanularia		3/4
Swan	River Daisy,	Blue	3/4
Swan	River Daisy,	Star Blue	3/4
Campanula	attica		1/2
Nemophila	insignis		1/2

PINK AND ROSE SHADES

TALL.			
Cornflower,	Pink		3
Larkspur,	Stock-flowered,	Rosy Scarlet	3
Lavatera	rosea	splendens	3
Lupinus	mutabilis,	Cream and Pink	3
Poppy,	Giant Double,	Chamois-rose	3
Scabious,	Pink		3
Nasturtium,	Salmon Queen		Cl.
Nasturtium,	Rosy Queen		Cl.
MEDIUM.			
Clarkia elegans,	Double Salmon		2
Clarkia elegans,	Double Delicate Pink		2
Godetia,	Double Rose		2
Jacobea,	Single,	Bright Rose	2
Poppy,	Pink Gem		2
Poppy,	Cardinal,	Salmon-pink	2
Poppy,	Shirley,	Single Rose-pink	2
Poppy,	Shirley,	Double Pink	2
Saponaria	Vaccaria,	Pink	2
Clarkia,	Double Rose		1-1/2
Hawkweed,	Pink		1-1/2
Jacobea,	Double,	Rose	1-1/2
Silene Armeria,	Rose		1-1/2
Statice	Suworowi		1-1/4
DWARF.			
Acroclinium,	Double rose		1
Acrolinium,	Single rose		1
Convolvulus minor,	Pink		1
Eschscholtzia,	Frilled Pink		1
Escholtzia,	Rosy Queen		1
Escholtzia,	Rose cardinal		1
Gypsophila elegans,	Delicate pink		1
Lupinus,	Dwarf	delicate pink	1
Nasturtium,	Dwarf,	Salmon Pink	1
Nasturtium,	Dwarf,	cæruleum roseum	1
Silene,	Double	Salmon Pink	1
Silene,	Double	Delicate Pink	1
Silene,	Bonetti		1
Silene,	Pseudo-Atocion		1
Statice	spicata		1
Viscaria,	Delicate Pink		1
Cornflower,	Victoria,	Dwarf rose	3/4
Godetia,	Dwarf Pink		3/4
Godetia,	Satin-rose		3/4
Abronia	umbellata		1/2
Candytuft,	Dwarf Pink		1/2
Saponaria	calabrica		1/2
Silene,	Double Dwarf	Delicate Pink	1/3
Silene,	Double Dwarf	Brilliant Rose	1/3
Silene,	Bonetti,	Dwarf Pink	1/3
Leptosiphon	roseus		1/4

CRIMSON AND SCARLET SHADES,

including Carmine and Ruby.

TALL.			
Coreopsis	atrosanguinea		3
Helichrysum,	Fireball		3
Poppy,	Giant Double,	Scarlet	3
Polygonum,	Ruby Gem		2-1/2
Malope,	Red		2
Nasturtium,	Tall,	Improved Lucifer	Cl.
Nasturtium,	Tall,	Black Prince	Cl.
MEDIUM.			
Chrysanthemum	atrococcineum		2
Clarkia elegans,	Salmon scarlet		2
Clarkia elegans,	Firefly		2
Godetia,	Double Crimson		2
Poppy,	Cardinal		2
Cacalia	coccinea		1-1/2
Coreopsis	cardaminigolia	Dwarf	1-1/2
DWARF.			
Candytuft,	Improved Carmine		1
Candytuft,	Dark crimson		1
Centranthus	macrosiphon		1
Godetia,	Crimson King		1
Godetia,	Scarlet Queen		1
Godetia,	Lady Albemarle		1
Linum	grandiflorum	rubrum	1
Nasturtium,	Dwarf,	Scarlet Queen	1
Nasturtium,	Dwarf,	King Theodore	1
Naturtium,	King of	Tom Thumbs	1
Viscaria	cardinalis		1
Collomia	coccinea		3/4
Coreopsis,	Dwarf Crimson		3/4
Eschscholtzia,	Ruby King		3/4
Godetia,	Afterglow		3/4
Godetia,	Lady Albemarle,	dwarf	3/4
Saponaria,	Scarlet Queen		1/2
Virginian Stock,	Crimson King		1/2
Viscaria,	Dwarf Carmine		1/2

Yet one other method of growing annuals calls for special mention. It is not fully recognised that a number of subjects, usually associated only with beds and borders, may also be flowered with the greatest ease under glass in winter and early spring. Those who have not hitherto attempted the culture of annuals in this way will be delighted with the charming effects produced. Among the subjects most suitable for the purpose are Alonsoa; the Star and Dunnettii varieties of Annual Chrysanthemum; *Clarkia elegans;* Dimorphotheca; *Gypsophila elegans*; Linaria; *Nemesia Suttoni*; Nicotiana, Miniature White and *N. affinis*; Phlox, Purity, one of the most lovely pot plants for the conservatory and of especial value for decorative work at Easter; Salpiglossis; and the pretty blue, Cineraria-like, Swan River Daisy. From the fact that these annuals are of the hardy or half-hardy types it will be readily understood that no great amount of heat is required to bring them to maturity; indeed, the more hardy the treatment the better for their well-doing. Seed should be sown during August or September in pots or pans placed in a cool frame, the seedlings being pricked off into other pots as soon as they have attained a suitable size. As colder weather approaches, transfer to the greenhouse or conservatory, and provided the night temperature is not allowed to fall below 45° all should be well. During the day give the plants the maximum of air whenever weather permits.

Hardy Annuals.—The seeds should be sown on a carefully prepared surface from which large stones have been removed, and the clods must be broken, but the soil should not be made so smooth as to become pasty under rain. Sow thinly, in rows spaced to agree with the height of the plant, cover with a very slight coat of fine dry earth—the smallest seeds needing but a mere dusting to cover them—and, from the first, keep the plants thinned sufficiently to prevent overcrowding. Spring-sown annuals are worthy of a better soil than they usually have allotted them, and also of more careful treatment. It is not wise to sow earlier than March or later than the middle of April. In the after-culture the most important matter is to keep the clumps well thinned. Not only will the bloom of crowded plants be comparatively poor and brief, but by early and bold thinning the plants will become so robust, and cover such large spaces of ground with their ample leafage and well-developed flowers, as really to astonish people who think they know all about annuals, and who may have ventured after much ill-treatment to designate them 'fugacious and weedy.' Although the sowing of hardy annuals direct on to beds and borders where the plants are wanted is economical in labour and avoids the check which transplanting occasions, the practice of raising annuals on specially prepared seed-beds and pricking out the plants to blooming quarters is sometimes followed. The soil into which they are transferred for flowering should be deeply dug, thoroughly broken up, and, if at all poor, liberally manured. It is an excellent plan also to sow hardy annuals outdoors in autumn, but it is needless to say more on this subject here, as it is dealt with fully at page 313.

Half-hardy Annuals.—Give these as long a period of growth as possible to insure a vigorous plant before the season of flowering. The best time for sowing is February, or the beginning of March; for although some kinds may with advantage be sown earlier, it is safer, as a rule, to wait for sunshine and full daylight, so as to keep up a steady and continuous growth. The soil for the seed-pans should be rich and fine. Good loam, improved by the addition of thoroughly decayed manure and leaf-mould, with sufficient sand to render the texture porous, will suit all kinds of annuals that are sown in pans under glass. Sow the seed thinly, cover very slightly, and lay squares of glass over to keep a uniform degree of moisture without the necessity of watering. Should watering become necessary, take care to avoid washing the seeds out. If the pans or pots are stood in a vessel containing several inches depth of water until sufficient has been absorbed, there will be no occasion to pour water on the surface. A gentle heat is to be preferred; when germination is too rapid it tends to the production of weak plants. As soon as the young plants appear, remove the glasses and place the seed-pans in the fullest light, where air can be given without danger to them. A dry east wind blowing fiercely over them will prove a blast of death. If they have no air at all, they will be puny, rickety things, scarcely worth planting out. Choice varieties should be carefully pricked out into pans and pots as soon as large enough; this will promote a fine, stocky growth and a splendid development of flowers. Take care not to plant out until the weather is favourable, for any great check will undo all your work, and make starvelings of your nurslings. If you cannot command heat for half-hardy annuals, sow in the first week in April, put the pans in a frame facing south, and the seeds will soon grow and do well. If that is too much trouble, sow in the open border early in May, making the border rich and friable, that they may have a good chance from the first.

Tender Annuals.—These require the same general treatment as advised for half-hardy annuals. But it is desirable to sow in a stronger heat than is necessary for annuals that are to be planted out. It is also requisite to be in good time in pricking out the seedlings, for if they get much drawn they cannot make robust pot plants. A light, rich, perfectly sweet soil, containing a fair proportion of sharp sand, is necessary to insure plants worth having. It is also important to get them into separate small pots as soon as possible, and to shift them on to larger and larger pots, until they have sufficient pot room for flowering, after which shift no more. As soon as these pots are filled with roots, give very weak manure water constantly until the plants are in flower, and then discontinue it, using instead pure soft water only.

Hardy Biennials and Perennials.—These are often sown in pans or boxes, and are pricked off when large enough into other pans or pots before they are transferred to beds or borders. The system has certain advantages in insuring safety from vermin and proper attention, for it is an unfortunate fact that too many cultivators consider it needless to thin or transplant sowings made in beds or borders. The plants are frequently allowed to struggle for existence, and the result is feeble attenuated specimens which, with trifling care and attention, might have become robust and capable of producing a bountiful bloom in their season. Still, it should be clearly understood that all the hardy biennials and perennials may be grown to perfection by sowing on a suitable seed-bed in the open ground, protecting the spot from marauders of all kinds, and by early and fearless thinning or transplanting. As a rule, we advocate one shift before placing the plants in final positions.

ABUTILON

Half-hardy greenhouse perennial

Handsome plants, two feet or more in height, can be produced from seed and flowered in a single season. They are useful for training to greenhouse walls, and they may also be transferred to open borders for the summer. When employed for the latter purpose, the plants should be lifted and put into pots about the end of August, after there has been a penetrating shower. In the absence of rain a soaking of water on the previous day will prevent the soil from falling away from the roots.

February and March are the right months for sowing seed, and for the pots any fairly light compost will answer. Prick off the seedlings when about an inch high, putting the plants in down to the seed-leaves. They must never be allowed to suffer for want of water, nor should they be starved in small pots. The growth had better not be hurried at any stage; the plants will then develop into shapely specimens with very little care.

ACHIMENES

Greenhouse or stove perennials

Although Achimenes can be propagated by division of the tubers, the simpler method of raising a supply from seed has become a common practice. During March or April sow in pots or pans, and while quite small transfer the seedlings to separate pots. It is important to insure free drainage, especially as frequent watering is a necessity while the plants are in active growth. Achimenes are generally kept in a high temperature; but they do not really need so much heat as Gloxinias, and in a warm greenhouse they can be flowered without the least difficulty. This is one of the finest subjects for growing in hanging-baskets.

ALONSOA

These popular half-hardy flowers are not only valuable for a summer display in borders, but they make charming subjects for the conservatory in the spring months. For blooming outdoors seed may be sown in pans in March and the plants treated in the manner usual for half-hardy annuals, or a sowing can be made in the open towards the end of April. Plants for flowering indoors in April and May should be raised from seed sown in the preceding August and September. Grow on the seedlings steadily in pots, but do not force them in any way. In fact, the treatment should be as nearly hardy as possible, a night temperature of 45° being generally sufficient to carry them through the winter.

AMARYLLIS

Hippeastrum

The majority of the named varieties are expensive, and a very considerable saving is effected by raising plants from seed. Thanks to the skill of the hybridiser, the seedlings not only compare favourably with flowers grown from costly bulbs, but they have been successful in winning certificates and awards of merit.

The germination is so irregular that it is well to put only one seed in each small pot. The most suitable soil is a mixture of two parts loam and one of leaf-mould, with sufficient coarse grit to insure free drainage. The proper temperature is about 65°. After the seedlings are established follow the treatment advised on page 340.

ANEMONE

The Windflower. Hardy perennial

The discovery that it is easy to flower the popular St. Brigid and similar Anemones from seed in about seven months from the date of sowing has given a great impetus to the culture of this plant, especially as it possesses a high value for decorating vases, in addition to its usefulness in beds and borders. From seed sown in February or March the plants should begin to bloom in September or October of the same year, and continue to flower until the following June, when it is unprofitable to retain them longer. No coddling of any kind is necessary. Dig a trench in a sheltered, sunny spot, and fill it with rich soil freely mingled with decayed cow-manure. If the land happens to be somewhat tenacious, Anemones will take kindly to it, but it should be well worked, and it may be needful to add a little fine sandy compost at the top as a preparation for the seed. The woolly seed should be rubbed with sand, and the two may be sown together thinly in lines. As a finish the ground should be lightly beaten with the back of a spade. Germination is decidedly slow, so that until the seedlings appear the removal of weeds requires care. The plants should be thinned until they stand six inches apart. Seed may also be sown in June or July for plants to flower in the following year, and the results will probably be even more satisfactory than from the spring sowing.

ANTIRRHINUM

Snapdragon. Hardy perennial

In bygone years Antirrhinums were seldom seen beyond the limits of old-fashioned cottage gardens. But even then the Snapdragon was a popular flower, and it was generally perpetuated by subdivision of the plants. Now, in common with a large number of perennials and biennials, the Antirrhinum is almost exclusively grown from seed. This altered method of culture has resulted in a marked advance in the size and colour of the spikes of bloom, and has also increased the vigour and floriferous character of the plants. In the process of raising, selecting and re-selecting the stocks, experts have found it possible to develop three distinct classes—Tall, Intermediate, and Dwarf—so that the value of the plant as an ornament in the garden has been advanced beyond the dreams of a former generation of gardeners. The Tall varieties attain a height of about three feet; the Intermediates generally range between twelve and eighteen inches, and the Dwarf or Tom Thumb section seldom exceeds six inches. All three classes have a distinct value for different positions in the garden.

Antirrhinums are not fastidious as to soil and may be relied on to give satisfaction in almost any spot chosen for them. Still, it must be admitted that they are conspicuously successful on dry soils and in sunny positions. This will account for the surprising displays occasionally seen on old walls and in large wild rockeries, where they are perfectly at home, apparently indifferent to the starving conditions in which their lot is cast.

The fact that the plant possesses such sturdy independence of character greatly enhances its value and usefulness. Nothing more handsome can be imagined in a border than the gigantic spikes of the Tall varieties, and they make a magnificent decoration for vases at a season when flowers suitable for cutting are much needed. The Intermediate Antirrhinums, like the Tall class, combine advantages for both bedding purposes and for cutting, perhaps in a still greater degree. The varieties are so numerous and charming that an enthusiast has suggested the desirability of devoting a garden to Antirrhinums alone. Although the Tom Thumb section is also frequently employed for bedding, these dwarf-growing varieties are better adapted for ribbon borders, or as an edging to carriage drives.

Antirrhinums may be grown as half-hardy annuals or as perennials, but the former is the simplest course for obtaining plants for summer bedding. Sow the seeds in pans or boxes from January to March, and prick off the seedlings as soon as large enough to handle. Grow on steadily and gradually harden off in readiness for planting out after the Wallflowers and other spring bedders have been removed. After flowering it will save trouble to consign the plants to the waste heap and again raise a sufficient supply to fill their places in the following spring. When grown as perennials, seed should be sown in July or August. Leave the plants in the seed-bed until ready for transfer to final positions. These will stand the winter and come into flower earlier than plants from spring-sown seed.

AQUILEGIA

Columbine. Hardy perennial

Since the introduction of the long-spurred hybrid varieties the Aquilegia has become exceedingly popular. Like the Nasturtium, it is particularly accommodating in character, and will thrive on poor soil and amid surroundings altogether uncongenial to many other subjects. Several of the fine varieties which have been recently introduced are, however, worthy of a place in the best of borders. Sow in February or early in March in a frame, and plant out when strong enough, or sow in June in an open border. If the season is favourable, those sown early may bloom the first year; the remainder will flower in the year following.

ASPARAGUS

Greenhouse foliage varieties. Half-hardy perennials

The finely laciniated foliage of *A. plumosus* is greatly prized for bouquets, and the plant invariably commands attention as a decorative subject on the table or in the conservatory. *A. decumbens* has long tremulous branches of elegant dark green foliage, and the plant is admirably adapted for hanging-baskets. *A. Sprengeri* is distinct from both, but is also very ornamental in baskets. Sow all three varieties in pans during February or March, in heat; prick off the seedlings immediately they are large enough to handle, and grow on in gentle heat until the beginning of June, when cool-house treatment will suit them.

ASTER

Callistephus sinensis. Half-hardy annual

In high summer so many flowers are available that no difficulty arises in making a varied display. The real trouble is in discarding, especially for a limited area. But when summer begins to merge into autumn the choice is not so extensive, and among the annuals which then adorn the garden Asters are indispensable. This superb flower has been developed into many forms, and each class affords a wide range of magnificent colours. Yet it must be admitted that in the majority of gardens Asters are seldom grown in sufficient numbers, and it is not unusual to find the flowers small in size and poor in colour. In many cases we believe the reason to be that the culture of Asters is often commenced too late. Preparations should therefore be made in good time, and apart from providing the requisite number of plants for filling beds and borders, and for supplying cut blooms, others should be raised for flowering in pots. For indoor decoration full use is rarely ever made of Asters, although the colours include many delightful shades which may be employed with most telling effect.

To secure a long-continued display of bloom there must be several sowings, and the earliest will need the aid of artificial heat. One secret of successful culture is to give no check to the plant from its first appearance until the time of flowering; and a suitable bed must be prepared, whether the seed be sown on the spot or plants are transferred from other quarters.

Asters do not readily accommodate themselves to violent alternations of heat and cold, particularly in the early stage of growth, and therefore the most sheltered position in the garden should be chosen for them; but avoid a hedge or shrubbery, where strong growing trees rob the soil of its virtue. Begin the preparation of beds during the previous autumn by deep digging, and incorporate a liberal dressing of well-rotted manure as the work proceeds. On light and shallow soils it will do more harm than good to bring the raw subsoil to the surface, but the subsoil may with advantage be stirred and loosened by the fork, and if a little loamy clay can be worked into it the land will be permanently benefited.

A very stiff soil will, however, present greater difficulties; but if by free working it can be made sufficiently friable, Asters will revel in it, and produce flowers of a size and colour that will reward the cultivator for all his trouble. Throw the ground up roughly in October. The more it is exposed to the action of wind, snow and frost, the more thoroughly will the winter disintegrate its particles and render it fertile. Early in spring give another digging, and then work in a good supply of decayed manure, together with grit, charcoal, wood ashes, or other material that will help to render the soil rich and free. Aim at inducing the roots to go down deep for supplies—there will then be a cool moist bottom even in dry weather, and these conditions will do much toward the production of fine stocky plants capable of carrying an imposing display of flowers.

For sowings from the end of March to the middle of April prepare a compost consisting principally of decayed leaf-mould, with sufficient loam to render it firm, and sharp sand to secure drainage. Either pots or seed-pans may be used. Place these in a cool greenhouse, or in a Cucumber or Melon pit, or even on a half-spent hot-bed. Sow thinly; a thick sowing is very likely to damp off. Just hide the seed with finely sifted soil, and place sheets of glass at the top to prevent rapid evaporation. Give no water unless the soil becomes decidedly dry, and then it is better to immerse the pot or pan for half an hour than to apply water on the surface. When the plants attain the third leaf they can be pricked off into shallow boxes or round the edges of 3-1/2 inch pots. From these they either may have another shift singly into small pots, or may be transferred direct to blooming quarters. A high temperature is not requisite at any stage of growth, indeed it is distinctly injurious. From 55° to 65° is the extreme range, and the happy medium should, if possible, be maintained. Give air on every suitable occasion, and as the time for transferring to the open ground approaches, endeavour to approximate nearly to the outside temperature. The plants will then scarcely feel the removal.

Another and simpler proceeding produces fairly good results, and we describe it for the benefit of those whose resources may be small, or who do not care to adopt the more troublesome method. In some spot shaded from the sun make a heap of stable manure, rather larger than the light to be placed upon it. Level the top, and cover with four or five inches of rich soil. Place a frame upon it with the light a trifle open. When the thermometer indicates 60°, draw drills at six inches apart; sow the seed, and cover with a little sifted soil. The light had better not be quite closed, in case of a rise of temperature. As the plants thrive, gradually give more air, until, in April, the showers may be allowed to fall directly upon them in the daytime. When the Asters are about three inches high they will be quite ready for the open ground, and a showery day is favourable to the transfer. After the bed has served its purpose, the manure will be in capital condition for enriching the garden.

In the event of there being no frame to spare, drive a stake into each corner of the bed. Connect the tops of the stakes, about one foot from the surface of the bed, with four rods securely tied, and upon these place other rods, over and around which any protecting material at command may be used. With this simple contrivance it is quite possible to grow Asters in a satisfactory manner.

The finest Asters are frequently grown in the open air, entirely without the aid of artificial heat, and indeed without any special horticultural appliances. Those who possess the best possible resources will find additional advantage in resorting also to this mode of culture. It gives another string to the bow, and prolongs the season of flowering. For open-air sowings in April make the soil level and fine, and about the middle of that month draw drills three inches deep. In these place an inch of finely prepared rich soil, and if it is largely mixed with vegetable ashes, so much the better. The distance between the drills should be regulated by the variety. For tall-growing Asters twelve to fifteen inches between the rows will not be too much. Ten inches will suffice for the dwarfs. Sow the seed thinly and evenly, and cover carefully with fine soil. Commence early to thin the plants, always leaving the strongest, and arrange that they finally stand at from eight to fifteen inches apart according to the sort.

Keep the ground clean, and before the flowering stage is reached gently stir the surface, but not deep enough to injure the roots. An occasional application of weak manure water will be advantageous, but it must not be allowed to touch the foliage.

For tall varieties it may be needful to provide support. If so, place a neat stick on that side of the plant towards which it leans, as this takes the strain off the tying material, and saves the plant from being cut or half-strangled. In a dry season, and especially on light soils, there must be a bountiful supply of soft water, alternated every few days with the manure water already alluded to. Evening is the best time to apply it.

For show purposes rather more room is required than we have stated. Only about five buds should be matured by each plant, and these, of course, the finest. To prepare flowers for exhibition is in itself an art, and each cultivator must be guided by his own resources and experience.

Asters in pots make excellent decorative subjects. It is only necessary to lift them carefully from the borders with balls of earth surrounding the roots, and pot them just before the buds expand, or they may be potted up while in full flower without flagging.

The plants are liable to the attacks of aphis, both green and black. While under glass the pests can be destroyed by fumigation; but in the open a solution of some good insecticide may be administered with the syringe at intervals of about three days, until a clearance is effected. Other foes are the various grubs which attack plants at the collar. On the first sign of failing vigour, gently remove with a pointed stick the soil around the plant, and in doing this avoid any needless disturbance of the roots. Do not be satisfied until the enemy is destroyed.

AUBRIETIA

Hardy perennial

In the early months of the year few subjects in the garden present so gay an appearance as Aubrietias, for with the first approach of genial weather the cushion-like plants burst into a mass of delightful blossom. For spring bedding, edgings, and the rock garden Aubrietias are indispensable, and they make a particularly effective show when grown in conjunction with Yellow Alyssum and White Arabis. Aubrietias are easily grown from seed sown in May and June. The plants are best raised in pans of light rich soil and may be put out in autumn where required to flower in the following spring.

AURICULA

Primula Auricula. Hardy perennial

Keen is the enthusiasm of the Auricula amateur. The only complaint we ever heard about the flower is that its most devoted admirer cannot endow it with perpetual youth and beauty.

It is well to bear in mind that seed from a worthless strain requires just as much attention as that which is saved with all a florist's skill from prize flowers. Some growers advocate sowing immediately the seed is ripe, but this intensifies the irregular germination that characterises seed of all the Primula species. Either February, March, or April may be chosen, and we give preference to the end of February. Use six-inch pots, and as there must be no doubt about drainage, nearly half-fill the pots with crocks, cover with a good layer of rough fibrous loam mingled with broken charcoal, and on the top a mixture of loam, decayed leaves, and sharp sand. Press the soil firmly down; sow thinly and regularly, putting the seeds in about half an inch apart; just cover them with fine soil, and place the pots in a cool frame or greenhouse, with sheets of glass over to prevent evaporation. Watering in the ordinary way is apt to wash out the seeds, and it is therefore advisable to immerse the pots in a vessel containing water until the soil has become saturated. Wait patiently for the plants. When they show four or six leaves, prick out into pans or boxes about two inches apart, and before the seedlings touch each other transfer to small pots. The surface soil in the pots may be lightly stirred occasionally to keep it free from moss. The plants must never be allowed to go dry, but as winter approaches water should be given more sparingly, and during sharp frosts it may be wise to withhold it entirely. There really is no need of artificial heat, for the Auricula is a mountaineer, and can endure both frost and snow. But we prize its beauty so highly that frames and greenhouses are properly employed for protecting it from wind, heavy rain, soot, dust, and all the unkind assaults of a lowland atmosphere, to which it is unaccustomed in a natural state. Still, the plants should be kept as nearly hardy as possible.

The Auricula is a slow-growing plant, and although there will probably be some flowers from seedlings in the second year, their value must not be judged until the following season. To the trained eye of the florist the Show Auriculas take precedence over the Alpine section; but for general usefulness the Alpines hold the first place. They may be fearlessly put into the open border, and especially the north border, where, with scarcely any care at all, they will endure the winter, and freely show their lovely flowers in spring.

BALSAM

Impatiens Balsamina. Half-hardy annual

The older methods of growing Balsams prescribed a false system, comprising disbudding, stopping, and other interferences with the natural growth of the plant. The rule of pinching back the leader to promote the growth of side shoots, and removing the flower buds to increase the size of the plants, was altogether vicious, because the natural growth is more elegant and effective. The finest flowers are produced on the main stem, and these are completely sacrificed by disbudding.

It is desirable to make two or three sowings of Balsam, say from the middle of March to the middle of May, the earlier sowings to be put on a sweet hot-bed, although March sowings will soon germinate in a frame, and the May sowing may be made in the open ground on a prepared bed. The soil at every stage should be rich and light, but not rank in any degree. Prick out the plants from the seed-pans directly the first rough leaves show, and soon after shift them again to encourage a stout dwarf habit. A sunny position should be chosen for the bed, in which they may be planted out about the first week of June, or earlier if the weather is particularly favourable. Heat, moisture, and a strong light favour a fine bloom, and, therefore, water must be given whenever dry weather prevails for any length of time. If kept sturdy while under glass, they will need no support of any kind, and although they are peculiarly fleshy in texture, it is seldom they are injured, even by a gale. When grown in pots throughout, the chief points are to shift them often in the early stages, to promote free growth in every reasonable way, and to cease shifting when they are in pots sufficiently large to sustain the strength of the plants. Generally speaking, eight-inch pots will suffice for very fine Balsams, but ten-inch pots may be used for plants from an early sowing. They will probably not show a flower-bud while increased pot room is allowed them; but as soon as their roots touch the sides of the pots the bloom will appear. It is occasionally the practice to lift plants from beds when pot Balsams are wanted. This method has the advantage of being the least troublesome, and as the plants need not be lifted until the flowers show, favourite colours can be chosen.

BEGONIA, TUBEROUS-ROOTED

Begonia hybrida. Half-hardy perennial

One of the most remarkable achievements in modern horticulture is the splendid development of single and double Tuberous-rooted Begonias from the plant as first introduced from the Andes. Originally the flowers were small, imperfect in form, and deficient in range of colour. But experts were quick in apprehending the capabilities of this graceful plant, and it proved to be unusually amenable to the hybridiser's efforts. Now the large symmetrical blossoms of both single and double flowers challenge attention for beauty of form and an almost endless variation of tints peculiar to the Tuberous-rooted Begonia. The plants are conspicuous ornaments of the conservatory and greenhouse for several months, and experience has proved that they make unique bedders, enduring unfavourable conditions of weather which are fatal to many of the older bedding subjects.

From the best strains of seed it is easy, with a little patience, to raise a fine stock of plants, possessing the highest decorative qualities. Under generous treatment the seedlings from a January or February sowing come into bloom during July and August. The seed should be sown in well-drained pots containing a good compost at the bottom, with fine sandy loam on the surface, pressed down. Before sowing sprinkle the soil with water, and sow the seed evenly, barely covering it with fine earth. A temperature of about 65° is suitable. Germination is both slow and irregular, and the plants must be pricked off into pans or small pots as fast as they become large enough to handle. This process should be followed up so long as seedlings appear and require transferring. They may be shifted on as the growth of the several plants may require. Begonias need more attention with reference to an even temperature during this stage than at any other period.

The merits of Begonias as bedding plants are now recognised in many gardens, and they deserve to be still more widely grown. It is wise to defer planting out until June. In the open ground they produce abundant supplies of flowers for cutting at the end of September and early in October, when many other flowers are over. The plants should be put out when they show themselves sufficiently strong, and it is better to be guided by the plants than by any fixed date. The beds must be freely enriched with well-rotted manure and decayed vegetable matter; it can scarcely be overdone, for Begonias are gross feeders.

The earliest plants to flower will often be retained in the greenhouse, as they follow in succession the Cinerarias and Calceolarias. Those that start later may be turned out as they come into bloom, which will probably be in June. By deferring the planting out until there is a show of bloom a selection of various shades of colour is possible, and this will greatly enhance the beauty of the beds. Begonias are hardier than is generally supposed; they need no protection, and require no heat, except in the stage of seedlings, when first forming their tubers.

For autumn decoration Begonias should be taken up from the beds during September and potted, when they will continue to bloom in the greenhouse or conservatory for a considerable time, and form a useful addition to the flowering plants of that period.

If not required for autumn decoration, let the plants remain out as long as may be safe; then pot off, and place in the greenhouse. Be careful not to hasten the drying of the bulbs. When the stems fall Begonias may be stored for their season of rest, allowing them to remain in the same pots. They can be put away in a dry cellar, or on the ground, covered up with sand, in any shed or frame where the bulbs will remain dry and be protected from frost. Both damp and cold are very injurious to them.

The temperature during their season of rest should be kept as near 50° as possible. When they show signs of growth in spring they must be put into small-sized pots, almost on the surface of the soil. As growth increases shift into larger sizes, inserting the bulb a little deeper each time until the crown is covered.

BEGONIA, FIBROUS-ROOTED

Begonia semperflorens. Half-hardy perennial

Fibrous-rooted Begonias are exceedingly valuable for either bedding in summer or greenhouse decoration during the autumn and winter. They produce a continual succession of flowers, rather small in size, but very useful for bouquets, and the plants are charming as table ornaments. The directions for sowing and after-treatment recommended for the Tuberous-rooted class will be suitable also for the Fibrous-rooted varieties, except that the latter must always be kept in a growing state, instead of being dried off at the end of the flowering season. Sow seed at the end of January or in February, and again at the beginning of March. Under fair treatment the first batch of plants will come into flower for bedding out in June.

CALCEOLARIA, HERBACEOUS

Calceolaria hybrida. Greenhouse biennial

The present magnificent race of Herbaceous Calceolarias, both as to constitution and the beauty of its flowers, is the result of much cross-fertilisation of the finest types, so that the best strains are capable of affording ever-new surprise and delight. The superb collections exhibited in recent years, which have made lasting impressions on the public by their form and brilliancy of colour, have invariably been raised from seeds of selected varieties, saved on scientific principles that insure vigour, variety, and splendour in the progeny.

Calceolarias thrive under intelligent cool-house culture, but it must be clearly understood that in every stage of growth they are quick in resenting neglect or careless treatment. The work must be carried out with scrupulous attention, and the result will more than justify the labour. Extreme conditions of temperature are distinctly injurious, and the plants are especially susceptible to a parched, dry atmosphere.

May is early enough to commence operations, and July is the limit for sowing. As a rule, the June sowing will produce the quickest, strongest, and most robust plants.

The soil, whatever its composition, should be rich, firm, and, above all, porous. Press it well into the pots or pans, and make the surface slightly convex and quite smooth. A compost that has been properly prepared will not need water; but should water become needful, it must be given by partially submerging the pans. The seed is as fine as snuff, and requires delicate handling. It is easily lost or blown away, and therefore it is wise not to open the packet until perfectly ready to sow. Distribute the seed evenly and sift over it a mere dusting of fine earth. Place a sheet of glass upon each pot or pan, and the glass must be either turned or wiped daily. This not only checks rapid evaporation, but prevents the attacks of vermin. Germination is always slower on an open than on a close stage. Perhaps the best possible position is a moist shady part of a vinery, if care be taken when syringing the vines to prevent the spray from falling upon the seed-pans.

Under favourable circumstances, from seven to nine days will suffice to bring the seedlings up in force, and very few will appear afterwards. When they are through the soil remove the sheet of glass, and give them prompt attention, or they will rapidly damp off. Immediately the second leaf appears, tiny as the plants may be and difficult to handle, commence pricking them off into other pots prepared to receive them, for it is unsafe to wait until they become strong. Allow about two inches between the plants. The occupants of each pan may generally be pricked off in about three operations, and there should be only the shortest possible intervals between.

With many subjects it is a safe rule to use the robust seedlings and throw the weakly ones away. This practice will not do in the case of Calceolarias, or some of the most charming colours that can grace the conservatory or greenhouse will be lost. The strongest seedlings generally produce flowers in which yellow largely predominates, a fact that can easily be verified by keeping the plants under different numbers. But it must not be inferred that because the remainder are somewhat weaker at the outset they will not eventually make robust plants.

Freely mix silver sand with the potting mould, and raise the surface higher in the centre than at the edge of the pot. From the first appearance of the seedlings shading is of the utmost importance, for even a brief period of direct sunshine will certainly prove destructive. Do not allow the plants to become dry for a moment, but give frequent gentle sprinklings of water, and rain-water is preferable. As the soil hardens, stir the surface with a pointed stick, not too deep, and give water a few hours after. About a month of this treatment should find each plant in the possession of four or five leaves. Then prepare thumb pots with small crocks, cover the crocks with clean moss and fill with rich porous soil. To these transfer the plants with extreme care, lifting each one with as much soil adhering to the roots as a skilful hand can make them carry. Place them in a frame, or in the sheltered part of a greenhouse, quite free from dripping water. Always give air on suitable days, and on the leeward side of the house.

Keep a sharp look-out for aphis, to the attacks of which Calceolarias are peculiarly liable. Fumigation is the best remedy, and it should be undertaken in the evening; a still atmosphere renders the operation more certain. Water carefully on the following morning, and shade from the sun.

By September the plants should be in large 60-pots, and it is then quite time to begin the preparation for wintering. Some growers put them in heat, and are successful, but the heat must be very moderate, and even then we regard the practice as dangerous. Place the plants near the glass, and at one end of the house where they will obtain plenty of side light, as well as light from above. During severe frosts it may be well to draw them back or remove them to a shelf lower down and towards the centre of the house, but they must be restored as soon as possible to the fullest light obtainable, as they have to do all their growth under glass. The more air that can safely be given, the better, and dispense with fire-heat if a temperature of 45° to 55° can be maintained without it.

When growth commences in spring, which will generally be early in March, give each plant its final shift into eight-or ten-inch pots. This must be done before the buds push up, or there will be more foliage than flowers.

The following is the compost we advise: one bushel good yellow loam, half-bushel leaf-soil, one gallon silver sand, a pound of Sutton's A 1 Garden Manure, and a pint of soot, well mixed at least ten days before use. Any sourness in the soil will be fatal to flowering. The compost must be carefully 'firmed' into the pots, but no severe pressure should be employed, or the roots will not run freely.

Neglect as to temperature or humidity will have to be paid for in long joints, green fly, red spider, or in some other way. But there are no plants of high quality that grow more thriftily if protected from cold winds and kept perfectly clean. A light airy greenhouse is their proper place, and they must have ample headroom.

After the pots are filled with roots, not before, manure water may be administered until the flower-heads begin to show colour, when pure soft water only should be used. About a fortnight in advance of the full display the branches must be tied to supports. If skilfully managed the supports will not be visible.

It may be that a few large specimens are required. If so, shift the most promising plants into 6-size pots. These large Calceolarias will need regular supplies of liquid manure until the bloom is well up, and if the pots are efficiently drained and the plants in a thriving condition, a rather strong beverage will suit them. For all ordinary purposes, however, plants may be allowed to flower in eight-or ten-inch pots, and for these one shift after the winter is sufficient.

New Types of Calceolaria.—There are now available a number of hybrid half-hardy perennial varieties, of which *C. profusa (Clibrani)* is the most popular, that bear the same relation to the Large-flowered Calceolaria as the Star Cineraria does to the Florist's Cineraria. In point of size the blooms produced by these new types are smaller than those of the Large-flowered section, but the tall graceful sprays are extremely beautiful and of the greatest decorative value. Except that seed should be sown earlier (February and March are the proper months), the plants should receive precisely the same treatment as that already described for Herbaceous Calceolaria.

CALCEOLARIA, SHRUBBY

Calceolaria rugosa. Half-hardy perennial

Notwithstanding the ease with which cuttings of the Shrubby Calceolaria can be carried through a severe winter, there is a growing disposition to obtain the required number of plants from seed sown in February; and seedlings have the advantage of great variety of colour. A frame or greenhouse, and the most ordinary treatment, will suffice to insure a large stock of attractive healthy plants for the embellishment of beds and borders.

CAMPANULA and CANTERBURY BELL

Hardy annual, hardy biennial, and hardy perennial

Among the numerous and diverse forms in the order Campanulaceæ are many flowers of great value in the garden, including Single, Double, and Cup and Saucer strains of the popular Canterbury Bell *(C. medium)*. The impression that some Campanulas are shy growers and require exceptionally careful treatment may arise from the frail habit of certain varieties, or from the fact that some of them occasionally fail to bloom within twelve months from date of sowing. The idea is not worth a moment's consideration. In moderately rich, well-drained soil the finest Campanulas not only prove to be thoroughly hardy, but they are most graceful in herbaceous borders or beds, and they may also be used alone in bold clumps with splendid effect. For instance, the handsome Chimney Campanulas *(C. pyramidalis* and *C. pyramidalis alba*) frequently attain a height of six feet or more, and sturdy spikes occasionally measure eight and even ten feet from base to tip. Such specimens are magnificent ornaments in conservatories and corridors, and cannot fail to arrest attention at the back of herbaceous borders, or when used as isolated plants on lawns. When grown in pots use a light rich compost, taking care to insure perfect drainage. The plants must never be allowed to become dry, as this not only checks growth but renders them liable to attack by red spider or green fly. Another distinctive subject for the decoration of the conservatory is *C. grandis*, which may be described as a dwarf Chimney Campanula. The freely branching plants, covered with attractive flowers, also form a striking group when grown in the open border.

Altogether different in character is *C. persicifolia grandiflora*, or the Peach-leaved Bell-flower as it is sometimes called. This plant is lighter and more graceful than the Canterbury Bell. It throws up handsome stems, two feet high, clothed from the ground with lance-like leaves and elegant bells which quiver in the slightest breeze. An interesting plant is the Giant Harebell, a dainty flower on a slender stem, resembling the wild variety in form, but larger, richer in colour, and a more profuse bloomer. *C. glomerata* is one of the hardiest plants that can be grown in any garden, and the large close heads of deep blue bells have long been familiar in herbaceous borders. For its very fine glistening, deep blue, erect flowers, *C. grandiflora* is also a great favourite.

Campanulas were formerly propagated by division, but this treatment has created the impression that they are unworthy to be ranked among the perennials. From seed, the plants are extremely robust. *C. persicifolia grandiflora* resents division, which frequently results in weakened growth and a tendency, especially in poor or badly drained soil, to dwindle away. The only satisfactory method of growing Campanulas is to raise plants annually from good strains of seed. If sown in gentle heat early in the year—February is the usual month—many of the varieties flower the same season. When they are well started, plenty of light and air must be admitted. Unless intended for potting they should be planted out in good soil where they will require no more care than is bestowed on the borders generally. Seed can also be sown in the open ground from May to July; transplant in autumn for flowering in the following season. During hot weather, particularly on light soil, the plants need to be well watered, but in retentive ground thorough drainage must be insured. Should signs of debility appear, transplant to rich soil, where they will soon regain vigour.

A popular half-hardy Campanula is *C. fragilis*, of trailing habit. The starry pale blue flowers are seen to most advantage in hanging-baskets. The charm of these flowers is wholly lost if they are placed on a stage in the greenhouse; and they are not entirely satisfactory in a window where the light is transmitted through the petals, as this robs them of colour and substance. But hanging in a conservatory with plenty of air and space their slender drooping stems are very graceful, and the light reflected from the flowers does full justice to their beauty. Sow in pans during February or March and pot on as required.

All the foregoing are perennials, but two little hardy annual Campanulas are *Attica* and *A. alba*, growing about six inches high. They make useful foreground plants, and are quite at home in rock gardens. Sow in April on light soil.

The Canterbury Bell has already been alluded to; it is a charming hardy biennial forming a valuable feature of the mixed border. The large semi-double blooms of the Cup and Saucer class and the double varieties are modern introductions which have become extremely popular; the range of colours now includes the most delicate shades of pink, mauve, and blue, in addition to pure white. Seed may be sown from April to July. When the seedlings are large enough transplant them where required for flowering in the summer of the succeeding year. But Canterbury Bells are also interesting in the greenhouse during spring; for this work pot them in October and on to December. So treated, they bloom even more generously than in the garden. There can be no more beautiful adornment for a hall or large drawing-room than a well-placed group of the fine white flowers, backed by a mass of dark-foliaged plants.

CANNA

Indian Shot. Half-hardy perennial

Cannas have ceased to be regarded simply as sub-tropical foliage plants, adapted only for the adornment of beds and borders. They have not lost their merits for this purpose, although in all probability the taller forms will be less grown than formerly, because the new dwarf varieties, which maintain a high standard of beauty in the foliage, include a diversity of rich tints previously unknown, and they possess the additional merit of producing flowers that have lifted the race into prominence as brilliant decorative subjects for the garden and the greenhouse.

The popular name is descriptive of the seed, which is almost spherical, black, and so hard that it has been used in the West Indies instead of shot. Hence it will occasion no surprise that the germs burst through the strong covering with difficulty, and that sometimes weeks elapse before the seedlings appear, one or two at a time. To facilitate germination some growers file the seed, others soak it until the skin becomes sufficiently soft to permit of the paring away of a small portion with a sharp knife. In either case caution must be exercised to avoid injuring the germ. A safer mode of attaining the object is to soak the seeds in water, placed in a greenhouse or stove, for about twenty-four hours before sowing. After soaking the seeds it is necessary to keep the soil constantly moist, or the germs will certainly suffer injury. The number of seeds sown should be recorded, so that it may be known when all are up. The first sowing should be made in January, in a temperature of about 75°, and as fast as the seedlings become ready transfer singly to small pots. As Cannas are gross feeders they must have a rich, porous compost, and an occasional dose of liquid manure will prove beneficial, especially when the pots are full of roots. If the seedlings from the January sowing are regularly potted on and properly managed they will begin to flower in June or July. Either the plants may be turned out into a rich soil, or the pots can be plunged, and after flowering in the open until late in autumn the plants can be lifted for another display of bloom in the greenhouse. In warm districts and in dry, sheltered situations, the roots may be left in the open ground all the winter under a covering of ashes; but they must be lifted from a damp, cold soil, and stored in a frame during the winter months. We have only mentioned January as the month for sowing, but seed may be put in up to midsummer, or even later, following the routine already indicated.

CARNATION

Dianthus Caryophyllus fl. pl. Hardy perennial

The Carnation belongs to the aristocracy of flowers and has attained the dignity of an exclusive exhibition. But in addition to their merits as show flowers, Carnations make conspicuous ornaments in the garden and the home, and it has been found that seed saved with skill from the finest varieties will produce plants yielding hundreds of flowers of which the grower need not feel ashamed. Since the introduction of the early-flowering class, which can easily be had in bloom within six months from date of sowing, an immense impetus has been given to the culture of Carnations from seed, and with judicious management it is not a difficult matter to insure a succession of these delightful subjects almost the year through. For the decoration of greenhouses and for providing cut flowers, seedling Carnations have a special value, which has only to be known to be universally appreciated. No trouble should be experienced with high-class seeds, which germinate freely and save much time and labour in comparison with the more tedious process of propagation; while an occasional new break may at times reward the raiser.

The proverb that what is worth doing is worth doing well is peculiarly exemplified in the cultivation of Carnations, the difference between the results of good and bad work being immense. We therefore advise the preparation of a compost consisting of about three parts of turfy loam, to one part each of cow-manure and sweet leaf-mould, with a small addition of fine grit. A compost that has been laid up for a year, according to the orthodox practice of florists, is very much to be desired; but it may be prepared off-hand if care be taken to have all the materials in a sweet, friable state, free from pastiness, and as far as possible free from vermin. By laying it in a heap, and turning two or three times, the vermin will be pretty well got rid of. Sow from April until August in 4-1/2 inch pots, which must be thoroughly drained. The seed must be very thinly covered, and sheets of glass should be laid over to check evaporation. Place the pots in a closed frame, or if the season be genial a sheltered border will suffice. Immediately the plants are large enough to handle, prick them off into seed-pans, or round the edge of 48-size pots. Place these in a cold pit or in the greenhouse. Give shade and water until the plants have formed six or eight leaves, and then choose a moist day for planting out.

To insure flowering plants in the following summer it is necessary to have them strong and robust before the winter sets in. As the blooming stems rise they must be carefully tied to tall sticks, stout enough to carry a cover for the bloom, if the plants are not flowered under glass. When the buds show they should be thinned, leaving as a rule the top, third, and fourth buds. The second is often too near the first, and some will not carry the fourth with vigour. When the petals nearly fill the calyx, each one must be carefully tied with a thin strip of material a little more than halfway down, to prevent the calyx from bursting, which disqualifies the flower for exhibition.

The early-flowering class is extremely valuable for the ease with which it can be grown. The seedlings offer the advantage of being far more floriferous than plants that have been propagated by the orthodox method, and they are quite immune from the disease which often decimates stocks raised from layers and cuttings. Two strains—Vanguard and Improved Marguerite—possess these characteristics in a very high degree. All the usual colours are included, and they not only make a very imposing display in the borders but are of great value for table decoration. Within about six months from the time seed is sown an admirable form of delightfully scented Carnation is at the command of every gardener, and a succession of these popular flowers is available long after the perennial varieties have ceased to bloom. Plants from seed sown in gentle heat in January or February will flower freely in the autumn of the same year, and if lifted and potted they will continue in bloom during the winter as ornaments of the greenhouse or conservatory. From another sowing in autumn there will be a display in the following spring.

CELOSIA PLUMOSA

Plumed Cockscomb. Greenhouse annual

The conditions which suit a liberally grown Cockscomb will produce long graceful plumes of *Celosia plumosa*, but the starving system will not answer with this plant. Sow in February or March, and by means of a steady heat, regular attention with water, and a rather moist atmosphere, the specimens should be grown without a check from beginning to end.

When they reach the final pots an occasional dose of weak manure water will help them, both in size and colour, but it must be discontinued when the flowers begin to show their beauty. As a rule it will be found more easy to manage this plant on a moderate-sized hot-bed than in a greenhouse. Repotting should always be done in time to prevent the roots from growing through the bottom of the pots.

CELOSIA CRISTATA—see COCKSCOMB, page 254

CHRYSANTHEMUM

Hardy perennial and hardy annual

The tedious method of propagating Begonias, Gloxinias, and Primulas by cuttings or layers has been replaced by the simpler and more satisfactory procedure of sowing seeds, which insures all the finest flowers in far greater variety than were obtained under the obsolete treatment. A similar revolution is now proceeding in the culture of Chrysanthemums. Many growers are relying entirely on seedlings raised from sowings early in the year for their autumn display. The culture of *C. indicum* from seed is as simple as that of Primulas or Stocks, and the variety and delicate charm of the seedlings far surpass the formal plants of years ago. Gardeners who require large numbers for decorative purposes may use seedling Chrysanthemums with excellent effect.

Seed should be sown in January or February, using a compost consisting of two parts leaf-soil to one part of loam. Place the pots or pans in a temperature of 65° to 70°. As soon as the seedlings appear they should be moved to a somewhat lower temperature—about 55° to 60°. When the young plants are large enough to handle, prick off into trays at about three inches apart, using a little more loam in the soil. The most convenient size for the purpose is fifteen inches long by nine inches wide and three inches deep. These trays produce a quicker root action than pots. After growth has started, place them in cold frames. Immediately the plants have made five or six leaves transfer singly to three-inch pots, and when nicely rooted they may be stopped once. About June shift into six-inch pots, adding a small quantity of coarse silver sand to the potting soil. Ten days later place them out of doors on a bed of ashes. Towards the end of July transfer to 9-1/2 inch pots for flowering, using soil of the composition already advised. Keep them standing on ashes or boards, if possible at the north side of a hedge or house. When thoroughly rooted a little manure water may be given once a week. In October stand the plants in a cool house, and in the first week of November move them to flowering quarters, keeping the temperature from 55° to 60°.

If required for blooming in the open, prick the seedlings off as soon as they will bear handling, and in May have them planted out in final positions, giving a little protection at first. They will yield a profusion of bloom which will prove invaluable for decorative purposes throughout the autumn months.

The Perennial Chrysanthemums include the well-known Marguerite, or Ox-eye Daisy *(C. leucanthemum)*, of which several new varieties have been introduced in recent years. Not only have these flowers been greatly improved in size and form, but there are now early-and late-flowering varieties which will give a succession of bloom from May until early autumn. The seed may be sown at any time from April to July on a carefully prepared bed of light fertile soil, and when the seedlings are large enough they should be transferred to permanent quarters for flowering in the following year. In the perennial border the plants make handsome specimens, and the long-stemmed flowers are also invaluable for vase decoration when cut.

Several of the Annual Chrysanthemums make superb displays in borders, especially when planted in large clumps, and they deserve to be grown extensively in odd corners to furnish a supply of charming flowers for bouquets and arrangement in vases. There is a considerable choice of colours, which come quite true, and the plants may be treated in all respects as hardy annuals. When grown in pots, the Star and Dunnettii varieties make most attractive subjects for the decoration of the greenhouse in winter and early spring. For this purpose seed should be sown in August and September.

CINERARIA

Greenhouse annual

The comparative ease with which the Cineraria can be well grown, together with the exceeding beauty and variety of its flowers, will always insure for it a high position in public favour. It is now so generally raised from seed that no other mode of culture need be alluded to. The plant is rapid in growth, very succulent, thirsty, requires generous feeding, and will not endure extremes of heat or cold. A compost of mellow turfy loam, either yellow or brown, with a fair addition of leaf-mould, will grow it to perfection. If leaf-mould cannot be obtained, turfy peat will make a fairly good substitute. Soil from an old Melon bed will also answer, with the addition of sharp grit such as the sifted sweepings from gravel walks; the disadvantage of a very rich soil is that it tends to the production of too much foliage.

The usual period for sowing is during the months of May and June, and, as a rule, the plants raised in May will be found the most valuable. A June sowing must not be expected to produce flowers until the following March or April. It is quite possible to have Cinerarias in bloom in November and December, and those who care for a display at that early period should sow in April.

Cinerarias grow so freely that it is not necessary to prick the seedlings off round the edges of pots or pans; but immediately the plants begin to make their second leaves, transfer direct to thumb pots, using rather coarse soil, and in doing this take care not to cover the hearts of the plants. Place the pots in a close frame; attend to shading, and sprinkle with soft water both morning and evening until well established. In the second week after potting, gradually diminish the heat and give more air. Too high a temperature, and even too much shade, will produce thin and weak leaf-stalks. If the plants are so crowded that they touch one another it will almost certainly be injurious, and render them an easy prey to some of their numerous enemies. It is far better to grow a few really fine specimens that will produce a handsome display of superb flowers, than to attempt a large number of feeble plants that will prove a constant source of trouble, and in the end yield but a poor return in bloom. Endeavour to grow them as nearly hardy as the season will allow, even admitting the night air freely on suitable occasions. Immediately the thumb pots are filled with roots, shift to a larger size, and it is important that this operation should not be delayed a day too long. To the practised eye the alteration of the colour of the leaves to a pale green is a sufficient intimation that starvation has commenced, and that prompt action is necessary to save the plants. It is the custom of some growers to transfer at once to the size in which they are intended to bloom. There is, however, some danger to the inexperienced in over-potting, and therefore one intermediate shift is advisable. As a rule 32-size pots are large enough, but the 24-or even the 16-size is allowable when very fine specimens are required. The seedlings should be in their final pots not later than the end of November.

It will help to harden and establish the plants if they are placed in the open air during August and September. A north border under the shelter of a wall or building is the most suitable spot, but avoid a hedge of any kind. Clear away suckers, and if many buds are presented, every third one may be removed when very fine blooms are wanted. From the first appearance of the buds, manure water can be given with advantage once or twice a week until the flowers show colour, and then it should be discontinued.

Although Cinerarias are thrifty plants, they are fastidious about trifles. If possible give them new pots, or see that old ones are made scrupulously clean. Even hard water will retard free growth, oftentimes to the perplexity of the cultivator.

A host of enemies attack Cinerarias; indeed, there is scarcely a pest known to the greenhouse but finds a congenial home upon this plant. Mildew is more common in some seasons than in others. As a rule, it appears during July and August, especially after insufficient ventilation, or when the plants have been left too long in one place or too near to each other. Obviously weakness invites attack, and the necessity of robust and vigorous growth is thus effectually taught. On the first appearance of a curled leaf, dust the foliage and soil with sulphur, and give no water overhead until a cure has been effected. The aphis is easily killed by fumigation carried out on a quiet evening. Some gardeners prefer to give an hour or two once a week to the removal of the pest by means of a soft brush. From three to four dozen plants are easily cleansed by hand in the time named.

Star Cinerarias *(C. stellata)* are grown under precisely the same conditions as the Florists' or Show Cinerarias, and this type of flower is highly valued for its singular gracefulness and beautiful decorative effect. In the conservatory and on the table it is an indispensable plant. The sprays admit of most charming arrangements in vases with any kind of ornamental foliage, and maintain their beauty for a long time in water.

Intermediate Cinerarias.—These new types of Cineraria, which in habit are intermediate between the Large-flowered and Stellata classes, make admirable subjects for table decoration, as well as for the adornment of the conservatory or greenhouse. In this class the Feltham Beauty strain undoubtedly has a great future before it. Originated at the Feltham Nurseries, this strain has attracted considerable attention at the numerous horticultural meetings where it has been exhibited, and since it passed into our hands a few years ago some very beautiful colours not to be found among the ordinary Stellata varieties have been added to it. The distinctive feature of the flowers is the white centre, which greatly enhances the vividness of the colouring of the petals. For the

Intermediate section the same methods of culture as advised for the other classes of Cineraria will apply.

CLARKIA

C. elegans. C. pulchella. Hardy annuals

The two distinct classes of Clarkia named above include several varieties that have long been freely grown in gardens as summer annuals. But the very beautiful recent introductions in the Elegans class have lifted these flowers to a higher plane of usefulness for producing brilliant sheets of colour in beds, borders, shrubberies, and beside carriage drives. Although all the Clarkias bloom profusely in ordinary garden soil they well repay liberal treatment. Seed may be sown from March to May, or in September if an early display is wanted. In good ground each plant of the Pulchella varieties should be allowed a space of eight or ten inches, but rather more room must be given to the Elegans class to do the plants justice.

The Elegans varieties are of special value when treated as pot plants for conservatory decoration in May and June. From seed sown in August or early in September the plants can be slowly grown into magnificent specimens four feet high and almost as much in diameter. Our own practice is to sow thinly in clean well-drained 48-size pots. These are placed in a temperature of from 50° to 55°, and when the seedlings are large enough to handle they are pricked off into shallow boxes about three inches apart, the base of the boxes being freely perforated to insure ample drainage. The most suitable soil is composed of equal parts of sound loam and leaf-mould, with the addition of a gallon of coarse sand to each bushel of the mixed soil. After the plants are well established, ventilate freely to secure robust growth. When three inches high pinch out the points, and a little later transfer separately to small pots, keeping them close for a few days and as near the glass as possible. As the roots develop, transfer again to larger pots, and then the second and final stopping of the shoots must be done. Should very large plants be wanted they can be flowered in 16-size pots, using a compost slightly heavier than that advised at a younger stage of growth. The night temperature during winter should be about 45°, giving air freely by day whenever possible to do so with safety. As the branches need support, sticks of a suitable length must be provided, and the stems tied out in good time to prevent them from breaking off.

CLERODENDRON FALLAX

Stove shrub

A very handsome erect shrub, which is extensively grown in tropical gardens. In this country it attains a height of about two feet, and is easily raised from seed in a warm greenhouse or conservatory, where it proves to be a really beautiful and striking plant.

Sow in pots or pans in March or April and transfer to single pots while small. From the commencement a very rich soil is necessary to insure robust growth and intense colour in the panicles of brilliant scarlet flowers. The plants bloom in August or September of the same year. When the leaves fall, if the intention be to store through winter, remove to a temperature of 55°; but raising plants annually is more satisfactory and entails less trouble than storing.

Like many other tropical plants, Clerodendron fallax is subject to attack by mealy bug, and this pest may be dealt with by hand picking or by washing the leaves with insecticide two evenings in succession. Aphis are also troublesome and should be cleared by fumigation.

COCKSCOMB

Celosia cristata. Tender annual

This fine old-fashioned flower has won renewed popularity of late years, probably as the result of a number of well-grown plants exhibited at horticultural shows. Those who can produce handsome Cinerarias, Balsams, and Calceolarias, will be likely to turn out grand Cockscombs, strongly coloured and on dwarf, leafy plants. Liberal culture is essential, and the first start should be made in a compost consisting mainly of rich light friable loam. Sow the seeds on a rather brisk heat in February or March, a newly-made but sweet hot-bed being the best place for the seed-pans. Prick out early into very small pots, and shift on so as to encourage growth without a check, and keep the plants on the hot-bed until the combs are formed. It is well not to shift beyond the 8-1/2-inch size; then, by allowing the roots to become pot-bound, the combs are soon produced. It matters not how select the seed, or how careful the culture, a certain proportion of unsymmetrical combs will appear; but these, if richly coloured, will be useful for decorative purposes, and should have all the attention needed to keep their leaves fresh and the combs pure in colour.

COLEUS

Stove perennial

There is so much difficulty in carrying Coleus through the winter in vigorous health that the modern plan of treating it as an annual is advantageous for the saving of trouble and fire-heat in winter, and also because it offers the charm of constant diversity. The fact is that our winter days are too short and gloomy to maintain the splendour of colouring which makes Coleus so attractive and valuable; and seed from a good strain may be relied on to produce plants which will delight the eye all through the summer and autumn. Some experienced men sow in February and succeed, but the majority of cultivators will show prudence by waiting until March, when increased daylight favours the rapid growth of the plants. Flowerpots are better than pans, as the greater depth affords opportunity of securing effectual drainage. The pots should be nearly half-filled with crocks, covered with a layer of moss to prevent the soil from being washed away. Fill them with light turfy loam, mingled with almost an equal bulk of sharp sand. Make an even surface, on which sow thinly, and shake over the seed a slight covering of fine soil. Place the pots in a temperature of not less than 65°. Watering needs particular care, because of the peculiar liability of the young plants to damp off, especially in dull weather. The strongest seedlings are pretty certain to be those in which green and black predominate, and they may without scruple be removed to make way for the slower-growing but better-coloured specimens. These should be transplanted round the edges of pots while quite small; and such as show delicate tints, especially those having pink markings on a golden ground, are worth nursing through the early stage with extra care. The pots must be shaded from direct sunshine, but should be kept near the glass. In May the plants will be large enough for 48-sized pots, beyond which there is no occasion to go. When the pots become full of roots the foliage increases in brilliancy, whereas larger pots encourage free growth to the detriment of colour. A dry atmosphere is particularly injurious, while an occasional dose of manure water will maintain the plants in health.

COLUMBINE—see AQUILEGIA

COSMEA

Cosmos. Half-hardy annual

Cosmeas make a striking show in the mixed border, and the flowers are also in large request for indoor decoration. Disappointment is often caused, however, through the plants failing to bloom until late in the season, and therefore it is important to grow an early-flowering strain in order to insure a long-continued display. The most successful method of raising plants is to sow the seed in pots during February, pricking off the plants as soon as large enough. When the first flowers appear in

May, transplant to positions in the open immediately danger from frost is past.

CYCLAMEN

Half-hardy perennial

Gardeners of experience will remember the time when the predominant colours of Cyclamen were purple and magenta, and it was impossible for the most friendly critic to feel enthusiastic concerning these flowers. But the new colours—Salmon Pink, Salmon Scarlet, the intense Vulcan, Rose Queen and Cherry Red, together with Giant White and White Butterfly—are now regarded as the brightest and most beautiful decorative subjects for the long period of dark winter days of which Christmas is the centre. As cut flowers for the dinner-table Cyclamens have no rival at that period of the year, and as specimen plants in the home they are delightful for their free-flowering habit, compact form, and elegant foliage.

Seed may be sown at any time during autumn or the early part of the year, and the plants will not only flower within twelve months, but if properly grown will produce more bloom than can be obtained from old bulbs. We do not advise more than three sowings, the first and most important of which should be made in August or the beginning of September. To obtain a succession of plants, sow again in October and for the last time early in the new year. Those who have not hitherto grown Cyclamen for midwinter blooming will be well pleased with the result. It is quite as easy to flower them in the winter as in the longer days, and this is more than can be said about most plants.

The best soil for Cyclamen is a rich, sound loam, with a liberal admixture of leaf-mould, and sufficient silver sand to insure free drainage. Press this mixture firmly into pots or seed-pans, and dibble the seed about an inch apart and not more than a quarter of an inch deep. Cover the surface with a thin layer of leaves or fibrous material to check rapid evaporation, and later on keep the soil free from moss. The autumn sowings may at first be placed in a frame having a temperature of not less than 45°. At the end of a fortnight transfer the pans to any warm and moist position in the greenhouse or propagating house.

Although the Cyclamen is a tender plant, it does not need a strong heat, and will not endure extremes of any kind. Sudden changes are always fatal to its growth. In winter the temperature should not be allowed to fall below 56°, or to rise above 70° at any time. The more evenly the heat can be maintained the better, and it is desirable to give all the light possible. In summer, however, although a warm and humid atmosphere is still necessary, the light may with advantage be somewhat subdued, but shading must not be overdone, or the constitution of the plant will suffer.

Cyclamen seed not only germinates slowly, but it also grows in the most capricious manner; sometimes a few plants come up long after others have made a good start. Do not be impatient of their appearance, but when some seedlings are large enough for removal transfer to thumb pots, taking care not to insert them too deeply. As the plants develop, shift into larger pots, ending finally in the 48-size. In the later stages mix less sand with the soil, and when potting always leave the crown of the corm clear. Keep the plants near the glass, and as the sun becomes powerful it will be necessary to provide shade and prevent excess of heat. Never allow the seedlings to suffer from want of water, or to become a prey to aphis. To avoid the latter, occasional, or it may be frequent, fumigations must be resorted to. About the end of May should find the most forward plants ready for shifting into 60-pots. Give all the air possible to promote a sturdy growth. In doing this, however, avoid draughts of cold air. From the end of June to the middle of July the finest plants should be ready for their final shift into 48-pots, in which they will flower admirably. The growth during August and September will be very free, and then occasional assistance with weak manure water will add to the size and colour of the flowers. As the evenings shorten, save the plants from chills, which result in deformed blossoms.

The whole secret of successful Cyclamen culture may be summed up in a few words: constant and unvarying heat, a moist atmosphere, and abundant supplies of water without stagnation; free circulation of air, avoiding cold draughts; light in winter, and shade in summer, with freedom from insect pests. These conditions will keep the plants in vigorous growth from first to last, and the result will be so bountiful a bloom as to prove the soundness of the rapid system of cultivation. This routine may be varied by the experienced cultivator, but the principles will remain the same in all cases, because the natural constitution of the plant gives the key to its management.

DAHLIA

Half-hardy perennial

Both the double and the single classes of Dahlia are increasingly grown as annuals from seed, and this practice has the great advantage of being economical in time and in the saving of space during winter. The seedlings grow freely and quickly, and will flower quite as early as those grown by the more lengthy and troublesome method from tubers. Even those who possess a stock of named sorts may with advantage raise a supply from seed, especially as there is a probability of securing some charming novelty, which is in itself no small incentive.

Although the Dahlia is a tender plant, it is easily managed in a greenhouse, or in a frame resting on a hot-bed. The seed may be sown as early as January, but unless sufficient space is at command to keep the plants stocky as they develop, it will be wise to wait until February. A sowing in the month last named will produce plants forward enough to bloom at the usual time. Even March will not be too late; but whatever time may be chosen, when the start has been made it must be followed up with diligence, so as to avoid giving any check from first to last. Sow thinly in pots or pans filled with ordinary light rich compost, and cover the seed with a mere sprinkling of fine earth. When the first pair of leaves attain the height of an inch, pot off each plant singly close up to the base of the leaves. It is not advisable to throw the weakly seedlings away; these are the very plants which are most likely to display new shades of colour and they are worth some additional trouble. Although weak at the outset, they may, by judicious treatment, be developed into a thriving and healthy condition.

When potted, place the plants in heat, giving a little extra care until growth is fairly started. In due time shift into larger sizes as may be necessary, and then it will be wise to consider whether there is space to grow the whole stock well. If not, do not hesitate to sacrifice the surplus, and in doing so reject the rankest-growing specimens, for these are least likely to produce a fine display of bloom. It is mistaken practice to take out the top shoot, as this checks the plant for no good end; but when about six inches high, each one will need the support of a stick. Give water freely, and air on all suitable occasions. The least tendency to curled leaves indicates something amiss, and demands immediate attention. A cold blast may have stricken the plants, or the soil may be poor; lack of sufficient water will produce the mischief, or it may arise from the presence of aphis. If the last-named assumption prove correct, fumigate on the first quiet evening, and omit watering on that day. The mere mention of the other points will be sufficient to show the remedy for them.

As the time for transfer to the open air approaches, all that is possible should be done to harden the plants for the change. They may be placed for a few days under the shelter of a wall or hedge, but on the least sign of frost be prepared to protect with hurdles or mats. Full exposure during genial showers and fair weather is advisable, and an occasional examination of the plants will prevent their rooting through the pots into the soil.

The border for Dahlias can scarcely be made too rich, for they are hungry and thirsty subjects, and will amply repay in a profusion of bloom the manure that may be lavished upon them. Slugs and snails are unfortunately too partial to newly planted Dahlias, but the vermin soon cease to care about them; therefore it is advisable to plant Lettuces plentifully at the same time, or previously, on the same ground, and to dust around the Dahlias with lime. Insert at least one stake, about a yard long, near each plant, to give support, and two or three others will have to be given before the branches spread far. Secure the first shoot when planting is completed, and follow up the tying as growth demands.

Dahlias bloom continuously for a long time, and appear to be especially at home in the shrubbery border, or in the centre of a bed. They are also valuable for training against buildings having a southern aspect, and here the flowering period is much prolonged, for an early frost will scarcely reach them. A light wall is an admirable background for deep-coloured varieties, and the white or yellow flowers are displayed to advantage against a dark building. Dahlias may be used either alone or in company with the climbing plants which are usual in such positions.

The flowers possess a special value for indoor decoration, and any odd corner of the garden can be utilised for producing a supply for this purpose. Cutting should invariably be done in the early morning, while yet the dew is upon them. They will then retain their beauty for a longer period than those taken at a later hour from the same plants. This remark is true of all flowers, but it applies with especial force to the Dahlia.

DAISY, DOUBLE

Bellis perennis fl. pl. Hardy perennial

The remarkable development of the Double Daisy in recent years has raised this simple garden subject to the foremost rank of spring bedding plants. So pronounced has been the improvement achieved in the size and form of the flowers, that plants raised from a reliable strain of seed will now produce blooms which may well be mistaken for specimens of finely shaped Asters. When massed in a large bed the flowers present one of the most striking sights to be seen anywhere in the spring garden. But apart from their use in formal beds and borders, Double Daisies make a pleasing break among Wallflowers, and are particularly attractive when grown as an edging to bulbous flowers and other spring-blooming subjects such as Polyanthus, Myosotis, &c. Plants from a sowing made in pans in April and put out when large enough, may be flowered in the autumn of the same year. But the method more generally practised is to sow on prepared beds in the open during June or July, and to transfer the seedlings when sufficiently developed to positions for blooming in the following season.

DELPHINIUM

Hardy perennial

Nearly all the perennial varieties may be raised from seed, and where large numbers are required this is the best method of obtaining them. They make handsome border flowers, and are extremely valuable during the early months of summer. Sow in May, June or July, in the open ground, and transplant in autumn. If mixed seed has been sown, it will not be wise to thin out all the weakly plants, or it may happen that some of the choicest shades may be lost. The first flowers will be over by midsummer, but if the stalks are promptly cut down instead of being allowed to seed, there will be a second display later in the year.

Three varieties, Queen of Blues, Dwarf Porcelain Blue, and Blue Butterfly, may be flowered as annuals, by sowing in pans in March and transplanting to the open as soon as the seedlings are ready. They also make particularly charming pot plants, for which purpose it is advisable to sow seeds in March.

The scarlet variety *(D. nudicaule)* is rather more delicate than the others, and it is wise to raise the plants in well-drained seed-pans, and to take care of them through the first winter in a cold frame; indeed, in a heavy soil there is a risk of losing them in any winter which is both cold and wet. It is not necessary to employ pots, but immediately after flowering take them up and store in peat until the following April, when they can be returned to the open ground.

D. sulphureum. The seed takes a very long time to germinate, and severely taxes the patience of the sower. But otherwise there is no difficulty in raising plants, and the long spikes of beautiful clear sulphur-yellow flowers are well worth the extra time the seedlings need. The best plan is to sow in autumn in the open ground, cover with a frame, and avoid disturbing the soil, except for weeding, until the next autumn, when the plants should be put into position for flowering in the following summer.

As slugs are exceedingly partial to Delphiniums, the crowns should be examined in spring, and the seed-beds may be dressed with soot and surrounded with ashes to save the seedlings from injury.

The annual Delphiniums are dealt with under Larkspur, page 274.

DIANTHUS

Pink. Biennials, hardy and half-hardy

Many varieties of Dianthus claim attention for their elegant forms and splendour of colouring. They have been so wonderfully improved by scientific growers that they almost supersede the old garden Pinks, and have the great advantage of coming true from seed. *D. Heddewigii* (Japan Pink) and its varieties, *D. chinensis* (Indian Pink) and *D. imperialis*, make interesting and sumptuous beds, and may all be flowered the first year from sowings made in heat in January or February. Immediately the seedlings are through the soil it is important to shift them to a rather lower temperature than is necessary for insuring germination, or the plants become soft and worthless. Be very sparing with water, especially if the soil is at all retentive. When two leaves are formed, transfer to pans, allowing about an inch between each plant, and place in a sheltered position. Gradually introduce to cool treatment, and when ready prick off again, allowing each plant more space. They will thus have a much better start, when planted out in May, than if taken from the seed-pans direct. Dianthus make a most attractive display in pots, and a number of seedlings should be potted on for flowering in this manner.

Where there are no facilities for raising Dianthus in heat, it is quite easy to grow plants in an open spot from a sowing in June or July, and they will flower freely in the following year. Prepare drills about six inches apart and line them with sifted soil; sow thinly, and carefully cover the seed with fine soil. Shade must be given during germination, but it should be gradually withdrawn when the seedlings are up. Transfer to final positions in August. Should this be impossible, prick the plants out, and shift them again a little later. It will only do harm to leave them crowded in the seed-bed, and the second move will better enable them to withstand winter frosts. The Dianthus thrives in a sandy or loamy soil, with full exposure to sunshine, and the plants scarcely need water or any attention the whole season through.

DIGITALIS

Foxglove. Hardy biennial

Besides the native Purple Foxglove, largely grown in gardens, there are several very handsome varieties that are valuable for adorning borders, shrubberies and woodland walks. Specially worthy of attention are Giant Primrose, a beautiful variety with rich cream or buff flowers; the Giant Spotted, which produces handsome flowers, rich and varied in colour; and the white variety with its abundance of charming ivory-white bells, which are occasionally slightly spotted.

Any deep rich soil suits Digitalis, and seed sown in May, June, or July will produce seedlings which, with very little attention, will yield a fine display of flowers in the following summer. Sow in the open in pans, or on a prepared border, and put the young plants into permanent positions, choosing showery weather in August or September.

DIMORPHOTHECA

Half-hardy annual

The Dimorphotheca, also called the Star of the Veldt, was introduced into this country from South Africa and, like the Nemesia, also a native of that Dominion, it has become one of the most valuable of our summer annuals. Under favourable conditions plants may be flowered in six weeks from time of sowing and they will continue to bloom in profusion until cut down by frost. In addition to the striking orange flower, *D. aurantiaca* (Orange Daisy), a wide range of colours, including many delicate tints, has been evolved by careful hybridisation.

Those who wish to obtain forward plants should sow during March or April in pans of light soil placed in a cold frame, and the seedlings will be ready for transfer to open quarters in May. Or seed may safely be sown in the open ground in May and June. As suggested by its native habitat, the Dimorphotheca loves a warm sunny position and grows to the greatest perfection in a light soil or a well-drained loam.

The practice of flowering half-hardy annuals in pots is rapidly increasing, and among this class of plants the Dimorphotheca has few rivals as a decorative subject for the conservatory. It is more effective to grow three or four plants in a pot than one only, and the best specimens are obtained by sowing direct into the pots and thinning the seedlings to the required number. Use a light rich compost containing a fair proportion of silver sand, and do not let the plants suffer for the lack of water.

ESCHSCHOLTZIA

Hardy perennial

A decade or so ago the predominant colours found in Eschscholtzias were yellow and orange, but in recent years a number of new and very attractive shades have been introduced, with the result that this plant is now regarded as indispensable for summer bedding and for borders. The modern practice is to grow Eschscholtzias as annuals, sowing in the open during March and April. As the seedlings do not readily transplant, the seed should be put in where the flowers are wanted. Thin out in due course, allowing each plant ample space for development. Sowings may also be made during September, from which the plants will bloom in advance of those raised in spring.

FREESIA

Half-hardy perennial

The Freesia is another of the bulbous flowers easily raised from seed, and it may be had in bloom within six months from date of sowing. Use a rich compost, and sow under glass in January, February, or March, as may best suit convenience. Seed should be sown again in August, to supply flowers in spring or summer of the following year. The brittleness of the roots makes re-potting a hazardous operation. It is therefore wise to sow in 48-pots and thin to four or five plants in each, thus avoiding the need for shifting until after flowering has taken place. When re-potting becomes imperative, it must be done with a gentle hand, and the bulbs ought to be carefully matched for each pot. The position chosen for Freesias should be light and freely ventilated in mild weather, but they will not endure a cutting draught. For further cultural notes see page 328.

FUCHSIA

Half-hardy perennial

To raise Fuchsias from seed will be new practice to many; but it is both interesting and inexpensive, and every year it secures an increasing number of adherents. Seed may be sown at almost any time of the year; if a start be made in January or February, the plants will bloom in July or August. Soil for the seed-pots should be somewhat firm in texture, but a light rich compost ought to be employed when the plants come to be potted off, and the final shift should be into a mixture containing nearly one-third of decayed cow-manure. For the early sowing we have named, a rather strong heat will be necessary to bring up the seed. When large enough to handle, prick off the seedlings round the edges of 60-pots, putting about six plants into each pot. Shade and moisture are requisite to give them a start after each transfer. Subsequently they must be potted on as growth demands, until the final size is reached; and flowering will not commence so long as increased pot-room is given. The growth must not be hurried, and the plants should at all times be kept free from vermin. Seedlings having narrow pointed leaves may be consigned to the waste heap without scruple; but plants with short rounded foliage, especially if dark in colour, are almost certain to prove of high quality.

GAILLARDIA

Half-hardy perennial

All the Gaillardias are most conveniently grown as annuals from seed. The plants remain in bloom for a long period, and for their gorgeous colouring the flowers are as highly prized for arranging in bowls and vases as for garden decoration. The best month in which to sow seed is March, and the plants will then be ready for putting out in May. Any good compost will answer, and only a moderate temperature is necessary to bring up the seedlings. The usual course of procedure in pricking off must be adopted to keep them short and stout.

GERANIUM

Pelargonium. Half-hardy perennial

Geraniums of all kinds are most valuable if treated as annuals. In their seedling state the plants are peculiarly robust and charmingly fresh in leafage and flowers, even if amongst them there does not happen to be one that is welcome as a novel florist's flower. When grown from first-class seed, however, a large proportion of fine varieties and a few real novelties may be expected. The seed may be sown on any day throughout the year, but February and August are especially suitable. Sow in pans filled with a good mixture, in a somewhat rough state. Cover with a fair sixteenth of an inch of fine soil. Put the seed-pans in a temperature of 60° to 70° if sown in February, but heat will not be necessary at all unless it is desired to bring the plants into flower early in the ensuing summer. We are accustomed to place the seed-pans on a sunny shelf in a cool greenhouse, and have fine plants by the end of June, many of which begin to flower in August.

GERBERA

Half-hardy perennial

The Gerbera, also known as the Barberton or Transvaal Daisy, is a native of South Africa. Under cool greenhouse treatment it may be grown to perfection in pots, and a charming display of bloom can also be obtained in the open border from plants put out in a well-drained sunny position and given slight protection in winter. The flowers somewhat resemble a Marguerite in form, having a number of long pointed petals radiating from a small centre. In addition to the brilliant *G. Jamesonii*, sometimes called the Scarlet Daisy of the Cape, many hybrid flowers having a wide range of delightful colours are also available. Although seed is often sown in spring, the best results are probably obtained from an August sowing, in pans placed in a gentle heat. Prick off the seedlings when large enough, and if required for the greenhouse or conservatory transfer to pots, or gradually harden off for planting in the open as soon as weather permits in the following spring.

GESNERA

Nægelia. Tender perennial

An extremely beautiful ornament for stove or conservatory. The new hybrids freely produce spikes of bright pendulous flowers of many charming colours. Although the Gesnera is a perennial, it is sound practice to treat the plant as an annual. Seedlings from a January sowing will commence flowering in about nine months. Very rich soil, a warm and even temperature, and plenty of water, are requisite to promote luxuriant growth. The culture advised for Gloxinias will exactly suit the Gesnera also.

GEUM

Hardy perennial

The introduction of the well-known double variety, Mrs. Bradshaw, which may easily be flowered from seed in the first season, has brought the Geum into prominence in recent years. Seed of the above-named variety should be sown in pans in March or April and the seedlings pricked off into boxes of rich soil when large enough. Put out in May or June and do not let the plants suffer for want of water. Geums may also be raised from sowings made in June or July, and transplanting in due course to permanent quarters, in the manner usual with hardy perennials.

GLADIOLUS

Corn Flag. Half-hardy perennial

Formerly the Gladiolus was seldom raised from seed, probably because the seed obtainable was not worth sowing. Now it is saved with so much care that it will give a splendid display of flowers, a large proportion of which will be equal to named sorts, and some may show a decided advance.

The use of large pots—the 32-size will answer—is advantageous for many reasons, and they should be either new or scrupulously clean, for they will have to remain unchanged for many months, so that a fair start is the more necessary. For the same reason special care should be taken to insure free drainage. Over the usual crocks place a layer of dry moss, and fill with a compost of fibrous loam and leaf-mould in equal parts, with sufficient sharp sand added to make it thoroughly porous. Press the soil firmly into the pots, making the surface quite even, and in February dibble the seeds separately about an inch apart, and half an inch deep. This will render it needless to disturb the seedlings during the first season. Put the seed-pots in a steady temperature not exceeding 65° or 70°. After watering, it will help to retain the moisture if the top of each pot is covered with a layer of *old* moss, until the plants show. When the seedlings are about an inch high remove to a lower temperature, and begin to harden off by giving air on suitable occasions. Take care, however, that in the process no check is given to growth. Soon after the middle of May the seedlings should be able to bear full exposure, and it will then be time to renew the surface soil. Gently remove the upper layer, and replace it with rotten cow-manure, or some other rich dressing. Water must be given regularly until about midsummer, when the pots may be plunged to the rim in a shady border, and this will keep them tolerably moist until, in September, the seedlings begin to ripen off, which they must be allowed to do. When the leaves have died down, shake out the bulbs and place them on a shelf to dry. A mixture of equal parts of peat and pine sawdust, placed in a box or seed-pan, will make the best possible store for them; the box or seed-pan to be kept in any spot which is safe from heat and frost. After about six weeks, each bulb should be examined, and decayed specimens removed. If any of them have commenced growing, pot them and place in a pit or greenhouse. In March take the bulbs out of store, pot each one singly, and prepare for planting out. The transfer to the open must not be made until the danger of frost is past, even though it be necessary to wait until the first week of June.

Further remarks on Gladiolus will be found at page 329, under 'The Culture of Flowering Bulbs.'

GLOXINIA

Tender perennial

Gloxinias can now be flowered in the most satisfactory manner within six months from the date of sowing seed. Hence there is no longer the least temptation to propagate these plants by the lengthy and troublesome method formerly in vogue, especially as seedlings raised from a first-class strain produce flowers of the finest quality, both as to form and style of growth. One great advantage to be obtained from seedlings is an almost endless variety of colour, for the careful hybridisation of the choicest flowers not only perpetuates those colours, but yields other fine shades also. Those who have never seen a large and well-grown collection of seedling Gloxinias have yet to witness one of the most striking displays of floral beauty.

Quite as much has been done for the foliage of the Gloxinia as for its flower, and the best strains now produce grand leaves which are reflexed in such a manner as almost to hide the pot, so that the foliage presents an extremely ornamental appearance.

By successive sowings and judicious management it is possible to flower Gloxinias almost the year through. The most important months for sowing seed are January, February, and March, and to secure an early display in the following spring some growers sow again in June or July.

The soil most suited to Gloxinias is a light porous compost of fibrous loam. If this is not obtainable, leaf-mould will answer, mixed with peat and silver sand in about equal parts. New pots are advisable, or old ones must be thoroughly cleansed, and free drainage is essential to success. Fill the pots to within half an inch of the top. Sow thinly, and slightly cover the seed with very fine soil. Place the pots in a warm, moist position, carefully shading from the sun. A light sprinkling of water daily will be necessary. Immediately some plants are large enough for shifting, lift them tenderly from the seed-pot, so as scarcely to disturb the rest, and prick off into large 60-pots in which the soil has a convex surface. Follow this process as plants become ready until all the seedlings have been transferred. When potting on allow the leaves to rest on the soil, but avoid covering the hearts. On the first warm day give air on the leeward side of the house, briefly at first, and increase the time as the plants become established. A clear space between the plants is necessary to prevent the leaves of neighbours from meeting. The final shift should be into 48-pots, unless extra fine specimens are required, and then one or two sizes larger may be used. An occasional dose of weak manure water will prove beneficial, taking care that the foliage is not wetted. A moist atmosphere, with the temperature at about 60° to 65°, greatly facilitates the growth of Gloxinias. With care, however, they may be well grown in greenhouses and pits heated by hot water. Although the plants love a humid atmosphere while growing, this ceases to be an advantage, and, in fact, becomes injurious when the flowers begin to expand. At that time, also, the manure water should be discontinued.

Under 'The Culture of Flowering Bulbs,' page 331, further instructions are given.

GODETIA

Hardy annual

So far as the culture of Godetias is concerned, the usual spring sowing and the regular treatment of hardy annuals will satisfy those who are content with a display entailing the least possible trouble. But the Godetia is no ordinary annual. The plants flower with such amazing profusion, and the colours are so magnificent, that those who wish to produce striking effects in beds or borders in July and August will find Godetias of the highest value. All the varieties come perfectly true to colour and admit of numerous contrasts and harmonies. As an example, we suggest the following combination for a long border, or beside a carriage drive. Sow two rows of Alyssum minimum, allowing twelve inches between the rows; one row of Dwarf Pink Godetia fifteen inches from the Alyssum; two rows of G. Dwarf Duchess of Albany eighteen inches apart; one row of G. Scarlet Queen eighteen inches from the preceding variety, and one row of Double Rose at the back. The result will astonish those who have not previously seen a really fine exposition of this flower. Many other combinations will occur to those who carefully study colour schemes.

There are few annuals more greatly valued for cutting than the taller varieties of Godetia. These mainly produce double flowers in sprays two feet or more in length which develop into full beauty after being placed in water.

March and April are the months for sowing seed in the open for a summer display, and September for spring flowering. Good effects, however, are obtained by raising a sufficient number of plants in boxes and pricking off in readiness for putting out after bulbs and spring bedders have been cleared away. Under this practice there need not be a blank or a defective specimen.

Dwarf Godetias make exceedingly symmetrical and attractive pot plants. For this purpose sow seed in October in pans and place them in a temperature of 55° until the seedlings appear, then remove to a cooler place. As soon as possible prick off three in each 48-pot and when established grow on during winter in cold frames, giving air daily except in frosty weather, when the frames must remain closed and can be protected with whatever covering may be at hand. Here it may be well to point out that even when touched by frost the plants will recover if they are shaded from the sun's rays until the pots are quite clear of frost. Godetias flowered in pots make bright groups in conservatories, and occasionally do good service where failures occur in beds.

GREVILLEA ROBUSTA

Australian Oak. Greenhouse shrub

In its native country, New South Wales, this is a stately tree. Here it is grown as a pot plant, and the finely cut, drooping, fern-like foliage produces one of the most graceful decorative subjects we possess. Its value is enhanced by the fact that it withstands the baneful influences of gas, dust, and changes of temperature better than the majority of table plants.

Seedlings are easily raised by those who can exercise patience; and afterwards the simplest cool culture will suffice to grow handsome specimens. But we do not know any seed—not even the Auricula—which takes more time and is so capricious in germinating. In all cases where seed is sown in fairly rich soil, which has to be kept constantly moist and undisturbed for a long period, there is a tendency to sourness, especially on the surface. Free drainage will do something towards preventing this. Another aid in the same direction is to cover the seed with a layer of sand, and the sand with a thin coating of ordinary potting soil. When the surface becomes covered with moss, the coating of soil can be gently removed down to the sand, and be replaced with fresh earth, without detriment to the seeds.

Sow at any time of the year, in 48-sized pots filled with rather firm soil; and as the seedlings straggle through and show two pairs of leaves, pot them off singly, and give the shelter of a close pit or frame until they become established. They must not be allowed to suffer for lack of water, but there is no necessity to give them manure water at any stage of growth. An occasional re-potting is the only other attention they will require until they reach the final size, and the pots need not then be large.

HOLLYHOCK

Althæa rosea. Hardy perennial

Generations of unnatural treatment had so debilitated the Hollyhock that disease threatened to banish it from our gardens. Just at the critical time it was discovered that the plant could be grown and satisfactorily flowered from seed. Florists at once turned their attention to the production of seed worth growing, and with marked success. The best strains may now be relied on to produce a large proportion of perfectly formed double flowers, imposing in size, colour, and substance. The seedlings also possess a constitution capable of withstanding the deadly *Puccinia malvacearum*, and there is no longer a danger that this stately plant will become merely one of the pleasures of memory.

In growing the Hollyhock it is necessary to remember that a large amount of vegetable tissue has to be produced within a brief period, so that the treatment throughout its career should be exceptionally liberal. Some gardeners are successful in flowering Hollyhocks as annuals. Where this course is adopted it is usual to sow in January in well-drained pots or seed-pans filled with rich soil freely mixed with sand, covering the seed with a slight dusting of fine earth. A temperature of 65° or 70° is necessary, and in about a fortnight the plants should attain a height of one inch, when they will be ready for pricking off round the edges of

4-1/2-inch pots, filled with a good porous compost. Put the seedlings in so that the first leaves just touch the surface. At the beginning of March transfer singly to thumb pots, and immediately the roots take hold remove to pits or frames, where they can be exposed to genial showers and be gradually hardened. Defer the planting out until the weather is quite warm and settled.

The shrubbery border is the natural position for the Hollyhock, but the regular occupants keep the soil poor, and for such a rapid-growing plant as we are now considering there is obviously all the greater need for deep digging and liberal manuring. If put out during dry weather, complete the operation with a soaking of water, and repeat this twice a week until rain falls. Give each plant a clear space of three or four feet to afford easy access for staking and watering. By midsummer offshoots will begin to push through the soil. The removal of these will throw all the strength of the plant into one stem. To insure its safety a strong stake will be required, which should be firmly driven into the ground, and rise six or seven feet above it. In case of an accident at any time to the central stem the hope of flowers for that year is gone, and it is therefore worth some pains to prevent a mishap. The tying must be done with judgment, and as the plants increase in size an occasional inspection will save the stems from being cut. Several inches of half-decayed cow-manure placed round the stems, with a saucer-like hollow in the centre to retain water, will be helpful to the roots, and if the flowers are intended for exhibition, the treatment can scarcely be too generous.

It is, however, easy to grow and flower Hollyhocks without the aid of artificial heat. On a south border in June prepare drills about two inches deep and a foot apart. Place an inch of rich sifted soil in each drill, and upon this sow the seed very thinly, covering it about a quarter of an inch. If the weather be dry, give a gentle soaking of water, and finish with a dusting of soot to prevent vermin from eating the seedlings. Thin the plants to six inches apart, and they may remain in the seed-rows until the end of September. Whether they are then transplanted straight to blooming quarters, or put into a cold frame for the winter, depends on soil and climate. In the southern counties, and on light land, it will generally be safe to winter Hollyhocks in the open, with merely a shelter of dry fern or litter. But in heavy loam or clay the risk is too great, and the cold frame must be resorted to. In this they will be secure, and can be ventilated as weather permits. As the season advances give more air, until they are planted out in May. Seed may also be sown in pans in July or August, the seedlings being transferred in due course to pots for the winter. The protection of a frame will suffice, provided that frost is kept away, and the plants may be put out in spring as already advised.

IMPATIENS

Sultan's Balsam. Tender perennial

Early sowing should be avoided for two reasons. The seed germinates but slowly in dull weather, and the seedlings when raised are almost certain to damp off. We do not advise a start before March, and not until April unless a steady heat of 60° or 65° can be maintained. Sow in well-drained pots, filled with soil composed of two parts of turfy loam and one part of leaf-soil, with very little sand added. The seedlings are exceedingly brittle at the outset, and re-potting should not be attempted until they are about an inch high. Even then they need delicate handling, and after the task is accomplished they should be promptly placed in a warm frame or propagating pit for a few days. In June or July the plants should reach 48-sized pots, but they must not be transferred to the conservatory without careful hardening, or the whole of the flowers will fall. *I. Holstii* also succeeds well when bedded out in summer in the same manner as Begonias.

JACOBEA—see SENECIO

KOCHIA TRICHOPHYLLA

Half-hardy annual

This remarkable variety of *K. scoparia* is a miniature annual shrub, which is also known as Summer Cypress, or Belvidere. It is singularly attractive, of rapid growth and graceful habit. In a very brief time the finely cut foliage forms a compact cylindrical plant, beautifully domed at the top, and the tender green changes to a rich russet-crimson in autumn.

Seed may be sown in slight heat during February or March to provide early plants for pots, or for setting out in the open immediately the bedding season commences. It is important not to crowd the seedlings, and every precaution should be taken to prevent them from becoming thin, leggy, or wanting in symmetry. Each plant must be allowed sufficient space to develop equally all round. An April sowing can be made in the open where the plants are intended to remain, and beyond regular thinning they will give very little trouble.

As a conspicuous dot plant in beds this Kochia is extremely useful, or it can be massed in borders, and it also forms an admirable dividing line in the flower garden. For the decoration of conservatories a number should be specially reserved. Specimens may be employed with striking effect on flights of steps, in halls, and many other positions where a plant of perfect outline will serve as an ornament. Height, 2 to 3 feet.

LARKSPUR

Hardy annual

The cultivation of the annual Delphiniums, more familiarly known as Larkspurs, is so simple in character that it calls for little comment. But these handsome subjects are so widely grown, and so greatly appreciated, that they are fully deserving of special mention here. The taller varieties, of which the Stock-flowered strain is the most popular, are best grown in large beds, borders and shrubberies, and the dwarfer kinds in small beds. Apart from their usefulness in the garden, however, the taller sorts of Larkspur are much in request for providing cut material, particularly for the decoration of the dinner-table, and a number of plants should always be grown in reserve for this purpose. It is usual to put in the seed where the plants are intended to stand, and March and April are the best months for sowing. Thin out the seedlings promptly, and give each plant ample room for development, especially when grown on good ground.

Larkspurs may also be sown in September for producing an earlier display in the following year than is possible from spring-sown seed.

LAVATERA

Mallow. Hardy annual and hardy perennial

Countryside gardens owe not a little of their floral brightness to the Mallows. The modern varieties of Lavatera, however, far surpass in effectiveness the flowers commonly met with and are regarded as among the finest subjects for creating an imposing display in tall borders and large beds. For this purpose the annual varieties, Loveliness, *Rosea splendens*, and *Alba splendens*, are the most popular. As transplanting is not to be depended upon, seed should be sown thinly in March, April or May where the plants are wanted to flower. If the ground has been generously prepared fine specimens will result, and each plant should be allowed a spacing of at least two feet for development.

The perennial variety, *L. Olbia*, makes a bold subject for herbaceous borders and shrubberies. Seed may be sown in pans any time from March to August, putting out the plants when large enough for flowering in the following season. Small plants of this variety may with advantage be potted for conservatory decoration.

LOBELIA

Annual and perennial; half-hardy

There are several distinct classes of Lobelia, differing materially in height and habit. For dwarf beds or edgings the *compact* varieties should alone be used. These grow from four to six inches high, and form dense balls of flowers. The *spreading* or *gracilis* class, including *L. speciosa* and *L. Paxtoniana*, is in deserved repute for positions which do not demand an exact limit to the line of colouring. The plants also show to advantage in suspended baskets, window boxes, rustic work, vases, and any position where an appearance of graceful negligence is aimed at. The *ramosa* section grows from nine to twelve inches high, and produces much larger flowers than the classes previously named.

All the foregoing can be treated as annuals; and from sowings, made in February or March plants may be raised in good time for bedding out in May. Use sandy soil, and place the seed-pans in a temperature of about 60°, taking care to keep them moist. By the end of March or beginning of April the seedlings will be ready for transferring to pots, pans, or boxes. The last named are very serviceable for this flower, for they afford opportunity of giving the seedlings sufficient space to produce a tufty habit of growth. A gentle heat will start them, and they will give no trouble afterwards, except on one point, which happens to be of considerable importance. It is that the plants should never be allowed to produce a flower while in pots or boxes. Pick off every bud until they are in final positions, and then, having taken hold of the soil, they will bloom profusely until the end of the season.

Lobelias make elegant pot plants, yet, with the exception of the *ramosa* varieties which are excellent for the purpose, they cannot be grown satisfactorily in pots. The difficulty is easily surmounted by putting them out a foot apart in a good open position, and if possible in a rather stiff soil. When they have developed into fine clumps lift them with care and place them in pots, avoiding injury to the roots. This method will produce a display of colour which cannot be attained by exclusive pot culture.

From the best strains of seed it is possible that a few plants may revert to long-lost characters. Florists are striving to obviate this, but it will require time. Meanwhile there are two ways of dealing with the difficulty. Some growers prefer to raise plants from seed, and take cuttings from approved specimens for the next season. This plan insures exactitude in height and colour, with almost the robust growth and free-flowering qualities of seedlings. But it necessitates holding a stock through the winter, and this may be a serious matter to many. The simpler proceeding, and one which answers well in practice, is to raise seedlings annually and to remove from the pans or boxes any plants which show the least deviation from the true type. A few kept as a reserve will replace faulty specimens which may be detected after planting out.

The handsome perennial section of Lobelias obtains less attention than it deserves, especially as the most ordinary routine culture will suffice for these plants. They are partial to moisture, and also to a deep rich loam. A sowing on moderate heat in February or March will secure plants fit for bedding out in May. They may also be grown entirely without the aid of artificial heat from sowings in June or July. Employ pots or seed-pans, and pot off singly immediately the plants are large enough to handle. The protection of a cold frame or hand-light is all that is necessary during winter, and the planting out may be done in May. These Lobelias reach two feet in height, and make excellent companions to such flowers as *Anemone japonica alba* and *Hyacinthus candicans*. The dark metallic foliage and dazzling scarlet flowers also have an imposing effect as the back row of a ribbon border.

LUPINUS

Lupine. Hardy annual and hardy perennial

Both the annual and the perennial Lupines are extremely valuable for garden decoration and for supplying an abundance of cut blooms. Each class includes a number of charming colours and many of the flowers are delightfully scented. Not the least of their merits is the fact that Lupines are not particular as to soil; indeed, the annual sorts will often thrive on ground that is too poor for other and more fastidious subjects.

The annual varieties should be sown where intended to flower, as they do not transplant well. Sow the seed in March, April, or May, and subsequently allow each specimen a space of about eighteen inches for development.

L. polyphyllus is a valuable race of perennial Lupines which, from a sowing made in March or April and treated as annuals, will produce a fine show in the following autumn. In order to insure a display earlier in the season, however, many growers of these flowers prefer to sow in June and July of the preceding year. Two varieties of *L. arboreus* form large bushes which are distinctly ornamental when in full bloom. The seed should be sown in June or July and the seedlings transplanted to flowering positions before they become very large.

MARIGOLD

Tagetes. Half-hardy annual

Marigolds of several classes are valued for the profuse display of their golden flowers in the later summer months. The choicest are the so-called French, or *Tagetes patula*, which have richly coloured flowers, and some of the varieties are beautifully striped. For their high quality these Marigolds are judged by the florists' standards. The African, or *Tagetes erecta*, make large bushy plants with flowers 'piled high' in the centre; the colours are intense orange and yellow. in various shades. The bedding section is represented by the dwarf varieties of *Tagetes patula*, or Dwarf French Marigolds; also by *Tagetes signata,* a very neat plant with fine foliage and rather small orange-coloured flowers, produced in great abundance. In hot seasons and on dry soils this proves an admirable substitute for the Calceolaria, which does not thrive when short of food, whereas the Tagetes bears drought, the shade of trees, and a poor soil with patience, and up to a certain point with advantage. Sow all these in March in a moderate heat, and prick the plants out in the usual way, taking care finally to allot them sunny positions. Seed may also be sown in the open ground at the end of April or early in May.

The section of Pot Marigolds, *Calendula officinalis*, includes two remarkably handsome varieties, Orange King and Lemon Queen; the flowers of both are large, double, perfectly formed, and are worth a place in the choicest garden. These may be sown on the open border in March, April, and May, and the best place for them is in the full sun on a rather dry poor soil, but they are not particular, provided they are not much shaded.

MARVEL OF PERU

Mirabilis Jalapa. Half-hardy perennial

This flower may be treated either as an annual or as a biennial. As an annual the plants are very compact and effective, the leaves and flowers forming round glittering masses in the late summer and autumn months. When the roots are saved through the winter and planted out in April larger plants are obtained, but there is no advance in quality over the very neat and sparkling specimens raised from seed in spring. Sow in heat in March and April, and treat in the same manner as Balsams until the time arrives for planting out. A rich sandy loam suits them, and they like full exposure to sunshine.

MIGNONETTE

Reseda odorata. Hardy annual

Mignonette is so much prized that we must devote to it a paragraph, although there is little to be said. In many gardens plants appear year after year from self-sown seeds, and it will therefore be evident that Mignonette may be grown with the utmost simplicity. As a border plant we have but to sow where it is to remain, at different times from March to midsummer; the one important point is to make the bed very firm; in fact the soil should be trodden hard. It is imperative to thin early and severely, for any one plant left alone will soon be a foot in diameter, and in some circumstances cover a much larger area. Where bees are kept and space can be afforded, seed should be sown in quantity, for Mignonette honey is of the finest quality in flavour and fragrance. In pot culture it should be remembered that Mignonette does not transplant well; therefore, having sown, say, a dozen seeds in each of a batch of 48-or 32-sized pots, firmly filled with rich porous soil to which a little lime or mortar rubble has been added, the young plants must be thinned down to five, or even three, in each pot, as soon as they begin to grow freely. If small plants are wanted early, leave five in a pot; if larger specimens are wanted later, leave only three, or even only one. For winter and spring, sow in August and September and keep them as hardy as possible until it becomes necessary to put them under glass for the winter. A further sowing for succession may be made in January or February. Several strains of different tints are now at the command of cultivators of this favourite flower.

MIMULUS

Monkey Flower. Hardy perennial

This flower will grow in almost any soil, although a moist retentive loam and a shady situation are best adapted for it. There are many varieties, differing in height, and all are worth growing, both in pots and borders. If sown in February or March, and treated as greenhouse annuals, they will flower in the first year. It is easy to raise a large number of plants in a cold frame, and they make a rich display in borders and beds later in the year. Sowings in the open ground during summer will supply plants for blooming in the following season, but the most satisfactory course is to grow them as annuals, and at the end of the summer consign them to the waste heap. The Mimulus is quite hardy, and the most ordinary care will suffice for it. Water in plenty it must have, or the flowering period will be curtailed.

The well-known Musk is a Mimulus *(M. moschatus)*, and is as easily grown from seed as other varieties. It makes a valuable pot plant.

MYOSOTIS

Forget-me-not. Perennials, hardy and half-hardy

AT one time an impression prevailed that all the varieties of Myosotis were semi-aquatic, and could only be grown satisfactorily in very damp shady places. And it is quite true that most of them bloom for a longer period in a moist than in a dry soil. Still, they all flower freely, and last a considerable time in any garden border.

The only half-hardy variety that need be referred to is Sutton's Pot Myosotis, which is a delightful subject for flowering indoors at Christmas time; and as Forget-me-nots are everywhere welcome, the practice of growing plants in pots is rapidly increasing. Seed should be sown in a cold frame in June, and the seedlings can be potted on as required, taking care from the commencement to avoid crowding as a precaution against mildew, to which the plants are very liable. The strain referred to produces fine free-growing specimen plants, and a batch should always be in reserve for cutting. For table decoration in winter Forget-me-nots are very telling.

All the hardy varieties may be sown from May to July for a brilliant display in the following spring. The seed should be put into a prepared seed-bed under the shelter of a wall or hedge; and in autumn the plants must be transferred to blooming quarters at the earliest opportunity.

Myosotis make an extremely effective groundwork for spring bulbs, for which purpose *M. dissitiflora* is the most valuable.

NASTURTIUM—see TROPÆOLUM

NEMESIA STRUMOSA SUTTONI

Half-hardy annual

THIS beautiful South African annual is remarkable for its floriferous character, long duration of bloom, and diversity of colour. Since we introduced it to this country in 1888 it has attained great popularity as a pot plant for table decoration, and some of the most resplendent bedding effects in public parks and gardens have been secured with this flower.

For an early show of bloom sow in pots or pans in March under glass, using a compost consisting largely of good fibrous loam, with the addition of a small proportion of wood ashes. No more heat than necessary should be used, and when the seedlings are large enough to handle prick them off and gradually harden for planting out in May. Other sowings may be made in May and June, and at this period of the year the seed germinates most quickly in boxes placed in a cool shady spot out of doors. In early summer seed may also be sown in the open border, and by thinning to a distance of six or eight inches sturdy plants will be secured, which will remain in bloom until quite late in autumn.

For winter and early spring flowering in pots seed should be sown in August or September. There must be no attempt at forcing, or attenuated worthless plants will result. A further sowing may be made in January for blooming in the later spring months.

Like the seed of Verbena, Furze, and some other subjects, the germination of Nemesia under artificial conditions is somewhat capricious, but no difficulty will be experienced with open-air sowings.

NICOTIANA

Tobacco. Half-hardy annual

The delicious fragrance of the Tobacco plant, especially during the morning and evening, has made it a great favourite in the greenhouse and conservatory, as well as in beds and borders near frequented paths.

As a pot plant too, the Nicotiana is exceedingly useful, the large sweet-scented white, soft pink, and rich red coloured flowers being very attractive. A group of plants placed in the porch will, in the earlier and later hours of the day, as the door is opened, fill the house with their delightful perfume. Seed may be sown from January to June, and a continuance of bloom may thus be secured during nearly nine months of the year. Prick off the seedlings as soon as they are fit to handle, for if sown too thickly they are liable to damp off rapidly. Gradually harden off if required for planting out in May or June. In some places, more especially in the South of England, Tobacco seed sown on an open sunny border early in May will produce fine plants that will flower freely in August.

PANSY

Viola tricolor. Hardy perennial

The popularity of this flower has been greatly extended and the culture simplified since it became the practice to raise the required number of plants every year from seed. For all ordinary purposes the trouble of striking cuttings and keeping stocks in pots through the winter is mere waste of labour and pit-room. The Pansy is a little fastidious, but not severely so. It thrives in a cool climate, with partial shade in high summer, and in a rich, moist, sandy soil. Notwithstanding all this, the Pansy will grow almost anywhere and anyhow; but as fine flowers of this old favourite are highly prized, the plant should be treated with reasonable care to do justice to its great merits.

A thick sowing is very liable to damp off: therefore sow thinly, either in pots or boxes, in February and March. The thin sowing, moreover, renders it possible to take out the forward plants without disturbing the remainder. In due course transplant into pans or boxes of good soil, and place in some cool spot where the plants may gradually harden off. When they have become stocky, remove to beds or borders, with balls of earth attached to the roots. Should the surrounding soil become set by heavy rain or by watering, a slight stirring of the surface will prove beneficial.

Seed sown in the open ground during the summer months will readily germinate, and the seedlings need no attention beyond thinning to about six inches apart until they are ready for transferring to their proper positions, where they will produce a mass of bloom in the following spring.

The Pansy puts forth its buds very early in the year. Whether they are particularly tasty, or the scarcity of young vegetable growth gives them undue prominence, we know not, but certain it is that sparrows show a marked partiality for them. And having once acquired a taste for the buds, these impudent marauders will not leave them alone; they evidently regard Pansies as the perfection of a winter salad. Their depredations can be prevented by an application of water flavoured with quassia or paraffin oil, which must be repeated after rain.

PELARGONIUM

Greenhouse perennial

All kinds of Pelargonium may be raised from seed with the certainty of giving satisfaction if the work be well done. An amateur, who contributed to the production of symmetrical flowers in the Zonal section, found that under ordinary treatment Zonals began to bloom in one hundred days from the date of sowing the seed, and some of those that flowered earliest proved to be the finest. The cultivator will soon discover that one rule is important, and that is to sow seed saved from really good strains. The simplest greenhouse culture suffices to raise Pelargoniums from seed. Some growers sow in July or August; others in January or February. The summer sowing necessitates careful winter keeping, and the flowers appear earlier than those from spring-sown seed. But the spring sowing is the easier to manage, and is recommended to all beginners. Any light, sandy loam will serve for these plants, and it is well to flower the principal bulk of them in 48-and 32-sized pots, for if grown to a great size the date of flowering is deferred without any corresponding advantage.

PENTSTEMON

Hardy perennial

Penstemons when grown as half-hardy annuals are a valuable addition to beds and borders, where they produce a brilliant effect in summer. In borders it is not advisable to plant singly, but they should be employed in groups of not less than one dozen. It is also important to sow a strain consisting principally of scarlet and pink shades with white markings, as well as white flowers; under fair conditions there will be a profusion of richly coloured blooms on stately spikes about two feet high. Sow in heat during February or March and plant out in genial weather. It is not necessary to keep them after flowering has finished, although seedling Pentstemons on comparatively dry soil in favourable districts scarcely feel the winter. Seed may also be sown in June, in the manner usual with hardy perennials, and the plants will bloom in advance of those which are spring-sown.

PETUNIA

Half-hardy perennial

The Petunia affords another example of the immense strides accomplished in the art of seed-saving. Formerly the colours were few, and the blossoms comparatively insignificant. Now the single strains produce large flowers, beautiful in form, including self colours and others which are striped, blotched, and veined, in almost endless diversity. Some are plain-edged, others elegantly fringed. The double varieties also come so nearly true to their types that there is little necessity for keeping a stock through the winter. Plants raised from seed of the large-flowered strain embrace a wide range of resplendent colours, and the doubles are perfect rosettes, exquisitely finished in form and marking.

The only way of obtaining double seedlings is to save seed from the finest single blooms fertilised with pollen of good double flowers. Plants raised from such seed may be relied on to produce a fair proportion of double flowers of great beauty, and those which come single will be of the large-flowered type.

The dwarf varieties attain the height of five to eight inches only, and make admirable edging and bedding plants. The taller strains range from one to two feet, and are handsome subjects for border and shrubbery work. Both dwarf and tall sections are sufficiently brilliant and free-flowering to produce a beautiful display as pot plants in the greenhouse and conservatory.

For indoor decoration, the third week in January will be early enough to commence operations. Two parts of leaf-mould, one of loam, and one of sharp sand, make an excellent soil for them. Fill the pots or seed-pans within half an inch of the rim, and press the soil firmly down. Sow thinly on an even surface, and cover the seed with almost pure sand. Keep the pots or pans uniformly moist with a fine rose and a light hand, and in a temperature of about 60°. Greater heat will render the seedlings weak and straggling. From this condition it will take some skill and much time to redeem them; indeed, they may not produce a good display of flowers until the season is well-nigh over. Just as the seed is germinating is a critical time for Petunias, and a little extra watchfulness then will be fully repaid.

In February the sun has not sufficient power to do mischief, so that shading is generally unnecessary. An even temperature and freedom from draughts should insure seedlings strong enough to prick off by the end of that month. Put the plants into seed-pans about an inch apart, so that the first leaves just touch the soil, still using a light compost.

In April they should be ready for transferring to small 60-pots. Subsequently they must be potted on as growth demands, until they reach the 48-or even the 32-size. After re-potting place the plants in a sheltered part of the house or frame, where shade can, if necessary, be given until the roots are established. Frequent sprinklings of water, and a temperature of 60° or 65°, will give them a vigorous start. The lights ought to be put down in good time in the evening, but this must be done with judgment, or the plants will lose their healthy colour and assume a yellowish tinge. Insufficient drainage has a precisely similar effect. In about ten days air may be given more freely, and then no suitable opportunity of exposure should be lost.

In raising Petunias for bedding, the same conditions are applicable; but as it is useless to put them into the open ground until the weather is warm and settled, the sowing need not be made until the end of February or the beginning of March. And for bedding there is no occasion to put the plants into larger pots than the 60-size. It will be necessary to give these seedlings shade in their young state, after they have been pricked off or potted.

The beds or borders intended for Petunias will be better without recent manure, for this tends to the excessive production of foliage and defers the flowering until late in the season. Do not be tempted by the first sunny day to put them out, but wait for settled weather. A cutting east wind, such as we sometimes have in May, will ruin them irretrievably. Each plant of the tall class will occupy a space of two feet, and the dwarfs may be one foot apart.

In potting Petunias, those which are weakly among the singles will probably produce the most valued colours, and from seed sown for doubles it may be accepted as a rule that from the feebler seedlings the finest rosette-shaped flowers may be expected.

All Petunias are impatient of being pot-bound, and this applies especially to the double varieties. They will, if treated generously, do ample justice to the 8-or even the 10-inch size. The growth should not be hurried at any stage, and if the foliage has a dark, healthy, green colour, free from blight, there will be magnificent flowers four or five inches across. The final shift should be into a sound compost, consisting, if possible, of good loam and leaf-mould in equal parts, with sufficient sand added to insure drainage. About a fortnight later commence giving weak manure water once a week instead of the ordinary watering, and as the buds appear it may be increased in strength, and be administered twice a week until the flowers expand.

Petunias are accommodating in their growth, and may be trained into various forms. The pyramid and fan-shape are most common, and the least objectionable. We confess, however, to a feeling of antipathy to fanciful shapes in plants, no matter what they may be. It is a necessity of our artificial conditions of culture that many of them should be trained and tied to produce shapely specimens, but the more nearly the gardener's art approaches Nature, the greater pleasure we derive from his labours.

PHLOX DRUMMONDII

Half-hardy annual

Those who are acquainted with the older forms of this annual might fail to recognise a friend under its new and improved appearance. There are now several beautiful types, each possessing characteristics of its own, and all producing flowers that are perfect in form and brilliant in colour. The large-flowered section produces splendid bedding plants, but the dwarf compact varieties are also highly prized for effective massing and general usefulness. The latter attain a height seldom exceeding six inches, and are therefore eminently suitable for edgings and borders, as well as for bedding. They bloom profusely for a long period, not only in the open ground, but also as pot plants in the greenhouse or conservatory, where they are conspicuous for the richness of their display.

For early flowering sow seed of all the varieties in February or March in well-drained pans or shallow boxes. Any good sifted soil, made firm, will suit them, and every seed should be separately pressed in, allowing about an inch between each; then cover with fine soil. This will generally give sufficient space between the plants to save pricking off; but if the growth becomes so strong as to render a transfer necessary, lift every alternate plant, fill the vacant spots with soil, and those left will have room to develop. Pot the plants that are taken out, give them a start in a frame, and shade from direct sunshine. Phloxes should not be coddled; the best results are always obtained from sturdy plants which have been hardened as far as possible by free access of air from their earliest stage of growth. This does not imply that they are to be rudely transferred from protection to the open air. The change can easily be managed gradually until some genial evening makes it perfectly safe to expose them fully. A space of about two feet each way is required for each plant of the large-flowered class, but a more modest allowance of nine or twelve inches will suffice for the dwarf varieties. Before they are put out the plants must be free from aphis; if not, fumigation should be resorted to once or twice until there is a clearance of the pest. Seed of the annual Phlox may also be sown in the open ground during the latter part of May, and the plants will flower abundantly from mid-August until frost destroys them.

The employment of Phlox as pot plants has already been alluded to, but special mention must be made of Purity, which is by far the most valuable of all the varieties for blooming indoors. The pure white flowers, which are sweetly scented, may be produced at almost any period of the year. They are, perhaps, more highly appreciated at Easter than at any other time, and to insure a display at that season seed should be sown in September or October. The plants will do well if grown on in a cold frame, the final shift being into pots of the 48-size. When grown under glass, Phlox should be given treatment as nearly hardy as possible, all that is necessary in regulating temperature being the exclusion of frost from the greenhouse or frame.

PHLOX, PERENNIAL

Hardy perennial

The seed of perennial Phlox is very slow and erratic in germinating, and from a sowing made in September the seedlings may not appear until the following spring. Seed may also be sown in the first week of March in shallow boxes, and put into moderate heat. In due time prick out into boxes filled with light rich soil, and having hardened them in the usual way, plant out a foot apart in a good bed, and help, if needful, with an occasional watering.

PICOTEE

Dianthus Caryophyllus fl. pi. Hardy perennial

Seedling Picotees are extremely robust and free-flowering, and seed saved from the best types will produce handsome specimens. The instructions for growing Carnation—sowing in pans from April to August and transplanting when large enough—are equally applicable to the Picotee.

PINK

Dianthus plumarius. Hardy perennial

This old English flower is valued in every garden. Both the double and single varieties are easily raised from seed and the plants bloom with the greatest freedom. Seed may be sown any time from April to August. Treat the seedlings in the manner advised for Carnations, and in due course transfer to open quarters. The foliage maintains its colour during the severest winter, and is therefore worth consideration for furnishing the border, to say nothing of the abundant display of perfumed flowers which the plants afford in early summer.

POLYANTHUS

Primula (veris) elatior. Hardy perennial

A sowing in February or March in pans will produce strong specimens for flowering in the following year. Or seed may be sown from May to July on a shady border. Prick off the seedlings when large enough to handle. The plants should never flag for want of water, and green fly must be kept down by syringing. Some good solution will be necessary against red spider if through starvation in a dry situation it has been permitted to gain a footing. All the varieties can be grown in a bed with a cool shaded aspect. They do not require a rich soil; a strong and fibrous loam with a little leaf-mould is sufficient. On passing out of flower the plants will split up into several heads, when they may be separated and potted singly. Exquisite colour effects can be created by planting Polyanthus in association with beds of Tulips for flowering in April.

POPPY

Papaver. Hardy annual and hardy perennial

The recent developments of this flower have brought it into great and deserved popularity, and it may be safely affirmed that few other subjects in our gardens afford a more imposing display of brilliant colouring during the blooming period. The delicate beauty of the Shirley Poppies is alone sufficient to create a reputation for the entire class, and the huge flowers of the double varieties make a gorgeous show. All the varieties are eminently adapted for enlivening shrubbery borders and the sides of carriage-drives.

Seeds of Annual Poppies should be sown where the plants are intended to flower, because it is difficult to transplant with any measure of success. During March or April sow in lines or groups, *and thin to about a foot apart.* Large clumps of some of the bolder colours should be sown in spots that are visible from a distance, and they will present glowing masses of flowers.

By sowing seeds of Perennial Poppies in pans in March, and putting out the seedlings when large enough, the plants will flower the same year. The more general practice, however, is to sow very thinly on a well-prepared border any time from May to August. Keep the seedlings free from weeds, and thin out if necessary. The plants may be transferred to permanent quarters early in autumn or in the spring months.

PORTULACA

Purslane. Half-hardy annual

This is a splendid subject when the weather favours it. In a dry hot season, and on a sandy soil, Portulacas can be grown as easily as Cress. Sowings are sometimes made early in the year in greenhouses or frames; but as a rule it is a vain attempt. Wait until May or June, when the weather appears settled; then put the seed into the open border, and the lighter the soil, and the hotter the season, the more brilliant will be the display of flowers. Sow on raised beds, in rows six or nine inches apart, and cover the seed with sand or fine earth. If the plants appear to be injuriously close they must be thinned. Should a period of rain ensue, the raised beds have a distinct advantage over a flat surface, and rows afford opportunity for stirring the soil and keeping down weeds.

PRIMROSE

Primula vulgaris. Hardy perennial

The mere name of this flower is sufficient to recall visions of spring and perhaps of happy visits to its haunts in days gone by. But many ardent lovers of the Primrose may not know that the strains which are now in favour embrace a wide range of colour, from pure white to deep crimson or maroon, various shades of yellow and orange, and rich blue. In fact, in a batch of seedlings nearly every plant may differ from its companions. They all agree, however, in possessing the delicate perfume which is characteristic of the hardy woodland favourite. Fancy Primroses are prized as pot and border flowers, and they fully reward florists for all the care which has been devoted to their improvement. They will bloom satisfactorily in any shady spot; but to grow them to perfection requires a stiff moist loam, on the north side of some hedge or shrubbery, where glimpses of sunshine occasionally play upon them. Here large flowers, intense in colour, will be abundantly produced far into the spring.

The finest plants are generally obtained from a February or March sowing made in pans or boxes. Seed may also be sown from May to July in carefully prepared ground in the open. If inclined to take some pains in raising the plants—and they are certainly worth it—make the summer sowings in seed-pans in ordinary potting soil; sprinkle a little sand over the seed, and as a finish press firmly down. Sheets of glass laid over the pans and turned daily will prevent rapid evaporation and help to keep the soil uniformly moist. The seedlings either may be potted once, and then be planted out, or, if strong enough, they may be transferred straight to flowering positions. Should this mode of procedure be considered too troublesome, prepare a shady patch of ground by deep digging; make it firm and level, and on this sow in shallow drills, covering the seed very lightly. A dressing of soot over the surface, and a cordon of ashes round it, will keep off slugs. Thin if necessary, and when the plants are strong enough, remove to their proper quarters. In February the buds will begin to show, and those intended for pots should be allowed to reveal their colours before they are taken up, so that a variety may be obtained. From a retentive soil each plant with its surrounding earth may be taken out almost exactly of the size required, and it should be rather smaller than the pot which has to accommodate it. A light soil must be watered the day before the operation, or the roots will be injuriously exposed. When potted, place the plants in a shaded cold frame or greenhouse, allowing them plenty of space, and withhold water until it is absolutely necessary. At first they should be kept close, but as the roots become established gradually give air more and more freely. Cool, slow treatment is all that is required. Any attempt to hurry the growth will only weaken the plants and ruin the colour of the flowers. Just before the buds open, one or two applications of manure water will be beneficial. When the display in pots is over, if the plants are put out in a shady border, they may flower again late in the season.

PRIMULA SINENSIS

Chinese Primrose. Greenhouse annual

The history of the Chinese Primula since it first reached this country has an almost romantic interest. As originally received the flower was, and now is, insignificant in size and miserably poor in colour. But florists at once perceived in it immense possibilities. The result of their labours, extending over many years, may be seen in the magnificent Single, Double, and Star Primulas which now adorn conservatories, greenhouses, and homes. From so small a beginning the range of colours is amazing; there are snowy-white flowers in several beautiful forms, a pure Cambridge blue, rich violet-blue, many shades of rose, pink, scarlet, and gorgeous crimson. Almost equally striking is the improvement in the foliage, especially the introduction of the fern-leaf, with its diverse shades of green and richly toned under-surface.

To enjoy the bloom for a long period make successive sowings in May and June. A further sowing may be made in July if necessary. Use new pots which have been soaked in water; but if these are not at hand, scrub some old pots clean, for Primulas are fastidious from the outset, and it is by apparent trifles that some growers produce plants so immensely superior to others treated with less care. Provide free drainage, and place a little dry moss over the crocks. Any fairly good rich soil will be suitable, but a mixture of equal parts of sound fibrous loam and leaf-mould, with a small addition of silver sand, is best. Press this compost firmly into the pots to within half an inch of the top. Water before sowing, and sprinkle sufficient sand over the surface to cover the soil. On this sand sow evenly and thinly, for it is well known that the finest new Primula seed comes up irregularly, and a thin sowing admits of the removal of plants that may be ready, without disturbing the remainder. Cover the seed with just enough fine soil to hide the sand, and gently press the surface. Place the pots in a sheltered part of the greenhouse, protected from draughts and direct sunlight; a small glazed frame will be useful for this purpose. While the seed is germinating the temperature should not rise above 70°, or fall below 50°. Immediately the plants are large enough, prick off round the rim of small pots, and if convenient place them in a propagating box. Water with care, and shade if necessary. When established give air, which should be daily increased until the plants will bear placing on the greenhouse stage. Transfer singly to thumb pots, and subsequently shift into larger sizes as may be requisite, but never do this until the pots are filled with roots, and always put the plants in firmly up to the collar. During July, August, and up to the middle of September expose freely to the air in any convenient position where shelter can be given in unfavourable weather.

Where there is no greenhouse, but only a hot-bed, it is still possible to grow good Primulas, with care and patience. The instructions given for treatment in the greenhouse may easily be adapted to the pit or frame, only there must be a little more watchfulness in affording shade on sunny days to prevent overheating.

Endeavour to give the plants a robust constitution from the first, for weak, rickety things cannot produce a satisfactory bloom. Primulas need a long period of growth before they flower; hence they should never be subjected to a forcing temperature. Sufficient heat must be provided to raise the plants, but afterwards the aim should be to render Primulas as nearly hardy as possible before cold weather sets in. There must, however, be ample protection against frost, damp, and cutting winds.

Primula stellata *(Star Primula)*.—This elegant strain of Primula, introduced by us in 1895, has attained a high position in popular favour. Although it is not intended to supersede or compete with the splendid strains of *P. sinensis*, it is a most valuable addition to the conservatory, and will be found indispensable for general decorative work. The plants are unusually floriferous and continue in bloom for a long time. When cut, the sprays travel well and remain fresh in water many days. For table adornment Star Primulas are unsurpassed by any other greenhouse flower at their own period of the year. The culture is precisely the same as for *P. sinensis*.

Half-hardy Primula.—This section, which embraces a number of very charming species, includes the well-known *P. obconica grandiflora,* which is almost perpetual-blooming under glass. Seed of this Primula may be sown from February to July, from the earliest of which the plants will flower in autumn and continue to bloom throughout the winter. In the early stages the seedlings may be managed as already directed for *P. sinensis*, bearing in mind that excessive watering should be avoided. Cool greenhouse treatment will suit the plants well.

Another half-hardy variety which has recently attained wide popularity is *P. malacoides.* The dainty flowers are produced tier upon tier to a height of about two feet and are very sweetly perfumed. For a winter display sow in February, and successional sowings may be made until July. *P. malacoides* especially resents a forcing temperature. Therefore the culture should be as nearly hardy as possible, and even in the seedling stage the plants must have free access of air on all suitable occasions, or they are very liable to damp off.

Hardy Primula.—A number of very elegant garden Primulas are worthy of attention. The majority answer well when grown in borders, but they are especially at home in rock or Alpine gardens. The family is now so large and so variable in time of blooming that it is possible to have the different species in flower during almost every month of the year. As a rule, it is advisable to raise the seedlings in pots or pans placed in a frame or greenhouse, and to transfer them to the open ground when thoroughly hardened off.

RANUNCULUS

Half-hardy perennial

The Ranunculus can be grown either from seed or from roots. The seed is thinly sown from January to March, in boxes four to six inches deep, filled with good soil. A cool greenhouse or frame is the proper place for the boxes until the spring is somewhat advanced. A little extra care is requisite to insure free growth and a hardy constitution, and the roots should not be turned out of the boxes until they have ceased growing and are quite ripe; then they may be stored for planting in November or February. For particulars on the treatment of roots, see page 348.

RICINUS

Castor-oil Plant. Half-hardy annual

Although this plant flowers freely, it is grown in the sub-tropical garden principally for its noble ornamental foliage, and also in the shrubbery border, either alone or in conjunction with other fine subjects, such as Canna, Solanum, Nicotiana, and Wigandia. Plants of the dwarfer varieties may also be used with very decorative effect in conservatories and greenhouses during the summer and autumn months.

To have plants ready for making a show in early summer they must be raised as half-hardy annuals in February or March. From the commencement a rich soil and abundant supplies of water are necessary for the production of stately specimens. The seed is large, and may be put singly into pots, or three or four in each, and the latter is the usual practice. A temperature of about 60° will bring them up. If several plants are grown in a pot, they must be separated while quite young, and put into small pots filled with very rich soil. It is almost impossible to have the compost too rich, so long as drainage is quite safe. When the pot is full of roots, shift to a larger size, and commence the process of hardening, in readiness for planting out in June. This is worth some care, for if the plant receives a check when put out, it may take a long time to recover, and then part of the brief growing season will be wasted. Many gardeners never raise Ricinus in heat, but trust entirely to a sowing in the open on the first day of May. The seeds are put in three inches deep, in groups of three or four, and finally the plants are thinned to one at each station.

Prepare the soil in advance by deep digging and the incorporation of an abundant supply of manure. The most effectual way of doing it is to take out the earth to a depth of eighteen inches or two feet, and fill in with decayed manure and loam, chiefly the former. Upon this put out the plant, or sow seed as may be determined. If this is too great a tax on resources, or the near presence of shrubs renders the proceeding impossible, drive a bar into the soil, which, if light, can be readily worked into a fair-sized hole. Fill this with rich stuff nearly to the top, and over it either put the plant or sow seed. A heavy top-dressing round each stem is also desirable, and the application of copious supplies of water will carry the nourishment down to the roots. Sub-tropical plants are only a source of disappointment under niggardly treatment, but they amply repay all the care and generosity which a liberal hand may lavish upon them. The plants will need the support of stakes to save them from injury in a high wind.

SAINTPAULIA

Greenhouse perennial

A very remarkable perennial, only four inches high, obtained from eastern tropical Africa. The plant has fleshy leaves, and the flowers, which are produced in clusters, somewhat resemble the Violet, but are much larger. Saintpaulia makes a beautiful table ornament, and a row of pot plants in full bloom forms a charming margin in conservatories, either for a stage or on the ground. The seedlings flower freely in about six months from date of sowing, and continue in bloom through the winter. Sowings may be made from January to March, in well-drained pots placed in a temperature of 60° to 65°. On no account should the soil be allowed to become dry. Subsequently the plants may be treated as recommended for Gloxinias.

SALPIGLOSSIS

Half-hardy annual

A highly ornamental half-hardy annual. The finest strains have large, open flowers, exhibiting extraordinary combinations of colours which range from the palest sulphur-white to orange, scarlet, and purple-violet, all being more or less pencilled and veined with some strong contrasting colour.

If an early display is wanted, a start should be made at the end of February or beginning of March, by sowing on a moderate hot-bed. In May the plants will be ready for flowering quarters. Or sow in April in the open ground where the plants are to remain, taking care to thin severely, and the thinnings will be useful for dibbling in out-of-the-way comers, where they will furnish acceptable material for table decoration, for which purpose this striking flower is well adapted.

Salpiglossis make charming pot plants for the greenhouse and conservatory. For this purpose seed should be sown in August or September, and under cool-house treatment the plants will bloom profusely in the following spring.

SALVIA

Hardy annual and half-hardy perennial

From a genus including 450 species a small number of Salvias have won deserved popularity for beds and borders. In summer and early autumn the long spikes of brilliant flowers produced by Fireball and Scarlet Queen make an extremely attractive display, and *S. patens* is one of the most superb pure blue flowers seen in gardens. As a bedding plant *S. argentea* is extensively grown for its silvery-white foliage, which completely covers the ground. These and other perennial varieties may be sown in pans during February and March for transfer to the open in May, and the plants need the usual treatment of half-hardy perennials.

A favourite annual variety is Blue Beard, growing eighteen inches high and presenting long spikes of bright purple bracts. The annual Salvias should also be sown in pans in February or March and transplanted in May; or seed may be sown in the open border during April.

SCHIZANTHUS

The Butterfly Flower. Half-hardy annual

At many leading horticultural displays in recent years masses of Schizanthus of extraordinary beauty have been exhibited with striking success. In conservatories, greenhouses, and on dinner-tables the plants form conspicuous ornaments and they should be freely grown for general decorative purposes. On special occasions the pots may be plunged to create a brilliant show of bloom as temporary beds and they are also extremely attractive in hanging-baskets.

The usual time for sowing seed to insure fine specimens is the end of August or early in September. Either well-drained pots or shallow boxes, filled with a good potting compost, will answer for raising the seedlings. Sow thinly, on a smooth surface, and cover the seeds with finely sifted soil. When the young plants appear place the pots or boxes near the glass where they can have abundance of light and air, so that from the start the plants may be short and healthy. Seedlings that are thin and drawn are never worth the space they occupy. Immediately they are large enough to handle, transfer to shallow boxes, allowing a space of three inches to each plant. The compost to consist of sound loam and leaf-soil in equal proportions, with the addition of sufficient coarse sand to render the mixture porous. For two or three days keep the boxes in a frame, which must remain closed and be shaded from sunshine until the seedlings are established, but remove the shading whenever possible;

then give air freely, and on attaining a height of three inches the first stopping may be done. A fortnight later the plants will be ready for pots of the 60-size. Treat them as nearly hardy as weather may permit. Stop the shoots a second time when about six inches high, with the object of forming bushy plants capable of yielding a bountiful bloom. When the 60-pots are filled with roots transfer to the 48-size, and in due time the final shift should be into pots of the 24-size. Larger pots may, of course, be employed for very fine specimens. The compost for this final shift ought to consist of two parts of rich loam, one part of leaf-soil, and one part of thoroughly decayed manure; the addition of sharp sand will be necessary. The stems to be tied out to stakes in good time to prevent injury. Just before the flowering period and while the plants are actually in bloom, weak liquid manure, instead of water, once or twice a week will be beneficial. A high temperature is not required, even in the winter months, to maintain Schizanthus in healthy condition. From 35° to 40° is all the heat they need; in fact, it is only requisite to keep frost at bay, and this near approach to hardy treatment will result in fine robust plants.

The Schizanthus may also be sown during March and April in pans placed in gentle heat, the seedlings being potted on for flowering in the conservatory or they may be put out in the open border. Towards the end of April or in May seed may be sown out of doors.

One point in the successful culture of Schizanthus should never be forgotten. The roots must not be allowed to become pot-bound. Where this is permitted at any stage of growth it is fatal to the development of a handsome show of bloom.

SENECIO ELEGANS

Jacobea. Hardy annual

Among the double varieties, the crimson, purple, rose and white Senecios take the lead for beauty and usefulness. They are remarkably accommodating plants, adapted for beds or the greenhouse. Sow early in pans or boxes, give the seedlings liberal treatment, and when bedded out the plants will produce myriads of bright flowers, until frost puts a stop to them. Any good soil which does not become pasty will suit, and full exposure to sunshine is essential to the production of a rich display of colour. In March or April seed may safely be sown in the open ground.

The Tall Single Bright Rose Jacobea is invaluable as a cut flower for table decoration under artificial light. It rivals the Star Cineraria in form and, being a hardy annual, it may be grown with the utmost ease.

SILENE

Catchfly. Hardy annual

Not one of the hardy annuals has established a better claim to be sown in autumn than the Silenes. Alone, they make a very attractive display, and they can be used with especial effect in beds planted with Daffodils, Hyacinths, and Tulips. While the Daffodils are in full beauty the Silenes clothe the ground with a carpet of green, and after the foliage of the bulbs has been cut off or pinned down the Silenes furnish a fresh display of floral beauty in advance of the summer bedders.

Silenes do not thrive on heavy damp soils, but the difficulty can be surmounted by keeping the plants in pans or boxes under a cold frame until growing weather sets in. The plants do very well in loam, and best of all in a dry sandy soil. The spring sowing should be made in March or April; the autumn sowing in August or early in September.

SOLANUM

Annual and perennial; half-hardy

Solanums are of importance, some as greenhouse plants, and others as sub-tropical bedders. They are somewhat tender in constitution, and must have good cultivation in a light rich soil. A sharp look-out for red spider is necessary, for this pest is very partial to Solanums. March is early enough to sow the seed, but for ordinary purposes April is to be preferred. By the middle of June the plants should be strong enough to put out, and with genial weather will make rapid progress. Those grown for their berries may be sown from February onwards, as it is important to secure bushy plants before they begin to flower, and an early start insures an early ripening of the bright, handsome fruit.

STATICE

Sea Lavender. Hardy and half-hardy annuals and hardy perennial

It would be difficult to decide whether the Sea Lavenders are more highly valued as border flowers or as cut material for use indoors. Certain it is that the light and graceful sprays of delicately coloured flowers are indispensable for house decoration, either when freshly cut or when dried for mixing with Helichrysums and other everlastings in winter. Yet Statice are very attractive when growing in the border, the varieties of branching habit giving a long-continued display of beautiful flowers.

The half-hardy varieties should be sown from January to March in pans placed on bottom heat. When large enough prick off the seedlings into boxes of good light soil, and gradually harden off in readiness for planting out in May. The hardy annual kinds also answer best when started in pans during March or April and transferred to the open in due course. Seed of the hardy perennial varieties should be sown in a nice light compost any time from April to July. Put out the plants into flowering positions when they have attained a suitable size.

When grown on in pots, the half-hardy sorts make exceedingly pretty subjects for house or conservatory decoration.

STOCK

Mathiola. Annual and biennial half-hardy

From the botanical standpoint Stocks comprise two main classes—the Annual and the Biennial. So accommodating as to treatment is this extensive family, however, that by selecting suitable sorts and sowing at appropriate periods, it is not difficult to obtain a succession of these delightful flowers the year through. With this object in view, our notes are divided into four sections covering the cycle of the seasons, as follows: Summer-flowering, or Ten-week; Intermediate varieties, for autumn-flowering; Winter-flowering; and Spring-flowering.

Summer-flowering, or Ten-week Stocks.—These annual varieties include a wonderful range of colours, as well as considerable diversity in the habit of growth. For their brightness, durability, and fragrance they are deservedly popular. It is usual to sow the seed under glass from the middle to the end of March. Pans or shallow boxes, filled with sweet sandy soil, make the best of seed-beds, and it may be well to say at once that no plants pay better for care and attention than the subjects now under consideration. Sow thinly, that the plants may have room to become stout while yet in the seed-bed, and from the very outset endeavour to impart a hardy constitution by giving air freely whenever the weather is suitable. This does not mean that they are to be subjected to some cutting blast that will cripple the plants beyond redemption, but that no opportunity should be lost of partial or entire exposure whenever the atmosphere is sufficiently genial to benefit them. If a cold frame on a spent hot-bed can be spared, it may be utilised by pricking off the seedlings into it, or the pans and boxes may simply be placed under its protection. The nearer the seedlings can be kept to the glass, the less will be the disposition to become leggy. In transplanting to the open ground, it is worth some trouble to induce each plant to carry a nice ball of soil attached to its roots.

On light, friable land, Ten-week Stocks can be successfully grown from sowings made in the open about the end of April. The character of the season must be some guide to the time chosen, and the sowing in this case should be rather thicker than in the seed-pans. Should the seed germinate well, severe thinning will have to be practised as growth demands. This method of culture entirely prevents loss by mildew, which so often proves fatal to young transplanted seedlings. It is difficult to make the soil too good for them, and there is no comparison between Stocks grown on a poor border and those grown in luxuriance. Some growers make a little trench for each row of seed, and this affords a certain degree of protection from cutting winds, and also forms a channel for water when there is a necessity for administering it. In a showery season, the plants will appear in about twelve days, but in dry weather it will be longer, and one or more gentle morning waterings may be necessary to bring them up. The distance between the rows must be determined by the variety. Nine inches is sufficient for the dwarf sorts; twelve or fifteen inches will not be too much for medium and tall kinds.

Slugs may be kept off by a dusting of soot or wood-ashes, and some precaution must also be adopted to prevent birds from disturbing the seed-bed.

Here it may be well to mention a fact which is not always remembered, although the knowledge of it is generally assumed. Seed can only be saved from single flowers, but those who have made a study of the business find little difficulty in selecting plants, and treating them in such a manner that seed obtained from them will produce a large percentage of double blossoms in the following generation. But the experience of the most skilled growers has not enabled them to save seed which will result entirely in double-flowering plants; and this is scarcely to be regretted, for the perpetuation of the race is dependent on single flowers. In keeping the various colours true there is one very awkward fact. Certain sorts invariably produce a difference in colour between the double and single flowers.

Intermediate Stocks form a valuable succession to the Summer-flowering, or Ten-week varieties. From seed sown in gentle heat in February or March, the plants usually commence flowering when the earlier varieties are beginning to fade, and will continue to bloom until winter sets in. It is also easy to grow the Intermediate section in pots for spring decoration, if the protection of a house or pit can be given during the winter to preserve them from frost. A simple plan is to sow in August or early in September five or six seeds in 48-sized pots. Thin to three plants in each, and of course a larger pot with more plants can be used when desirable. Give air whenever possible, and water regularly. There is no need for artificial heat; indeed, it is not well to hurry the plants in any way. A good top-dressing of rich soil is advisable before flowering, and as the buds appear, manure water, weak at first, but gradually increased in strength, may be given once a week until in full bloom.

Winter-flowering Stocks.—During the winter months Stocks afford an immense amount of pleasure. They are particularly welcome at Christmas, and to insure flowering plants at that season of the year suitable varieties, such as Christmas Pink or Beauty of Nice, should be selected, and a start made in June. As soon as the first leaf is attained, prick off three seedlings in a three-inch pot; place in a cool frame under a north wall, keeping the light off all day until they are ready for another shift into six-inch pots. Use three parts of good yellow loam and one part of leaf-soil—no sand. Pot firmly and restore to the frame until the plants start growing, when they may be removed to the greenhouse. Manure water, not too strong, once a week is beneficial, and pure water should be given sparingly. Keep near the glass and ventilate freely. Further sowings made in July and August will extend the supply of flowers.

Spring-flowering Stocks, which include the popular Brompton strain, come into flower in spring and early summer. Although in some seasons it may answer to sow where the plants are required to bloom, the practice is too precarious to be risked generally. A safer method is to sow in seed-pans in June or July. Place these under shelter until the plants are an inch high, then stand them in the open for a week before transplanting. Have ready a piece of freshly-dug soil, and on a dull day put them out at eight to twelve inches apart. If the growth is too rapid during September, it may be advisable to lift them and plant again, for the winter must not find them soft and succulent. There should be hard stems and sturdy growth to carry them through the cold weather. In districts that are specially unfavourable it may be necessary to pot each plant singly in the 60-size, and plunge these in ashes in a cold frame, or under the shelter of a south wall, until severe weather is past, and they can then be turned out into the borders.

STREPTOCARPUS

Cape Primrose. Tender perennial

The hybrids are a very striking race, invaluable for greenhouse and conservatory decoration, producing a continuous succession of large trumpet-shaped flowers, embracing colours ranging from pure white, through lavender, purple, violet, rose, and red, to rich rosy-purple. Sow very thinly from January to March in well-drained pots, and a dusting of fine soil will sufficiently cover the seed. Place the pots in a temperature of 60° to 65°, and take care that the soil is not allowed to become dry. Prick off the seedlings when large enough to handle, keeping them in the temperature named until the final potting. When established they thrive with ordinary attention in a greenhouse, and they winter well in a temperature ranging between 40° and 50°. Seed sown in January and February will produce plants which will come into bloom during the following June and July.

Streptocarpus Wendlandii is a singularly interesting variety. Only one immense leaf is produced, which frequently attains a width of two feet, with a proportionate length. This leaf is reflexed, completely hiding the pot on one side, and from its midrib scapes of elegant violet-blue flowers with white throat are thrown up to a height of eighteen inches. The seeds should be sown in a warm greenhouse early in the year. The plants will begin to flower in the winter and continue in bloom for about six months. The temperature which is suitable for Gloxinias will answer for this plant also.

SUNFLOWER

Helianthus annuus. Hardy annual

The utility of the Sunflower has been alluded to in a former page. Here we have only to regard the plant in its ornamental character, as an occupant of the shrubbery or flower border.

In addition to the common species, there are several strains which are adapted for special purposes. The dwarf varieties grow about three to four feet high, and produce fine heads of bloom. The 'giant' attains the enormous height of eight or ten feet in a favourable season, and the flowers are of immense size. The double strain generally reaches six feet in height, and is valuable for its fine show of colour and enduring quality. There is no difficulty, therefore, in making a selection to suit the requirements of any border. The Sunflower can also be employed in one or more rows to make a boundary or to hide an unsightly fence, and some growers use it as a screen for flowers which will not bear full sunshine.

Seed may be sown very early in the season, and the plants can be brought forward in the manner usual with half-hardy annuals, but there is no necessity for this mode of growing them. Sow in April or May where the plants are to flower, on soil which has been abundantly manured to a depth of eighteen inches, and they will bloom in good time. To maintain the rapid growth, water must not be stinted in dry weather.

SWEET PEA

Lathyrus odoratus. Hardy climbing annual

The history of the Sweet Pea can be traced back for more than two hundred years; and it is almost as fascinating as an exhibition of the flowers. Recent improvements in this highly popular subject include an amazing diversity of colours, a marked increase in the number of flowers on each stem, and an extraordinary enlargement in their size. A modern list may run into hundreds, but those who grow every known variety find that there are many close resemblances, arising no doubt from simultaneous introductions by hybridists who have experimented on similar lines. Enthusiastic growers of Sweet Peas are no longer content with a limited number of named varieties, for it is obvious that in competitions where fifty or a hundred bunches have to be staged for certain prizes, a large and representative collection must be grown. For general garden decoration, however, and to provide sprays for the adornment of homes, the Giant-flowered class, offered under colours only, will continue to be extremely popular.

The change in character and the increased usefulness of Sweet Peas have necessitated a revolution in the methods of culture. The freer growth and more robust habit demand greater space than was formerly allowed. Instead of crowded rows of attenuated plants, producing a meagre return of small flowers, poor in colour, it is now the practice to prepare the ground by deep trenching and liberal manuring, and to give every plant ample space for full development both in rows and in clumps. In the ensuing paragraphs we outline the cultural routine which should be followed as nearly as possible by those who desire to insure a long-continued supply of the very finest flowers. But where circumstances do not permit of these recommendations being adopted in full, the details may be modified according to the materials at command and the requirements of the cultivator.

It is usual to commence the preparation of the ground in autumn. Trenching is of paramount importance, for the roots of the Sweet Pea require a considerable depth of good soil in which to ramify for the support of robust healthy plants capable of producing handsome flowers over a long season. Where the surface soil is shallow, care must be exercised to avoid bringing uncultivated subsoil to the top, and it is well worth incurring a little extra trouble to provide a sufficient depth of fertile material for full root development. Therefore dig out a wide trench and place the good top soil on one side. Then remove and discard the subsoil to a depth of twelve inches and, after breaking up the bottom of the trench with a fork or pickaxe, replace with an equal quantity of decayed manure, leaves, old potting soil or any other suitable stuff that may be on hand. Finally return the top soil to its original position.

The use of manure needs discrimination, and in fixing the quantity, as well as in selecting the most suitable kinds, due consideration must be given to the character of the soil. For light land, four barrow-loads of well-rotted farmyard manure per square pole will make an excellent dressing, but a rather smaller amount will suffice for heavy ground. In place of farmyard manure an unlimited quantity of leaf-soil, if obtainable, may be used, and it is also a good plan to dig in any available green refuse. Garden ground which for some years previously has been kept in a state of high cultivation by the liberal use of natural manure will not, as a rule, need further help in this direction, but it should receive a good dressing of lime. Indeed, any soil in which Sweet Peas are to be grown should contain not less than two per cent. of lime. The employment of artificial, as well as organic, manures is essential in any first-class scheme of cultivation. But here a word of warning is necessary. Nitrogenous manures in any form are harmful to the plant when applied in large quantities, and are liable to predispose it to disease, except on extreme types of sandy soil. Heavy ground should be dressed with seven pounds of basic slag in autumn and two pounds of sulphate of potash in spring. On light soils apply in spring four pounds of superphosphate of lime and two pounds of sulphate of potash. The quantities stated in each case are sufficient for a square pole of ground. Wood ashes (in a dry state) are also of great value, and these should be raked in a little in advance of planting out.

The special preparation of the soil just described entails the raising of plants in pots or boxes in readiness for transfer to the open as early as weather permits in spring. The finest flowers are undoubtedly obtained from an autumn sowing, and about the middle of September may be regarded as the best period for putting in the seed. This early commencement possesses the advantage of allowing ample time for the development of sturdy, well-rooted plants, which will not only bloom in advance of those sown in spring but will remain in flower for an unusually long period. Sow in light porous soil, and either three-inch pots, pans or boxes may be used. Place in a cold frame and keep the lights down until the seeds have germinated, but afterwards the frame should never be closed except during severe weather. There must be no misunderstanding on the question of air-giving. The Sweet Pea is almost hardy, and robust healthy seedlings, grown as nearly as possible under natural conditions, are wanted. Therefore to subject the plant to artificial heat will only defeat the object in view. A current of air should be admitted to the frame day and night, and the lights may be entirely removed on all favourable occasions. But the seedlings will need protection from excessive moisture, for if too wet at the roots they are liable to injury from frost. When four pairs of leaves are formed, stop each plant once, and after a little further progress has been made transplant singly into three-inch pots. Keep the pots in the frame, giving only such protection from hard weather as may be absolutely necessary, and plant out on the first suitable opportunity. In the South transplanting may be possible late in February or at the opening of March, but a month later will be safer in districts north of the Trent.

Those who for any reason do not find it convenient to sow in autumn may start the seed early in the year—from mid-January onwards, according to the district. The general principles described in the preceding paragraph apply equally to spring sowings, but it may be well to say that there must be no attempt to hasten growth by the application of a high temperature. A frame will afford all the protection necessary, and even a box covered with glass and placed in a sheltered spot will be found serviceable for raising seedlings.

Before planting out, the top soil of the ground prepared in autumn must be well worked and made friable. The disposition of the plants, and the method adopted for staking them, will, to a great extent, depend on the precise purpose for which the flowers are required. For garden decoration single rows answer well, and the plants should be spaced one foot apart. Or, if preferred, put out in clumps of three to five plants, allowing a diameter of from nine to fifteen inches. Carefully remove the plants from the pots or boxes in which they were raised, disentangle the roots and shake them quite free from soil. Make a hole of the necessary depth, and allow the roots to descend into the ground to their full extent, which may be as much as two feet in the case of well-grown specimens from autumn-sown seed. Give support immediately with well-branched twigs, and it is important that the plants be kept perfectly upright. Finally stake with bushy hazel sticks eight to ten feet in height, or taller still where the ground has been generously prepared.

Long-stemmed flowers free from blemish are essential for show work and for the highest forms of house decoration, and to insure an adequate supply over an extended period the following method, which is adopted by some of the most successful exhibitors, is strongly to be recommended. The plants are put out in double rows one foot apart, and spaced a foot apart in the lines. Each plant should carry two shoots only, both of which must be provided with a rod of bamboo, ash, or hazel, ten to twelve feet in length. For this double cordon system the rods will stand six inches apart in the rows, and it is desirable to make them secure against damage from high winds. Insert a stout pole at each end of the row, and about seven feet from the ground-level fix to each pole a substantial wooden crosspiece a little more than a foot in length. From these cross-pieces tightly stretch strands of wire, to which securely tie the rods. As growth develops commence disbudding promptly, regularly remove all laterals and tendrils, and tie each cordon to its supporting rod with raffia as often as may be necessary.

After transfer to the open ground the plants must never be allowed to become dry at the roots. Keep the hoe going between the rows, especially after the soil has been beaten down by rain.

The blooming period can be prolonged by the simple expedient of daily removing the dead or faded flowers. The ripening of only a few seed-pods speedily puts a stop to flowering.

In the open ground seed may be sown in spring from February to May, and successional sowings at intervals of a fortnight will extend the supply of flowers far into autumnal days. Even where a few clumps only can be grown it is unwise to depend on a single sowing. Autumn sowings outdoors are often made in September or October where a warm soil and favourable situation can be insured.

Sweet Peas have two principal foes, the slug and the sparrow. Against the former the usual precautions, such as ashes, old soot, lime, and various traps, are available; and the latter must by some means be prevented from doing mischief. After the buds show through the soil, it is generally too late for the adoption of remedies. Nearly all the heads will be found nipped off and laid ready for inspection. One could almost forgive the marauders were food the object, but the birds appear to commit havoc from pure wantonness, and whole rows are sometimes destroyed in a single morning.

Early sprays are so much prized that the practice of flowering Sweet Peas in pots under glass is yearly increasing, and for this purpose seed must be sown in August or September; the plants to be kept slowly moving during the dark days. In February the growth will be more rapid, but it is important to give the plants the hardiest possible treatment. In April, if properly managed, there will be a brilliant display.

The winter-flowering race blooms freely at a still earlier period, although the plants are less vigorous than other varieties.

SWEET WILLIAM

Dianthus barbatus. Hardy biennial

Sweet William belongs to the same genus as the Pink. The finest strains produce superb heads of flowers, some of them intensely rich in colour, while others have a contrasting edge. The new varieties are so marked an advance on older colours that they have created a fresh interest in this favourite garden flower.

In several instances we have advised that biennials and perennials should be treated as annuals, both on the ground of economy and for the excellent results obtained by this practice. But the Sweet William is not amenable to any treatment which reduces the natural period of growth.

Seed may be sown in May, June or July for transplanting in autumn, and the numerous colours afford opportunity of obtaining a great diversity of splendid effects in beds and borders.

TOBACCO—see NICOTIANA

TORENIA

Greenhouse annual

Sow in a warm temperature in March or April. Prick off while small into pots, and subsequently pot the seedlings singly. Any fairly good compost will suit them. The branches need support, and the plants must be kept free from green fly. The Torenias make very elegant pot plants, and they are also well adapted for hanging baskets and other ornamental contrivances.

TROPÆOLUM

Nasturtium, or Indian Cress. Hardy and half-hardy annuals

The *Tropæolum tuberosum* is treated under the 'Culture of Flowering Bulbs,' so that here we have only to consider the varieties that are grown from seed. There are two distinct classes, both widely cultivated, for the seed is inexpensive, and the plants extremely showy durable, and easily raised.

Tropæolum majus is the climbing Nasturtium, or great Indian Cress. The flower as originally obtained from Peru was a rich orange, marked with deep reddish-brown, but it has been developed into various shades of yellow and red, culminating in a tint which is almost black. The leaves are nearly circular, and are attached to the long footstalks by the centre instead of at the margin. Loudon fancifully compares the leaf to a buckler, and the flower to a helmet. The Lobbianum section is close in habit, with smaller foliage borne on somewhat woolly stems. All the varieties bloom freely, and constitute a brilliant class of climbers of great value for brightening the backs of borders or hiding unsightly objects. After the seeds have been dibbled about an inch deep in either April or May, the only attention the plants require is to nip out a straggling shoot occasionally, or prevent a stray branch from reaching over and smothering some plant which will not endure its embraces.

The well-known Canary Creeper *(T. canariense)* is a perfectly distinct variety, and as a half-hardy annual should be raised under protection and planted out in May, although sowings in the open ground in April and May often prove satisfactory. Unlike the others, it needs a rich soil to insure vigorous growth. When liberally treated the entire plant will be covered with its bright fairy-like flowers, until frost ends its career.

Tropæolum majus nanum.—The Tom Thumb, or Dwarf varieties, make excellent bedding plants, blooming far on into the autumn after many of the regular bedders have faded and become shabby. There is an extensive choice of colours in reds, yellows, and browns, which come perfectly true from seed, and all possess the merit of flowering freely on very poor soil. They grow luxuriantly on rich land, but then the foliage becomes a mere mask under which the flowers are concealed. There is not one of the Tom Thumb class that may not be treated as a hardy annual, and all afford opportunity of making a gorgeous show of colour at a cost ridiculously disproportionate to the effect obtained. They are also admirably adapted for pot culture, making shapely plants covered with bloom for a long period.

Many of the later introductions in Nasturtium are notable for their refined and delicate colouring, and are extremely desirable subjects for the decoration of the dinner-table and small vases in the drawing-room.

As the flavour of the flowers and leaves somewhat resembles that of common Cress, they are frequently used in salads, and are accounted an excellent anti-scorbutic. The flowers are legitimately employed in decorating the salad-bowl, because they are not only ornamental but strictly edible.

In a green state the seeds of both tall and dwarf varieties make an excellent pickle which is occasionally used as a substitute for capers.

VERBENA

Hardy and half-hardy perennials

VERBENAS raised from the best strains of seed come true to colour and the plants are models of health and vigour, and make resplendent beds. It is of the utmost importance to remember that the Verbena requires very little of the artificial heat to which it is commonly subjected, and which fully accounts for the frequency of disease among plants propagated from cuttings. Seed may be sown in boxes in January, February, and March, the earlier sowings naturally requiring more heat than the later ones. As the seedlings become large enough, they should be potted on and planted out in May, when they will flower throughout the summer, and far into the autumn.

Verbenas may also be sown in March or April in boxes, put into a frame, and if kept moist a lot of plants will appear in about a month. When large enough these must be carefully lifted and potted. A rich, mellow, and very sweet soil is needed by the Verbena. Many of the failures that occur in its cultivation are not only traceable to the coddling of the plant under glass, but also to the careless way in which it is often planted on poor worn-out soil that has been cropped for years without manure, or even the sweetening effects of a good digging. Raising Verbenas from seed has restored this plant to the list of easily grown and thoroughly useful flowers for the parterre.

The hardy perennial *V. venosa* also comes perfectly true and uniform from seed.

VIOLA

Tufted Pansy. Hardy perennial

This plant well merits its popularity for use in beds and borders. It is perfectly hardy, the habit is good, and it continues in bloom for several months in the year. The treatment prescribed for Pansy is also suitable for Viola.

WALLFLOWER

Cheiranthus Cheiri. Hardy biennial

Wallflowers are often sown too late. As a result the growth is not thoroughly matured, and the plants present but a feeble show of bloom. They should in their season be little mounds of fire and gold, exhaling a perfume that few flowers can equal in its peculiar freshness. Sow the seed in May or June, in a sunny place, on rather poor, but sweet and well-prepared soil favourable to free rooting. When the plants are two inches high, transplant into rows six inches asunder, allowing three inches apart in the row, and as soon as the plants overlap transplant again, six or nine inches apart every way, aiding with water when needful to help them to new growth. Or lift every other row and every other plant, leaving the remainder untouched to supply flowers for cutting. When the beds are cleared of their summer occupants, they may be filled with the best plants of Wallflower, to afford cheerful green leafage all through the winter and a grand show of bloom in the spring, as frost will not hurt the single varieties; but the doubles will not always endure the rigours of a severe winter.

Early-flowering Varieties.—By selection and cross-fertilisation an early-flowering race of Wallflowers has been obtained, and it is now possible to enjoy for many months of the year a fragrance which has hitherto been associated exclusively with spring. From a sowing made in May or June the plants commence flowering in autumn and continue throughout the winter, unless checked by frost. With the advent of spring weather, however, they burst into full bloom, making a delightful display in advance of the ordinary varieties.

WIGANDIA

Half-hardy perennial

This plant is grown for its foliage, and is extensively used in sub-tropical gardening. The instructions given for raising Ricinus in heat apply equally to this subject; but it is not wise to rely on an open-air sowing for a supply of Wigandias.

ZINNIA

Zinnia elegans. Half-hardy annual

THE double varieties of Zinnia have entirely eclipsed the single form of this flower. They grow to an immense size and are extremely valuable for beds and borders, the plants remaining in bloom for a considerable period. Double Zinnias are so varied in colour and beautiful in form that they deserve to take high rank as exhibition flowers.

The Zinnia is delicate, and should not be sown too soon. March is quite early enough to commence operations, and the first week in April will be none too late for sowing. A compost that suits Asters will answer admirably for Zinnias. Sow in 4-1/2 inch pots, which should have very free drainage, and cover the seed thinly with fine soil. Plunge the pots at once in a temperature of about 60°, when the seed will germinate quickly, and the plants on attaining one inch in height can be potted off separately. Place them in a close frame, shade from sunshine, and when well established, gradually give air and harden off. It will not be safe to transfer to the open until the first week in June, unless the position is exceptionally sheltered and the soil very dry. A shrubbery border is a suitable spot, and the more scorching the season the finer will be the flowers. There must, however, be shelter from the wind, for the stems of Zinnias are hollow, and easily damaged by a storm.

A satisfactory display of this flower may be obtained without the aid of heat by sowing in the open ground about the middle of May. Select a sunny sloping border or bed for sowing, enrich the soil, and make it fine. Press this down rather firmly, then drop three or four seeds at intervals of from fifteen to eighteen inches between each group, and lightly cover them. In due time thin to one plant at each station. If they thrive the branches will not only meet but overlap, and produce a grand display. In the event of very dry weather at sowing time the ground may be watered before the seed is put in, and then be covered with dry fine soil.

Zinnias do not transplant well, except as small seedlings. When it is necessary to undertake the task, choose, if possible, a showery day, and shade each plant with an inverted flower-pot for a few days, but take off the pots in the evening.

Zinnias intended for exhibition must be treated in a more generous fashion than plants that are grown for border decoration, or for the sake of yielding cut flowers. The seed may be raised in heat as already directed, but the border will need to be prepared with special care and liberality. Should the soil be heavy, it must be reduced to a friable state during winter. Before the plants are put in, raise the land into ridges about four or five inches high. Plant on the top of the ridge, and then an application of soot or lime (not too near to inflict injury) may be used as a precaution against slugs. In a wet season the plants will stand a better chance than if put on the flat, and if a scorching summer comes they will be none the worse for it. As the flowering time approaches mulch the ground with well-decayed manure.

The plants must be carefully staked and tied out. It is not merely necessary to secure the main stem, but the branches should also be supported, or when weighted with flowers they will be very liable to give way under a moderate wind. Superfluous branches may be removed, but not so severely as to start new growth to the detriment of the flowers. Disbudding also will have to be practised for the highest class of flowers. Only one bloom should be allowed to develop on each branch at a time, and this must be protected from sun and rain after it is about half grown.

SPRING FLOWERS FROM SEEDS

It is the spring flowers that perhaps give the greatest charm and interest to the English garden. Commencing with the flowering trees, the Almond, Double Peach, *Prunus Pissardi*, and many others, we soon have the Daffodils, Wallflowers, and Pansies, making the ground bright and gay after the long dreary winter. It may promote economy in the production of these brilliant and charming displays if we offer a few remarks on the employment of spring-flowering plants which can easily be raised for the purpose from seeds. It will, of course, occur to the reader that a considerable proportion of the annuals that are usually sown in autumn are particularly adapted for producing rich and varied displays in spring. A type of this class is found in the well-known Erysimum, Orange Gem, one of the cheapest, hardiest, and most resplendent plants of the kind, cheap enough for the humblest amateur to employ freely in his borders and beds, and at the same time so effective in its colouring as to be adapted for the most complex and highly finished examples of geometric work. Another striking subject is the Siberian Wallflower *(Cheiranthus Allionii*), so nearly allied to the Erysimum, Orange Gem, the gorgeous orange flowers adding a fresh colour to the many new shades given us in recent years by the old English Wallflower. Among the annuals are several valuable spring flowers—such as, for example, *Nemophila insignis*, well known for its lovely blue blossoms, and the white variety, *alba*, of the same; *Saponaria calabrica,* exquisite rosy pink; Silene, rose, dwarf rose, and dwarf white; Virginian Stock, of which the distinct varieties are remarkably well adapted to form bands and masses of red, white, and yellow, and also to make a delightful groundwork for enhancing the splendour of late Tulips; and clumps of Aubrietia, Yellow Alyssum, and other of the more distinctive plants that are employed in high colouring in first-class geometric gardening. A list of such plants will at once indicate that there is a field of enterprise for the practitioner of spring flower gardening; and while cheap and effective materials are thus brought into the service, there is no interference with the later summer bedding, because, if the annuals are well managed, they will give their plentiful bloom when the garden is most in need of colour, and may be cleared off in time to make way for the plants that are generally employed in the summer display and which are known as ' bedding plants' *par excellence.*

In the management of annuals for an early bloom, it is of great importance to sow them at a proper time, so that they will be strong enough to perform what is required of them, and yet not so forward (or 'winter proud') as to suffer from the severity of the weather. In the North the middle of August is none too early for a general sowing in beds, and in the South the middle of September is none too late. In some few sheltered spots in the extreme South-West seed may be got in at the middle of October. As a rule, however, the sowing should be made as late as those familiar with the soil and climate of the place may deem safe, the main point being to have the seedlings in a short-jointed condition, close to the ground, in which state they are least likely to be injured by frosts. We prefer sowing in drills on a rather poor soil well broken up to a kindly state, and if the weather happens to be dry, the drills should be freely watered before the seed is sown, and there will be no more watering needed. The after-management is extremely simple: the plants must be kept clear of weeds, and be slightly thinned out if much crowded, for a few sturdy specimens are of more value than any number that have run up weak and wiry through overcrowding.

In sheltered gardens, having dry chalk or sandy soils, the greater part, or perhaps the whole stock, might be transplanted from the seed-beds to the flower-beds and borders as soon as sufficient growth has been made; but on heavy soils and in exposed places it will be advisable to delay the removal until March. This part of the work must be nicely done, the plants being lifted in clumps and no attempt made to single them, and they must be carefully pressed in and aided with water, if necessary, to promote a quick 'taking hold' of their new quarters. Those planted out in October on a dry soil will not only bloom early and gaily, but will be beautiful in their different tints of green all the winter through.

But we are not restricted to annuals in seeking for spring flowers from seeds. With very few exceptions, *all* the favourite plants of the spring garden may be grown from seeds at a cost almost infinitesimal as compared with the raising of named varieties from cuttings and divisions.

Daisies, some of them now almost as large as Asters, are not only suited to the ribbon border, but make an amazingly brilliant show when the white, pink, and crimson are planted in masses or in separate beds. Seedlings flower with far greater freedom and produce much larger blooms than divided plants, and even after the first few weeks, when the later flowers become smaller and less perfect in form, a brilliant display is maintained till late in the summer if the beds are not wanted for other things. Pansies, which are still unsurpassed for beds and borders, are easily raised from seed. What is more interesting than a long row of plants of Perfection Pansy beside the pathway? every step brings one to a flower of perfect charm, quite different in marking or colour from any other. The several species and varieties of Arabis, Alyssum, Aubrietia, Viola, Polyanthus, Iberis, and Forget-me-not also come quite true from seed. The precision of style and colouring that results from raising these from cuttings is, of course, admitted; but in forming masses and ribbon lines, minute individual characters are of less consequence than a good general effect, and this may be insured by raising the plants from seed in a manner so cheap and expeditious that we feel assured spring bedding would be more often seen in its proper freshness and fulness were the system we now recommend adopted in place of the tedious one of multiplication by offsets and cuttings.

Wallflowers cannot be grown in too great numbers in any garden, for either their delightful perfume or charming colour effect. The striking displays to be seen in some of our public parks and on seaside fronts have done much to popularise this old favourite flower. Since the first edition of this book was issued, many new and remarkable colours in Wallflowers have been introduced, among the last, but by no means least, being the Fire King and Orange Bedder. It is by the blending of the colours that the most telling effects can be produced. Probably Blood Red, a very inadequate name, and Cloth of Gold will always be the most favourite combination, and when planted together one sets off the other to a degree little thought of when these varieties are grown separately.

Purple and the other yellows (Faerie Queene and Monarch) also make a pleasing bed. Fire King and Orange Bedder should be grown in masses, separately or together, and when seen in the late afternoon or early evening their vivid and gorgeous colouring is almost unsurpassed by any other flower. The early-flowering Wallflowers will, in mild winters, bloom from January till April, or even as early as Christmas.

It should not be forgotten that these biennial and perennial plants require more time to prepare themselves for flowering than do the annuals. If sown in August they may not bloom at all the next season, or the bloom may be late and insignificant. But if sown in May and June they have a long season of growth before winter sets in, and at the turn of spring the plants will be matured and strongly set for bloom.

The sowing of biennial and perennial plants for a display of spring flowers must be carefully done. The ground should be moderately rich and quite mellow through being well broken up; in other words, a good seed-bed must be prepared. If the weather is dry, the drills should be watered before the seed is sown; and in the event of a drought, the young plants must have the aid of water to keep them going through the summer. The seed should be sown thinly, and, as soon as the plants are large enough, they should be thinned out if at all crowded, and the thinnings can be planted in rows and shaded for a while. As a rule, the whole of the work will be comprised in sowing, thinning, and weeding. In average seasons they will not require watering, and in this matter alone will be seen the advantage of raising from seeds instead of cuttings.

Ordinary care, with such plants as we have named, will insure a splendid display of spring flowers; and they are worth whatever attention may be necessary to promote complete and early development. It may happen that plants from early sowings will show a few flowers in autumn if neglected. This is easily prevented, to the great advantage of the plants, by the simple process of 'stopping' or nipping out the points of the leading shoots to cause the production of side shoots. If a sturdy growth is thus secured, and the plants are transferred to the flower-beds in October, the result will justify the labour.

Practical gardeners will not need to be informed that the system we now propose is capable of many applications and expansions; but it may be suggested to amateurs who lament the dreary aspect of their beds and borders in the month of May and early part of June, that the plants we recommend for the formation of masses in the geometric garden are equally well adapted to form beautiful clumps and sheets on borders, banks, and rockeries, as well as in many instances to serve as a groundwork to Hyacinths, Tulips, Narcissi, and other splendid hardy spring flowers.

Sweet Peas deserve to be considered separately. These flowers are now so varied and exquisitely beautiful that they never appear in the garden too early. From autumn sowings not only are the most forward blooms obtained, but for size and intensity of colour the flowers are unsurpassed by the later displays from spring sowings.

THE CULTURE OF FLOWERING BULBS

Our popular flowering bulbs are obtained from many lands; they are exceedingly diversified in character, and they bloom at different periods of the year. Each variety has a value of its own, and answers to some special requirement in its proper season under glass or in the open ground. In the darkest winter days we prize the glow of Tulips and Hyacinths for brightening our homes. And bleak days are not all past when Aconites and Snowdrops sparkle in beds and borders. The Anemones follow in March, and during the lengthening days of spring there are sumptuous beds of Hyacinths, Narcissi, and Tulips. When high summer begins to decline we have stately groups of Gladioli and many beautiful Lilies in the shrubbery borders.

Not least among the merits of Dutch Bulbs is the ease with which they can be forced into flower at a period of the year when bright blossoms are particularly precious, and they are equally available for the grandest conservatory or the humblest cottage window. They are attractive singly in pots or vases, or they can be arranged in splendid banks and groups for the highest decorative purposes. Another advantage is that bulbs endure treatment which would be fatal to many other flowers. They can be grown in small pots, or be almost packed together in boxes or seed-pans; and when near perfection they may be shaken out and have the roots washed for glasses, ferneries, and small aquaria; or they can be replanted close together in sand, and covered with green moss. Their hardiness, too, permits of their being grown and successfully flowered without the least aid from artificial heat. Small beds and borders may be made brilliant with these flowers, and the number of bulbs that can be planted in a very limited space is somewhat astonishing to a novice. Unlike many other subjects, bulbs may be rather crowded without injury to individual specimens.

For the decoration of windows no other flowers can compare with Dutch Bulbs in variety and brilliancy of colour. Some of them are not particularly long-lived, and this need occasion no regret, for it affords opportunity of making constant changes in the character and colour of the miniature exhibition, which may easily be extended over many weeks. And a really beautiful display is within reach of those who have not a scrap of garden in which to bring an ordinary plant to perfection. Unused attics and lead flats can, with a little skill and attention in the case of bulbs, be made to answer the purpose which pits and greenhouses serve for many of our showy plants. Some of the most attractive flowering plants cannot be successfully grown in large centres of population, but bulbs will produce handsome blossoms even in smoky towns.

We do not recommend the attempt to grow bulbs in the actual window-boxes. It is seldom entirely satisfactory. They should be treated in the manner advised under the several varieties in the following pages, and just as the colours are becoming visible, a selection can be made from pots or boxes for crowding closely in the ornamental arrangements for the window. When the first occupants show signs of fading, others can be brought forward to fill their places, and this process may be repeated until the stock is exhausted. Winter Aconites, Snowdrops, Squills, and Glory of the Snow furnish the earliest display; these to be followed by Crocuses, Tulips, Hyacinths, and the many forms of the great Narciss family, until spring is far advanced.

The secret of their accommodating nature lies in the fact that within the Hyacinth or Tulip every petal of the coming flower is already stored. During the five or six years of its progressive life the capacities of the bulb have been steadily conserved, and we have but to unfold its beauty, aiming at short stout growth and intensity of colour. Of course there is an immense difference in the quality of bulbs, and they necessarily vary according to the character of the season. The most successful growers cannot insure uniformity in any one variety year after year, because the seasons are beyond human control. But those who regularly visit the bulb farms can obtain the finest roots of the year, although it may be necessary to select from many sources.

Such bulbs as Lilies, Iris, Montbretia, Hyacinthus, and Alstroemeria suffer no deterioration after the first year's flowering. Indeed, it will be the cultivator's fault if they do not increase in number and carry finer heads of bloom in succeeding years. As outdoor subjects some of them are not yet appreciated at their full value. Magnificent as *Lilium auratum* and *L. lancifolium* must ever be in conservatories, they exhibit their imposing proportions to greater advantage, and their wealth of perfume is far more acceptable, when grown among handsome shrubs in the border. Very little attention is needed to bring them up year after year in ever-increasing loveliness.

Growing Bulbs in Moss-fibre.—A most interesting method of growing bulbs is to place them in bowls and jardinières filled with prepared moss-fibre, and far better results for home decoration may be obtained in this way than by using ordinary potting soil in vases, &c. For this system of culture no drainage is necessary, and the bowls and vases which are specially made for the purpose are not pierced with the usual holes for the escape of water. The receptacles are non-porous and may be placed on tables and columns, or they can be employed in halls and corridors without the slightest risk of injury. The fibre is perfectly clean to handle, odourless, and remains sweet for an indefinite period.

Vases of any kind may be used, provided they are non-porous, but the bulbs to be planted in them should be of a suitable size. For quite small jardinières, white and purple Crocuses, Scillas, Snowdrops, and Grape Hyacinths are available, also the smaller varieties of Narcissi. Larger vases will accommodate Hyacinths, Narcissi, Tulips, &c. It is better not to mix different kinds of bulbs in one bowl unless simultaneous flowering can be insured. The specially prepared fibre needs only to be moistened before use. Having selected suitable receptacles for the bulbs to be grown, place a few pieces of charcoal at the bottom of each bowl. Then cover the charcoal with one to three inches of moistened fibre according to the depth of the bowl, placing the bulbs in positions so that their tips reach to within half-inch of the rim. The spaces between and around the bulbs to be filled with moistened fibre, carefully firmed in by hand. The bulbs will require practically no attention for the first few weeks and may be stood in a warm, airy position, but on no account must they be shut up in a close cupboard. If the fibre has been properly moistened there will be no need to give water until the shoots are an inch or so long, but the fibre must not be allowed to go dry, or the flower-buds become 'blind.' The surface of the fibre should always look moist, but if too much water has been given the bowl may be held carefully on its side so that the surplus water can drain away. As the growth increases more water will be required and all the light possible must be given to insure sturdy foliage. This fibre also answers admirably instead of water for Hyacinths grown in glasses, but care should be taken to fill the glasses as lightly as possible with the compost; if crammed in tightly the root growth is liable to lift the bulbs out of position.

ACHIMENES

Showy stove bulbs remarkable for their beauty. Given a sufficiency of heat, the cultivation is of the easiest nature, for they grow rapidly and flower freely, if potted in sandy peat, and kept in a warm greenhouse or the coolest part of a stove, in a somewhat humid atmosphere. It needs only the simplest management to have these plants in bloom at almost any season of the year, for the bulbs may be kept dormant for a considerable length of time without injury, and may be started into growth as required to keep up a long succession of flowers. They are occasionally well grown in common frames over hot-beds. For suspended baskets Achimenes are invaluable.

AGAPANTHUS

In favoured districts on the South coast this noble plant succeeds admirably if planted out between September and March in a rich, deep, moist loam, either in full sun or in partial shade. When grown in pots it requires a strong loamy soil, with plenty of manure, and throughout the summer the pots should be allowed to stand in pans of water. As the Agapanthus is a gross-feeding plant, it should be re-potted annually in autumn, and be wintered in a cool pit or frame. In transferring to new pots a little care must be taken to avoid injuring the mass of fleshy roots.

ALLIUM

The *Allium neapolitanum* is the finest white-flowered variety, and is exceedingly valuable for bouquets and vase decoration. The large umbels of blossoms are of the purest white. It is one of the earliest spring-flowering bulbs, and, although quite hardy, it comes forward quickly and easily in a cool house.

ALSTROEMERIA

An elegant plant which belongs to the nearly hardy group referred to in the notice of Ixia. In autumn it may be safely planted out in almost any part of the United Kingdom, provided it is planted nine inches deep, and can have a sunny position on a dry soil, for damp is more hurtful to it than frost. As a pot plant it is comparatively useless, but if allowed to remain several years in a dry border, a large clump of any of the varieties presents a brilliant appearance when in flower.

AMARYLLIS

See remarks under Lilies at page 340.

ANEMONE

Windflower

Our observations on this flower will be limited to the tuberous varieties; but even with this restriction, the range of form and colour is exceedingly wide. The Anemone is an accommodating plant, and can be successfully flowered either in pots or in beds, at the option of the cultivator.

The most natural place for it is near shady woodland walks, where it can be seen to the greatest advantage. But it is also a splendid subject for masses in the mixed border, or in front of shrubberies; and alone in beds it makes a brilliant and lasting show. For all the purposes of garden decoration to which the Crocus, Hyacinth, and Tulip are applied, the Windflower is equally well adapted. We do not advise planting singly, but the Anemone answers admirably in lines, groups, or beds, and the colours admit of numberless harmonies and contrasts.

The commoner Anemones need only to be planted about three inches deep, with the eyes upwards, at any time between September and March, and they will require little or no attention afterwards. Under trees, instead of planting in a formal pattern, it is worth while to put them in with some attempt at natural grouping, and not too close together—say from six inches to a foot apart. In such positions they may be left undisturbed for years; and if the soil happens to be a good sandy loam, they will thrive and increase. In masses or beds within the garden, however, a richer effect is wanted, and the distance between the roots should not exceed from four to six inches.

A choice collection of roots is worth more care, and florists are accustomed to prepare the beds for their reception with fastidious exactness. The soil, if not considered suitable, is taken out to the depth of two feet, and is replaced by a rich and specially prepared compost. Although the individual flowers produced by this method are generally very fine, and the total effect of the bed is exceedingly beautiful, yet the truth must be confessed that for ordinary gardening the system is extravagant and unnecessary. As a hobby, it is, of course, justifiable enough; but Anemones of high quality can be grown by a much simpler mode of procedure. One deep digging there certainly should be, and a layer of manure at the bottom of each trench is sound treatment, for it supplies the roots with food and a cool subsoil. Poor land should also be enriched by incorporating a dressing of decayed manure as the work proceeds. Subsequently one or two light surface forkings will help to make the bed mellow. A rough plan, showing the name and position of every root, will be a safer record than labelling in the usual way, and it also prevents the disfigurement of the bed. There should be a distance of six inches between the roots; and they may be put in singly by means of the trowel, or in drills drawn three inches deep. The former method is generally adopted for groups; but to insure regularity in flowering the planting must be uniform in depth. For beds, drills are more reliable, and they are speedily made.

The time of planting determines to a considerable extent the date of flowering; and, as the roots may be put in during autumn, winter, and early spring, it is easy to secure a succession of Anemones from January until May. But this flower is of so much more value early in the year than at a later period, when many other subjects brighten the garden, that it is scarcely worth while to plant so late as March.

The Anemone is well worth growing in pots, both for its foliage and flowers. It does not resent forcing to the same extent as the Ranunculus; nevertheless, cool treatment is almost essential to do it full justice. The potting should be done in batches to insure a succession of flowers, and the first lot may be put in at the end of August, or beginning of September. They should have the benefit of really good soil; a mixture of leaf-mould and loam, with the addition of a little powdered charcoal, will suit them exactly. In preparing the pots, place a layer of light manure above the crocks, which will assist the drainage and benefit the plants. Then fill with compost to within two inches of the top, and lay in the roots; add soil to a level with the rim, and press lightly down. The strongest roots should, of course, be selected for potting, and it will need more than a hasty glance to put them in with the eyes upwards. One or more roots may be planted in each pot, according to the size of the latter.

The early plantings can be placed in any warm position out of doors, such as under a south wall; but after the middle of October remove to a cold pit, or on to the greenhouse stage. Watering is all the attention they will require, and of this there must be no stint, especially during the blooming period. A high temperature at any stage is needless, and if they are just kept out of the reach of frost they will take excellent care of themselves.

Anemones are adapted for many decorative purposes; they make capital window plants, and their sharply cut foliage is very ornamental in the drawing-room or on the dinner-table.

BABIANA

Babianas are delicately constituted, but extremely elegant plants when well grown. Though far from showy, they appeal to the educated eye for appreciation of their blue and purple oculate flowers. The culture is the same as for the Ixia, and we incline strongly to the practice of keeping the bulbs at least two seasons in the same pots.

BEGONIA, TUBEROUS-ROOTED

Few flowers have a greater claim on the attention of horticulturists than the Tuberous-rooted Begonia, either for the ease with which it can be grown, or for the many valuable purposes to which the plant may be applied. It can be flowered at any time from February until October, and is available for all kinds of indoor decoration, and also for growing in the open ground during the summer months.

Instead of allowing the plants to be rudely dried off, it is worth a little trouble to reduce them slowly to the dormant state by gradually withholding water. They should still be retained in pots, which may be stored under a thick layer of ashes or dry peat in any cellar, frame, or shed where the thermometer stands pretty uniformly at about 50°. The store should also be dry, for damp is quite as injurious to these roots as cold. Generally speaking, it may be said that any store which is safe for Dahlias will also preserve Tuberous-rooted Begonias.

After the winter's rest the bulbs are invariably saucer-shaped, and in the event of their being watered before growth has commenced, sufficient water will remain in the hollow to destroy the bulb. This peculiarity makes it dangerous to start the plant before activity is evident. In January or February, as the bulbs show signs of life, pot them almost on the surface of a rich loamy soil, and employ the smallest pots possible. Nurse them with a little care in a warm place for about ten days, and they should then be very gradually hardened. A regular system of potting on will be necessary until the final size is reached; and at each operation the plants should be inserted rather deeper than before. If re-potting is deferred too long, the foliage will turn yellow—a sure sign that the plant is starving. No flowers should be allowed in the early stages of growth, and this rule is imperative if fine specimens are wanted; but when the plants are transferred just as the pots are full of roots, there will be little disposition to bloom prematurely. While growing, the Tuberous Begonia delights in a humid atmosphere, but this should be avoided after flowering has commenced. When sticks are inserted for tying out the flowers, the bulbs must not be wounded.

The erect-growing varieties are valuable for low conservatory stages, and they form splendid groups in corners of drawing-rooms. The drooping kinds are seen to advantage on brackets, shelves, and in suspended baskets; and the short-jointed plants of the drooping class are specially adapted for rockeries and beds. They must not be put into the open until the danger of a nipping east wind is past. The early part of June is generally about the right time.

In the autumn it is usual to lift and pot the plants, although in mild districts, and in a light soil, they may safely be left out all the winter under the shelter of a heap of ashes or decayed manure. In beds this plan is scarcely worth adoption, because it leaves the ground bare for several months; but where Begonias are grown in the reserve border to furnish a supply of flowers for cutting, it may be a considerable advantage to leave them until the following year.

A word is necessary as to soil. The Begonia is a gross feeder, and to develop its fine qualities there must be a liberal employment of manure. As a matter of fact, it is scarcely possible to make the soil too rich for this flower.

CHIONODOXA

Glory of the Snow

The varied blue tints of the Chionodoxa, its more open blossoms, and larger size, distinguish this flower from its older and justly prized rival, the Scilla. Indeed, the Chionodoxa is exquisitely beautiful, and of great value for pot culture, beds, or borders. Five bulbs may be grown in a 48-sized pot, and in the border not less than half a dozen should be planted in a group. Employed as a single or double line, it also produces a striking bit of colouring. The bulbs should be planted in autumn four inches deep, the distance between being not more than three inches. Any ordinary garden soil will grow this flower, and it is advisable to allow the bulbs to remain undisturbed for several years, as the effect will be the greater in each succeeding spring.

CROCUS

This brilliant harbinger of spring will thrive in any soil or situation, but to be brought to the highest possible perfection it should be grown in an open bed or border of deep, rich, dry sandy loam. The bulbs should be planted during September, October, and November. If kept out of the ground after the end of the year they will be seriously damaged, and however carefully planted, will not flower in a satisfactory manner. Plant three inches deep in lines, clumps, or masses, as taste may suggest, putting the bulbs two inches apart. If convenient, let them remain undisturbed two or three years, and then take them up and plant again in well-prepared and liberally manured soil. A bed of mixed Crocuses has a pleasing appearance, but in selecting bulbs for the geometric garden it is more effective to employ distinct colours, reserving the yellow for the exterior parts of the design to define its boundaries, and using the blue and the white in masses and bands within. In districts where sparrows attack the flowers, they may be deterred from doing mischief by stretching over the beds a few strands of black thread, which will not interfere with the beauty of the display, and will terrify the sparrows for a sufficient period to save the flowers.

The named varieties are invaluable for pot and frame culture, and to force for decorative purposes; for though the individual flowers are short-lived, the finest bulbs yield a long succession of bloom, and in character Crocuses are quite distinct from all other flowers of the same early season. When grown in pots and baskets, the bulbs should be placed close together to produce a striking effect. A light, rich soil is desirable, but they may be flowered in a mixture of charcoal and moss, or in fibre, or moss alone. When required in quantity for ornamental baskets and similar receptacles, it is wise to plant them in shallow boxes filled with rotten manure and leaf-mould, and to lift them out separately, and pack them when in flower in the ornamental baskets. A perfect display of flowers in precisely the same stage of development can thus be secured, and successional displays may follow as long as supplies remain in the boxes.

CROWN IMPERIAL

Fritillaria imperialis

A noble plant which needs a deep, rich, moist soil, and an open situation, to insure the full degree of stateliness, but it will make a very good figure in any border where it can enjoy a glimmer of sunshine. There are several distinct varieties, the flowers of which range in colour from palest yellow to the deepest shade of orange and reddish buff, and there are others which have variegated leaves. They should be planted in autumn eighteen inches apart, allowing from four to six inches of soil above the crowns.

CYCLAMEN

Although it is advisable to raise Cyclamens from seed every year, occasions arise when it is necessary to store the bulbs for a second season, and the best method of treating them during the period of rest must be considered. As the production of seed weakens the corms, preference should be given to those which have not been subjected to this tax on their energies.

At the close of the flowering season the bulbs should be gradually reduced to a resting state by withholding moisture. When the foliage turns yellow the pots may be laid on their sides in a cold frame, if available, or in any other convenient place where they will not be forgotten, until about the middle of July. They should then be placed upright, and have a supply of water. After fresh growth has fairly commenced, shake the bulbs out of the pots, remove most of the old soil, and re-pot in a compost consisting of mellow turfy loam and leaf-mould, with a sufficient admixture of silver sand to insure drainage. The corm should be so placed in the pot as to bring the crown about level with the rim, and every care must be taken to avoid injuring the young roots. Place the pots in a close frame for a few days, after which ample ventilation should be given to maintain a robust condition. The lights may remain constantly open until there is danger from autumn frosts. Specimens that show a great number of flower-buds should be assisted occasionally with weak manure water.

C. Coum and *C. europæum* are rarely well grown, for although quite hardy, the climate of this country does not suit them in their season of flowering, which is the early spring. The cool greenhouse is the safest place for them, except in sheltered spots, where they may be planted out on a border of peat, or amongst ferns in a rockery. When grown in pots, light turfy loam and peat in equal quantities, with a fourth part of cow-manure and a liberal addition of sand, will form an excellent compost for them. The pots should never be exposed to the drying action of the sun or wind, but should be plunged to the rim in coal-ashes. The best time for potting or planting them is September or October.

Instructions on raising Cyclamens from seed will be found at page 256.

DAFFODIL—*see* NARCISSUS, *page* 344

DOG'S-TOOTH VIOLET

The red and white varieties are as hardy as any plant in our gardens, and by their neat habit and elegant leaves and flowers they are admirably adapted to plant in quantities in the front of a rockery, in either peat or sandy loam and leaf-mould. They are equally suitable for edging small beds in gardens where spring flowers are systematically grown; in fact, they are true 'spring bedders.' Autumn is the proper time to plant the bulbs. But Dog's-tooth Violets are also worth growing in pots, especially where an unheated 'Alpine house' is kept for plants of this class. Several bulbs may be put in a pot of the 48-size.

FERRARIA—*see* TIGRIDIA, *page 350*

FREESIA

The singularly graceful form, wide range of beautiful colours, and delicious perfume of this flower have made it an immense favourite; and happily there is no Cape bulb which can be grown with greater ease in the frame or cool greenhouse. One characteristic is very marked, and it is the disproportion between the small bulb and the fine flowers produced from it.

Procure the bulbs as early in the autumn as possible, and lose no time in potting them. Any light rich soil will answer, but that which suits them best is composed of two parts of loam, one of leaf-mould, and one of peat, with enough sand or grit added to insure drainage. Commence with pots of the right size, for the roots are extremely brittle, and there must be no risk of injuring them by re-potting. The 48-size will accommodate several bulbs. Place under a south wall, and cover with leaf-mould until top growth commences, and then remove the covering.

At the end of September transfer the pots to a cold frame, and when the plants attain a height of four inches, support them with neat sticks, which should not be inserted too near the bulbs. Watering will require judgment, for too much moisture turns the foliage yellow. When the pots are full of roots, liquid manure twice a week will be helpful.

After the blooming season has passed, encourage the foliage to wither by withholding water. The roots may be stored away in their own pots until the following August.

FRITILLARIA

Fritillarias produce bell-shaped flowers, varying in colour, but generally of a purplish tint, and beautifully spotted. They thrive in a good deep loam, but may be grown in almost any soil, and do well under the shade of trees. They are quite hardy, and, like most other bulbs, should be planted in autumn. Fritillarias are occasionally grown in pots kept in a cold frame, but they will not endure forcing in the least degree, and the mixed border is the best position for them. These flowers make a charming ornament when grown in bowls filled with moss-fibre.

GLADIOLUS

The Gladiolus is adapted for many important uses and it associates admirably with Dahlias, Hollyhocks, Pyrethrums, and Phloxes in the furnishing of clumps on the lawn and in the mixed border. It is perfectly in harmony with surroundings when planted in American beds or in the shrubbery. For supplying cut flowers it is invaluable, as they retain their freshness in a vase for many days, and a plentiful supply should be grown in reserved spots expressly for this purpose.

Culture in Pots.—The early-flowering varieties are of especial value for decorating greenhouses and conservatories during spring and early summer. The corms of these Gladioli are small, and a 32-sized pot will accommodate several. The soil should be decidedly rich, and it must be porous, because water has to be given freely when the plants are in full growth. Pot the corms in autumn, and cover with leaf-mould until the roots are developed, when successive batches can be brought forward and gently forced for a continuous supply of elegant flowers during April and May. A mild temperature of about 55° is quite sufficient for them.

Culture in the Open Ground.—The autumn-flowering Gladioli are grown in the open ground, and preparations should begin well in advance of planting time. Almost any soil can be made to answer, but that which suits them best is a good medium, friable loam with a cool rich subsoil, and each grower must decide for himself how far this is within reach naturally, or can be secured by resources at command. Thus, a light soil may be made suitable by placing a thick layer of rotten cow-manure a foot below the surface, and a heavy, retentive loam can be reduced to the proper state by the admixture of lighter material. On the surface spread a liberal quantity of manure and dig it in, leaving the soil in a rough state to be disintegrated by frosts. Before the planting time arrives it is worth some trouble to free the ground from wire worms, or they will play havoc with the growth just as it is appearing above ground. Potatoes serve admirably as traps for these pests.

Gladioli are peculiarly liable to injury from wind, so that a sheltered, but not a shaded, position should, if possible, be chosen for them. The time of planting depends partly on the district, partly on the season; but the soil must be in suitable condition and fine weather is necessary. From the middle of March to the middle of April should afford some suitable opportunity of getting the bulbs in satisfactorily. Give the land a light forking, not deep enough to bring up the manure, and make the surface level. The rows may be twelve or eighteen inches apart; we prefer the greater distance, because of the convenience it affords in attending to the plants when growing; nine inches is sufficient space in the rows.

There are two methods of putting in the bulbs, each of which has advocates among practised growers. One is to take out the soil with a trowel to the depth of six or seven inches for each corm, then insert about two inches of mixed sand and powdered charcoal or wood ashes; lay the root upon it, and carefully cover with fine soil. If that process is considered too tedious, draw a deep drill with a hoe, and at the bottom put the light mixture already named; place the roots at regular distances upon it, and lightly return the top soil. The operation should be so performed as to leave the crown of the corm four inches below the surface. When planting is completed, give the bed a finishing touch with the rake.

An eminent grower strips off the outer coat or skin of each bulb before planting to ascertain that there is no disease; and this cannot otherwise be discovered. No doubt the procedure prevents the bed from showing blanks, but that object can be more safely attained by growing a reserve in pots. There is, however, another practice which possesses very decided advantages, and it is to break the skin at the crown of the bulb to allow the foliage free exit. The skin is so tough that it is frequently the means of distorting the plant in its attempt to force an opening.

The bed for a time needs little attention, except to keep it free from weeds, and this is best done by hand. When the shoots reach about a foot high, tying must be resorted to in earnest. The most effectual plan, of course, is to put a separate stake to each plant, and for exhibition specimens this is certainly advisable. But rows can be secured by a stake at each end, with two or three strands of strong material carried across, to which each flower must be tied. Whatever method is adopted, care should be taken to avoid cutting the plant, while holding it secure from damage in a high wind. Let the material which is placed round the flowering-stem be soft and wide, such as list, which answers admirably.

Water must be freely and regularly given during dry weather, either in the morning or in the evening; and a mulch of old manure spread over the bed will prevent evaporation, and save the ground from caking hard.

Another important matter is shading. For ordinary purposes this is not essential; but as it very much lengthens the duration of the flower, it is worth attention on that ground alone, and for exhibition it is indispensable. Whether shading is provided by separate protectors made expressly for the purpose, or by home-made contrivances of canvas or wood, the point to be quite certain about is security, or an accident may wreck well-grounded hopes.

The lifting and storing of the corms affect the quality of the next year's flowers so much that it is important to accomplish lifting at the most suitable time, and the storing in the best manner. By the middle or end of October, on some fine day, take up the roots, even if the foliage be still green; tie a label to each variety, and hang them in some airy place until they can be cleared of soil and leaves. Remove each stem with a sharp knife, and lay out the bulbs to dry for another fortnight. They can then be stored in paper bags or in boxes on any dry shelf which is safe from vermin and frost.

An article on the culture of the Gladiolus from seed will be found on page 267.

GLOXINIA

Gloxinias may be had in bloom almost all the year by judicious management. When required for early flowering, those that start first should be selected and carefully shifted into other pots, and be kept near the glass, as they depend much on light for rapid and luxuriant growth. A moist atmosphere, with the temperature about 60° to 65°, greatly facilitates the growth of Gloxinias, but they may be grown well in greenhouses or in pits heated by hot water. The most suitable soil is a light fibrous loam, combined with a little peat and silver sand. Manure water during the growing period twice a week is helpful, but it should be discontinued when the flowers show colour. The plants love shade, and at no time should suffer from drought. Storing Gloxinias for their season of rest, *i.e.* the winter, must be carefully attended to, as losses frequently occur during this stage. It is also important that the plants should not be 'dried off' too quickly; place them in a light, airy position, and by a gradual reduction of moisture the leaves will fall off naturally. The bulbs may then be stored away on a shelf, in an even temperature of about 50°, each bulb being closely surrounded by cocoa-nut fibre and peat in equal parts to prevent excessive dryness, which, like too much damp, often causes the loss of the bulb.

Besides growing the same plants from year to year, it is always desirable to have a fresh stock coming on, as the old bulbs may deteriorate after two or three years. This can easily be managed by successive sowings of seed, as advised at page 268.

HEMEROCALLIS—*see under* LILIES, *page 343*

HYACINTH

One of the most valuable characteristics of the Hyacinth is the ease with which it can be flowered in a variety of ways by very simple modes of treatment. It may be employed as a hardy, rough-weather plant for the garden border, or as a grand exhibition and conservatory flower. The bulbs may be planted at any time from September to the middle of December, with the certainty of their blooming well, if properly cared for; but the prudent cultivator will plant them as early as possible in the autumn, and so manage them afterwards as to secure the longest period of growth previous to their flowering. They can be forced to flower at Christmas, but the more slowly the flowers are developed the finer in the end will they be. To obtain good bulbs is a matter of the utmost importance, and it may be useful here to remark that the mere size of a Hyacinth bulb is no criterion of its value—nor, indeed, is its neatness of form or brightness of appearance. The two most important qualities are soundness and density. If the bulbs are hard and heavy in proportion to their size, they may be depended on to produce good flowers of their kind. The bulbs of some sorts are never large or handsome, while, on the other hand, many others partake of both these qualities in a marked degree.

One other matter in general relating to the treatment of Hyacinths needs to be referred to. Harm has often been done by the practice of massing the flowers, whether in pot groups or in garden beds, without consideration of colour harmonies. Yet no other bulbous flower offers such a wide choice of delightful colours, or is so eminently adapted to artistic blending, as the Hyacinth. By eschewing the dull blues and allied shades and by bringing into association exquisite tones of mauve, pink, apricot, salmon, pale yellow, rich lilac, bright red, &c., it is easy to demonstrate that there are possibilities in Hyacinths which may never have been suspected before. The following are a few of the charming blends which may be made, and will especially appeal to those who grow Hyacinths indoors: (i) Apricot, cream, and pale blue; (2) cream, pale pink, and rose-pink; (3) bright pink and pale blue; (4) bright red, rich blue, and pure white; (5) rose-pink and rich blue; (6) pale yellow and rich blue; (7) deep mauve and pale mauve; (8) cream and pale blue; (9) bright blue shades (dull, washy, and nondescript blue, purple, and violet tints must be avoided); (10) blush pink and rose-pink; (11) apricot and cream; (12) pale lavender, cream, and apricot. These examples will show that charming effects can be secured either with two or with three varieties. Colour-grouping may also be carried out in the garden, but in this case great care must be exercised to get varieties of clear, bright hues which flower at the same time, such as Inimitable Bedding Hyacinths. Modern taste further dictates that the bare soil shall be hidden, and this end is best served by providing a groundwork of dwarf plants, such as Daisies, Forget-me-nots, double white Arabis, and mauve Aubrietia. Another course is to mix Hyacinths with Daffodils of the Chalice or Star section; there is no better variety than Sir Watkin, but others may be used.

Culture in Pots.—It is not necessary to use large pots, or pots of a peculiar shape, for Hyacinths. There is nothing better than common flower-pots, and in those of 60-size single bulbs may be flowered in a most satisfactory manner. The pots usually employed are the 48-and 32-sizes, the last-named being required only for selected bulbs grown for exhibition. We advise the use of small pots where Hyacinths are grown in pits and frames for decorative purposes, because they can be conveniently placed in ornamental stands, or packed close together in baskets of moss, when required for the embellishment of the drawing-room. As the use of new pots for Hyacinths is often the cause of failure, they should not be employed if well-cleansed old pots are available. The tender roots of the bulbs frequently become too dry owing to the absorbent nature of the new pots. A rich, light soil is indispensable, and it should consist chiefly of turfy loam, with leaf-mould and a liberal allowance of sharp sand. The mixture ought to be in a moderately moist condition when ready for use. In small pots one hollow crock must suffice, but the 48-and 32-sized pots can be prepared in the usual way, with one large hollow crock, and a little heap of smaller potsherds or nodules of charcoal over it. Fill the pots quite full of soil, and then press the bulb into it, and press the soil round the bulb to finish the operation. If potted loosely, they will not thrive; if potted too firmly, they will rise up as soon as the roots begin to grow, and be one-sided. In large pots the bulbs should be nearly covered with soil, but in small pots they must be only half covered, in order to afford them the largest possible amount of root-room. When potted, a cool place must be found for them, and unless they go absolutely dry, they should not have a drop of water until they begin to grow freely and are in the enjoyment of full daylight. The pots may be stored in a dark, cool pit, or any out-of-the-way place where neither sun, nor frost, nor heavy rains will affect them; but it is advisable to plunge them in coal-ashes and also to cover them with a few inches of the ashes. As to their removal, they must be taken out as wanted for forcing, and certainly before they push up their flower spikes, as they will do if they remain too long in the bed. The cultivator will be guided in respect of their removal from the bed by circumstances; but when they are removed, a distinct routine of treatment must be observed, or the flowering will be unsatisfactory. For a short time they should be placed in subdued daylight, that the blanched growth may acquire a healthy green hue slowly; and they need to be kept cool in order that they shall grow very little until a healthy colour is acquired. The floor of a cool greenhouse is a good place for them when first taken out of the bed and cleaned up for forcing. Another matter of great importance is to place them near the glass immediately their green colour is established, and to grow them as slowly as the requirements of the case will permit. If to be forced early, allow plenty of time to train them to bear a great heat, taking from bed to pit, and from pit to cool house, and deferring to the latest possible moment placing them in the heat in which they are to flower. Those to bloom at Christmas should be potted in September, those to follow may be potted a month later. If a long succession is required, a sufficient number should be potted every two or three weeks to the end of the year. Those potted latest will, of course, flower in frames without the aid of heat. In any and every case the highest temperature of the forcing-pit should be 70°; to go beyond that point will cause an attenuated growth and poverty of colour. If liquid manure is employed at all, it should be used constantly and extremely weak until the flowers begin to expand, and then pure soft water only should be used. No matter what may be the particular constitution of the liquid manure, it must be weak, or it will do more harm than good. The spikes should be supported by wires or neat sticks in ample time, and a constant watch kept to see that the stems are not cut or bent, as they rapidly develop beyond the range allowed them by their supports.

Culture in Glasses.—It is of little consequence whether rain, river, or spring water be employed in this mode of culture, but it must be pure, and in the glasses it should nearly but not quite touch the bulbs. Store at once in a dark, cool place, to encourage the bulbs to send their roots down into the water before the leaves begin to grow. When the roots are developed, bring the glasses from the dark to the light, in order that leaves and flowers may be in perfect health. Let them have as much light as possible, with an equable temperature, and provide supports in good time. Hyacinths are often injured by being kept in rooms that are at times extremely cold and at others heated to excess. Those who wish to grow the bulbs to perfection in glasses should remove them occasionally as circumstances may require, to prevent the injury that must otherwise result from rapid and extreme alternations of temperature. It is not desirable to introduce to the water any stimulating substance, but the glasses must be kept nearly full of water by replenishing as it disappears. If the leaves become dusty, they may be cleansed with a soft brush or a sponge dipped in water, but particular care must be taken not to injure them in the process.

Culture in Moss-fibre.—While Hyacinths, differing from Daffodils and Tulips, are perhaps relatively better in pots of soil than in bowls of moss-fibre, they may still be grown successfully in bowls provided a fairly deep receptacle is chosen and care is taken to avoid making the fibre hard. With a shallow bowl and very firm fibre it may be found that the roots strike upward and the plant does not get that abundant supply of moisture which is essential to its welfare. For this method of culture preference should be given to the Roman, Giant Italian, Christmas Pink,

Miniature and Grape Hyacinths, which look particularly charming in bowls and similar contrivances. Detailed directions are given on page 319.

Culture in Beds.—The Hyacinth will grow well in any ordinary garden soil, but that which suits it best is a light rich loam. The bed should be effectually drained, for though the plant loves moisture it cannot thrive in a bog during the winter. It is advisable to plant early, and to plant deep. If a rich effect is required, especially in beds near the windows of a residence, the bulbs should be six inches apart, but at a greater distance a good effect may be produced by planting nine inches apart. The time of blooming may be to some extent influenced by the time and manner of planting, but no strict rules can be given to suit particular instances. Late planting and deep planting both tend to defer the time of blooming, although there will not be a great difference in any case, and as a rule the late bloom is to be preferred, because less liable to injury from frost. The shallowest planting should insure a depth of three inches of earth above the crown of the bulb, but they will flower better, and only a few days later, if covered with full six inches of earth over the crowns. The Hyacinth is so hardy that protection need not be thought of, except in peculiar cases of unusual exposure, or on the occurrence of an excessively low temperature when they are growing freely. Under any circumstances, there is no protection so effectual as dry litter, but a thin coat of half-rotten manure spread over the bed is to be preferred in the event of danger being apprehended at any time before the growth has fairly pushed through.

The bulbs may be taken up as soon as the leaves acquire a yellow colour, so that the brilliant display of spring may be immediately followed by another, equally brilliant perhaps, but in character altogether different. When grown in beds, Hyacinths do not require water or sticks; all they need is to be planted properly, and they will take care of themselves.

Miniature Hyacinths.—These charming little sparkling gems are invaluable for baskets, bowls and other contrivances which are adapted for the choicest decorative purposes. In quality they are excellent, the spikes being symmetrical, the flowers well formed, and the colours brilliant. But they are true miniatures, growing about half the size of the other kinds, and requiring less soil to root in. They will flower well if planted in a mixture of moss-fibre and charcoal, kept constantly moist, and covered with the greenest moss, to give to the ornament containing them a finished appearance.

Feather and Grape Hyacinths will grow in any good garden soil, and are admirably adapted for borders that are shaded by trees. They should be planted in large clumps, and be allowed to remain several years undisturbed. Both classes are beautiful—the Feather Hyacinth emphatically so; indeed, numerous as beautiful flowers are, this, for delicacy of structure, has peculiar claims to our admiration, when presenting its feathery plumes a foot or more in length, all cut into curling threads of the most elegant tenuity. Grape Hyacinths make a charming ornament for the drawing-room when grown in bowls of moss-fibre.

Roman Hyacinth.—This flower is particularly welcome in the short, dark days of November, December, and January. For placing in glasses to decorate the drawing-room or dinner-table the spikes of bloom are largely grown; and the separate flowers, mounted on wire, form an important feature in winter bouquets, for which purpose their delicious perfume renders them especially valuable.

The bulbs can be grown with the utmost ease. Pot them immediately they can be obtained in August or September, and stand them in some spare corner in the open ground, where they can be covered with a few inches of leaf-mould. This will encourage the roots to start before there is any top growth. In October remove the covering, and transfer the pots to a pit or frame, or they may be placed under the greenhouse stage for a time, provided they will not be in the way of dripping water. A little later, room should be found for them upon the stage, or the foliage may become drawn. When the buds are visible, plunge the pots in a bottom heat of 65° or 70°, and in a week the flowers will be fit for use. Like its more imposing prototype, the Roman Hyacinth may have its roots gently freed from soil for packing in bowls or vases filled with wet moss or sand; but they ought not to be subjected to a violent change of temperature. If wanted in glasses, they can be grown in water after the usual fashion, but the flower is scarcely adapted for this mode of treatment. They will, however, grow well in bowls filled with moss-fibre.

Italian Hyacinth.—Although rather later in flowering than the Roman variety, the Italian Hyacinth deserves to be grown as a pot plant, especially for its more lasting quality. The graceful flowers are carried on long stout stems which are most effective for the decoration of vases. The bulbs are perfectly hardy, and may be planted in clumps in the open border, where they will bloom in April and afford abundant sprays for cutting. The habit is less formal than that of the Dutch Hyacinth and the flowers exhale a sweet delicate perfume. As previously stated, the Italian Hyacinth is especially suitable for growing in moss-fibre.

HYACINTHUS CANDICANS

An excellent companion to Delphiniums, Salvias, and perennial Lobelias in the mixed border. The stately spikes of this flower also associate well with shrubs, and help to enliven a bed of Rhododendrons at a period of the year when the latter is uninteresting. Roots may be planted in any soil from November to March; and, as they are perfectly hardy, they can be left in the open ground all the year without the least misgiving as to their safety. A strong root will produce a succession of flower-spikes, and this tendency will be assisted by cutting off each spike immediately it has ceased to be attractive.

IRIS

The common varieties of Iris are well-known favourites of the border, and the whole family have claims on the attention of amateurs, on account of their excellent faculty of taking care of themselves if properly planted in the first instance. The tuberous or bulbous rooted kinds do not require a rich soil; a sandy loam suits them, and they thrive in peat. Such beautiful species as Reticulata, the Chalcedonian, and the Peacock are worth growing in pots placed in frames or in a cool greenhouse. The English, Dutch, and Spanish varieties should be planted in clumps in front of a shrubbery border, where they may be seen to advantage. The crown of the bulb must not be more than three inches below the surface. From September to December will answer for planting, and the roots may be taken up when the flowering period is over, or if the space is not wanted they can be allowed to remain for the following season. Bulbs of the English class should never be kept out of the ground longer than can be helped, but they ought not to be grown in one spot for more than three years; after that time the clumps must be divided and a fresh position found for them.

IXIA and SPARAXIS

These attractive Cape bulbs are hardy in favoured districts, and may be left out for years in a sheltered border. In places where none but the hardiest plants pass through the winter safely, they must be grown in the greenhouse or the frame, and any good sandy soil will suit them, whether peat or loam. They should be potted early in the autumn, and have plenty of air at all times when the weather is favourable, especially when they are growing freely in spring. If carefully managed, they may remain two seasons in the same pots. Use the 48-size, and plant four or five bulbs in each. A dry, deep, sandy border under a wall in any of the warmer western and southern districts might be furnished with such plants as Ixias, Sparaxis, Alstroemerias, Oxalis, Tritonias, Babianas, and the choicest of the smaller kinds of Iris. It would constitute a garden of the most interesting exotics.

JONQUIL

For its delicious fragrance and exquisite beauty the Jonquil has long been considered one of the most valuable of the Narciss family for cultivation in pots, and it is also a first-rate border and woodland flower. When forced, the treatment should agree as nearly as possible with that prescribed for the Narcissus. Four or five bulbs may be planted in one pot.

LACHENALIA

An elegant plant which is not quite hardy enough to be trusted in the open ground; but it is the easiest matter possible to grow it well in the greenhouse. The bulbs should be potted as soon as they begin to grow in the autumn, and several bulbs may be put into each pot. There can be no better soil than turfy loam, without manure or sand. It is of the utmost importance that the plants should have abundance of water, when they will produce leaves two inches across, and spikes of flowers fully double the size of those commonly met with. An admirable use for these bulbs is to insert them all over the outside of hanging-baskets, which they will cover with the most graceful display of aërial vegetation imaginable, the flower-spikes turning upwards, and the leaves hanging down.

LEUCOJUM

The Spring Snowflake *(L. vernum)* blooms as early as February or March, and the Summer Snowflake *(L. æstivum)* comes into flower in May and June. They closely resemble the Snowdrop, but are much larger than that well-known spring favourite. The bulbs are perfectly hardy, and will grow in any garden soil. Plant in clumps three inches deep, any time from the end of September until the middle of November.

LILIES

Hardy border Lilies are among the most useful garden plants known. They are peculiarly hardy and robust, requiring no support from sticks or ties; several of them remain green all the winter, and are capable of resisting any amount of frost. If left alone, they increase rapidly, and become more valuable every year. We will say nothing of their beauty, for that is proverbial; but it may be useful to observe that many of the most lovely Lilies, usually regarded as only suitable for the greenhouse, and grown with great care under glass, are really as hardy as the old common white Lily, and may be grown with it in the same border. To grow Lilies well requires a deep, moist, rich loam. A stubborn clay may be improved for them by deep digging, and incorporating with the staple plenty of rotten manure and leaf-mould. They all thrive in peat, or rotten turf, or, indeed, in any soil containing an abundance of decomposing vegetable matter. The autumn is the proper time to plant Lilies, but they may be planted at any season, if they can be obtained in a dormant state or growing in pots. They should be planted deep for their size, say, never less than six inches. After they have stood some years it is necessary to lift and part the clumps, when the borders should be deeply dug and liberally manured before replanting. If the stems of Lilies become leafless and unsightly before the flowers are past, it is a sign that the roots are too dry, or that the soil is impoverished; and therefore, as soon as the stems die down, they should be lifted, and perhaps transferred to a more favourable spot.

Amaryllis.—These magnificent plants do not require the high temperature in which they are usually grown, nor should they be allowed to remain for a great length of time dust-dry, as we sometimes find them. It is important to remember that they have distinct seasons of activity and rest, but must not be forced into either condition by such drastic measures as are occasionally resorted to. The proper soil for them is turfy loam, enriched with rotten manure, and rendered moderately porous by an admixture of sand. The light soil in which many plants thrive will not suit them; the soil must be firm, and somewhat rough in texture. When first potted, give them very little water, and promote growth by means of a bottom heat of 65°. Increase the supply of water as the plants progress, and shift them into 6-inch pots for flowering. While they are in flower they may be placed in the conservatory, or wherever else they may be required for decorative purposes. When the flowers have faded take them to the greenhouse to complete their growth, after which dry them off slowly, but with the clear understanding that they are never to be desiccated. They may be wintered in the greenhouse, and should certainly be placed where they will always be slightly moist, even if a few leaves remain green throughout the winter. Frequent disturbance of the roots is to be particularly avoided in the cultivation of Amaryllis, and therefore it is desirable to allow them to remain in the same pots two or three years; or if they are shifted on, it should be done in such a way that the roots are scarcely seen in the process. Top dressing and liquid manure will help them when they have been some time in the same pots.

Lilium auratum.—This magnificent Lily has proved to be as hardy as the white garden variety, and is now freely planted in borders and shrubberies where the noble heads of bloom always command admiration. But the splendour of the flower will continue to insure for it a high degree of favour as a decorative subject for the conservatory. When grown in a pot the best soil is sandy peat, but it will flower finely in a rich light mixture, such as Fuchsias require. It is advisable to begin with the smallest pot in which the bulb can be placed, and then to shift to larger and larger sizes as the plant progresses, taking care to have the bulb two inches below the soil when in their flowering pots, because roots are thrown out from the stem just above the bulb, and these roots need to be carefully fed, as they are the main support of the flowers that appear later. When the flower-buds are visible, there should, of course, be no further shifting. In respect of temperature, this is an accommodating Lily; but as a rule a cool house is better for the plant than one which is maintained at a high temperature. The supply of water should be plentiful during the period of growth and flowering, but afterwards it can be reduced.

Lilium Harrisii *(The Bermuda, or Easter Lily)* is of the *longiflorum* type, but the flowers are larger, and are produced with greater freedom than by the ordinary *L. longiflorum*. Moreover, the Bermuda Lily flowers almost continuously. Before one stem has finished blooming another shoots up. This perennial habit gives it a peculiar value for the greenhouse, and renders forcing possible at almost any season.

Immediately the bulbs are received they should be potted in rich fibrous loam—the more fibrous the better—and be placed in a cold frame. They need little water until growth has fairly commenced, after which more moisture will be necessary. So far as safety is concerned, they only require protection from frost; but for an early show of bloom artificial heat is imperative. The temperature should, however, be very moderate at first, and rise slowly. When the buds show, a top-dressing of fresh loam and decayed manure will be helpful, and to allow for this the soil must be two inches from the tops of the pots when the bulbs are first potted. After producing two or three flowering stems, it will be wise to place the pots out of doors and give less water, or the bulbs will be exhausted. But they must never be allowed to become quite dry, and after a partial rest of six weeks or two months they may be re-potted in fresh soil and started for another show of bloom.

We do not recommend the planting of this Lily in open borders during autumn, for growth will commence immediately, and a severe frost will cut it down; but if planted in spring, it succeeds admirably, and will produce a long succession of its handsome trumpet-shaped flowers. For the following winter it can be either protected, or lifted for storing in a frame.

Lilium lancifolium.—A graceful and highly perfumed Lily, which is perfectly hardy, and will grow in good loam, though peat is to be preferred for pot culture. To produce handsome specimens the same routine must be followed as directed for the cultivation of *L. auratum*. It scarcely need be added that, instead of growing the bulbs separately in pots, several may be grown in a large pot to produce a richer effect. But it is not advisable to place the bulbs in a large mass of earth in the first instance. It is better that they should commence their growth in small pots, and be shifted on as they require more room. Aphis is extremely partial to these Lilies, particularly if they are badly grown and allowed to suffer for the want of water. The simplest way to remove the pest is to dip the plants in pure water, taking care, of course, to prevent them from falling out of the pots in the operation.

Lily of the Valley.—The popular name of this native plant is a misnomer. Botanically it is known as *Convallaria majalis*, and structurally the roots differ from those which are characteristic of the whole tribe of Liliums. However, we have no quarrel with a charming name for a most dainty flower of fairy-like proportions. The sprays of pure white pendulous bells have captivated the popular fancy, and they are in public demand from the moment florists are able to place them on the market.

Whether for early or late spring forcing, or for planting in the open ground, the most vigorous strain should be chosen, and there is one which is incomparably superior to all others, producing finer spikes and larger individual flowers. As a rule these roots are obtainable in November, but, if necessary, it is far better to wait a week or two than attempt to force such as have been lifted prematurely.

The crowns may be potted, and where few are grown this is the usual course. The large growers pack them in boxes, with a little fine soil, and cover the tops with about four inches of cocoa-nut fibre. For the earliest supply a temperature of 90° is necessary, accompanied with plenty of moisture. After the spikes of bloom show, slightly reduce the temperature, and remove the fibre to afford the leaves an opportunity of maturing. When sufficiently advanced transfer the plants to pots for the conservatory or the decoration of windows. Successive supplies can be brought forward with less heat.

In the open, Lily of the Valley require a partially shaded position. The soil must be freely manured, and a good proportion of leaf-mould worked in. Plant single crowns at a distance of six inches from each other, and supply them with liquid manure during the growing period. After four, or at most five years, they will become too crowded, when they should be lifted, and the largest and finest crowns be selected for the formation of a fresh bed.

Japanese Day Lily *(Hemerocallis Kwanso fl. pl.)*.—Admirably adapted for pot culture to decorate the conservatory, the rich variegation of its graceful curling leaves affording an elegant display of colour in the early months of the year, and its fine double flowers being extremely showy during their short blooming season. As this variety is quite hardy, it may be planted in the select border with perfect safety, and, in common with other Day Lilies, it bears the shade of trees remarkably well. This is certainly one of the handsomest hardy plants in cultivation.

MONTBRETIA

Of this useful autumn-flowering bulb there are several varieties, *M. crocosmiflora* probably being the most popular. In the warm and sheltered gardens of the South and in light well-drained soil the roots pass the winter safely. But where frost prevails some protection, such as a small mound of litter, must be provided; the covering to be removed immediately the danger of frost is past. The most favourable time for planting is the autumn, but during open weather the roots may be put in up to the end of March. It is usual to plant in clumps at a depth of about three inches, allowing a distance of six inches between the corms. As they may remain undisturbed for several years the spacing will permit them to spread and produce masses of their graceful flowers.

NARCISSUS

Narcissi and Daffodils differ from Hyacinths, Tulips, and some other bulbs in one particular which is important, because it furnishes the key to the management of these flowers. The rootlets do not perish during the season of rest, and this fact clearly indicates that the bulbs should not remain out of ground for a day longer than is necessary.

Culture in Pots.—All the Polyanthus class, and almost all the Garden varieties, thrive in pots, and can be forced with extreme ease. Pot them early in any rich, porous compost, and put them into the soil a little deeper than is usual for Hyacinths. For a few weeks keep them in a cool spot in the open ground under a thick covering of ashes to promote root-growth without prematurely starting the tops. With all bulbs this is an important point, especially for such as are intended to be brought forward in heat. When the pots are full of roots, leaf-growth will commence, and the covering should be removed. A cool pit is then the best place for them. The after-treatment will depend entirely on the date the flowers are wanted. A low temperature, long continued, means late flowering, so that within reasonable limits the grower can control the time of their appearance. For the earliest display select the Roman and Paper White, which are naturally early-blooming varieties. After a few days in a cool pit, transfer to the greenhouse, and about a week or ten days before they are needed in flower plunge them in a brisk bottom heat, and give plenty of water of the proper temperature. The forcing should not begin until the plants are sufficiently advanced, or it will injure the flowers in both size and colour. Weak manure water will be beneficial occasionally, but when the blossoms begin to open this must be discontinued, and at the same time the heat should be diminished.

A succession of Narcissi for indoor decoration can be secured by starting batches at intervals of two or three weeks; and by moderating the treatment as the season advances, the last lot will flower naturally without artificial stimulus. Large bulbs should be potted singly, but several roots of the smaller sorts may be put into one pot. Heavy heads of bloom will need support, and there is nothing neater than the wires which are made expressly for the purpose.

Culture in Moss-fibre.—The lightsome charm of Narcissi and Daffodils is never seen to greater advantage than when these are grown in bowls of fibre for the decoration of rooms. Well-filled bowls of Daffodils are as delightful indoors as are sturdy clumps nodding over grass or Polyanthuses in the open air. The cultural routine is clean, pleasant, and full of interest. The bowls are chosen with care, the fibre is well saturated by repeated turning and moistening (this is essential to success), enough crushed oyster shell is incorporated to make the compost glisten brightly through and through, the mixture is pressed into the bowl until it is firm without being hard, the bulbs are half embedded, a few pieces of charcoal are pushed in here and there, the bowls are put in a dark place for six weeks or so, and the rest is merely to see that the fibre never gets dry.

Culture in Water.—For growing in glasses no other bulbous flower is equal to the Narcissus. Darkness at the outset is not essential to it, and therefore the gradual development of the roots may be observed from the time they start; and contact with water will do no harm to the bulb. The glasses should, however, be kept in a low and fairly uniform temperature, to discourage the growth of foliage until the bulbs have fully formed their roots. Pure rain water is desirable, but it is not actually necessary; and for the sake of appearances, as well as on the score of health, it should be changed immediately it ceases to be quite transparent. Those who do not care to observe the growth in glasses, but like to have the plants in water during the blooming period, may grow the bulbs in pots in the usual way, and wash off the soil when wanted. In this case the roots will not be quite so regular as those which have been wholly grown in water. Perhaps we need scarcely say that it is possible to utilise this flower in many other ways—such, for instance, as in decorating épergnes, glass globes, and fancy vases. They may also be made to float on a small fountain or aquarium; indeed, it is surprising to what varied and effective purposes a little ingenuity will adapt them.

Culture in Open Ground.—For this purpose the Narcissus will always command attention for its graceful appearance; and this observation applies with as much force to the Polyanthus section, when thus used, as to the varieties which are specially recognised as Garden Narcissus. The latter class includes many old favourites, among which is the Pheasant's Eye—one of the most exquisite flowers grown in our gardens.

The Narcissus is often used for bedding with superb effect. The graceful habit, which is one of its principal charms, is very striking in large masses, and its elegant appearance in the positions for which it is naturally suited cannot fail to arrest attention. Beneath trees, by the side of a shady walk, in front of shrubberies, or in the mixed border, the Narcissus is thoroughly at home.

If possible, choose a position where the bulbs need not be disturbed for several years, and plant them early. When the spot they are to occupy happens to be full, pot the bulbs until the ground is vacant, and in due time turn them out. A southern or western aspect is desirable, but the nature of the soil is comparatively unimportant, provided it is dry when the bulbs are in their resting state. In sour land or in stagnant water they will certainly rot, but a touch of sea spray will not injure them. If the soil needs enriching, there is no better material than decayed cow-manure, which may be incorporated as the work goes on, or it can be applied as a top-dressing. Those which are evidently weak may be assisted with a few doses of manure water, not too strong.

In planting groups, put the smaller bulbs four or five inches, and the larger sorts from six to nine inches apart; depth, six to nine inches, according to size. Where exposed to a strong wind, it may be necessary to give the flowers some kind of support to save them from injury.

The Double and Single Daffodils are now in marked public favour and their bright colours make them extremely useful for beds and borders. For planting under and among trees they are invaluable, and a sufficient number should always be put in to produce an immediate effect. They thrive in damp, shady spots, and every three or four years it will be necessary to divide and replant them.

The Chinese Sacred Lily (*Narcissus Tazetta*).—The popular name of this flower is misleading. It is not a Lily, but a Narcissus of the Polyanthus type, and, like others of the same class, the bulbs may be successfully grown in soil or in water. But *Narcissus Tazetta* has proved to be singularly beautiful in water, and the management of it entails very little trouble. A wide bowl of Japanese pattern is appropriate for the purpose, and to obtain the best effect the bowl should be partially filled with a number of plain or ornamental stones, with a few pieces of charcoal to keep the water sweet. On the top, and so that they will be held by the stones, place one or more bulbs: pour in water until it covers the base of the bulbs. Store in a dark cool cellar until the roots have started and the leaves begin to appear; then remove to the room where the ornament is wanted. Occasionally the water must be replenished. The development of the flower-heads is surprisingly rapid, and a large bulb generally produces several clusters of sweetly scented flowers. But if the bulbs are forced too quickly the blossoms are sometimes crippled.

ORNITHOGALUM

Star of Bethlehem

During the month of June *O. arabicum* produces heads of pure white fragrant flowers, each having a green centre. The roots are large and fleshy, and should be planted in the autumn six inches deep. A sheltered position, such as under a south wall, is desirable for them, and some protection in the form of dry litter, or a heap of light manure, will be necessary to carry the roots safely through severe winter weather. The bulbs are frequently potted for indoor decoration. Another variety, *O.*

umbellatum, with pure white starry flowers, makes an attractive show in May, and is valuable for naturalising in clumps or masses in the border.

OXALIS

These frame plants are suitable for the cool greenhouse or for forcing, and they are adapted also for the open border in peculiarly favourable districts. They are particularly neat and cheerful, flowering abundantly, and requiring only the most ordinary treatment of frame plants. In winter they should be kept dry. The 48-sized pot is suitable, and about five bulbs may be planted in each, using light soil freely mixed with sand.

RANUNCULUS

To maintain a collection of named Ranunculuses demands skill and patience, but a few of the most brilliant self-coloured, spotted and striped varieties may be easily grown, if a cool, deep, rich, moist soil can be provided for them. The best soil for the Ranunculus is a loam or clay in which the common field Buttercup grows freely and plentifully. The situation should be open, the bed well pulverised, and the soil effectively drained, both to promote a vigorous growth and, as far as possible, to save the plants from injury by wireworms, leather-jackets, and other ground vermin. Elaborate modes of manuring, such as mixing several sorts of manure together in mystical proportions, are altogether unnecessary, but a good dressing of rotten manure and leaf-mould should be dug in before planting, and if the soil is particularly heavy, sharp sand must be added. The roots may be planted in November and December in gardens where vegetation does not usually suffer from damp in winter; but where there is any reason to apprehend danger from damp, the planting should be deferred until February, and should be completed within the first twenty days of that month, if weather permit. Prepare a fine surface to plant on, and draw drills six inches apart and two inches deep, and place the tubers, claws downwards, in the drills, four inches apart, covering them with sifted soil before drawing the earth back to the drill. Rake the bed smooth, and the planting is completed. To keep free from weeds, and to give plentiful supplies of water in dry weather, are the two principal features of the summer cultivation. When the flowers are past, and the leaves begin to fade, take up the roots, dry them in a cool place, and store in peat or cocoa-nut fibre.

Turban Ranunculus.—This class is remarkably handsome, of hardier constitution and freer growth than the edged and spotted varieties. For the production of masses of colour, and to form showy clumps in the borders, the Turban varieties are of the utmost value. They require a good loam, well manured, and the general treatment advised for the named varieties; but as they are not so delicate they will thrive under less congenial conditions.

SCILLA

The Blue Squill may be grown in exactly the same manner as the Roman Hyacinth for indoor decoration, and it makes a charming companion to that flower. It is perfectly hardy, and for its deep, lovely blue should be largely grown in the open border, where it appears to especial advantage in conjunction with Snowdrops. It is also valuable for filling small beds, and for making marginal lines in the geometric garden.

The *Scilla præcox*, or *sibirica*, thrives on the mountains of North Italy, where masses of it may be seen growing close to the snow, and in this country it withstands wind and rain which would be the ruin of many another flower. Still we like to see it in a sheltered border, where it has a fair chance of displaying its beauty without much risk of injury. In such a position it will flower in February, and in the bleakest quarter it will open in March. It is not at all fastidious as to soil, but when planted will give no further trouble until the foliage withers, and it is time to lift the bulbs to make way for other occupants. If convenient, the roots may remain for years in one spot.

The *Scilla campanulata* deserves more attention than it has hitherto received. After almost all other spring-flowering bulbs are over, it makes a beautiful display, which lasts until nearly the end of May. It somewhat resembles the wild Blue-bell, but is much larger than that woodland flower.

SNOWDROP

Snowdrops are among the hardiest flowers known to our gardens, and are invaluable for their welcome snow-white bells in the earliest days of the opening spring. They should be planted in clumps, and left alone for years. The double-flowering variety is exquisitely beautiful: we might, indeed, speak of it as a bit of floral jewellery. The flowers are bell-shaped, closely packed with petals, like so many microscopic petticoats arranged for the 'tiring' of a fairy: they are snow-white and sometimes delicately tipped with light green. This variety is as hardy as the single, and the best for growing in baskets and pots. When employed in lines the planting ought to be very close together, and the line should be composed of several rows, making, in fact, a broad band. Such a ribbon when backed with *Scilla sibirica* is very beautiful. The best way of displaying the Snowdrop alone is in large groups densely crowded together. The effect is much more telling than when the same number of bulbs is spread over a larger area. Put the roots in drills, two inches deep, and if possible in a spot where they need not be disturbed for two or three years. Snowdrops may be grown in pots, and be gently forced for Christmas. But unless wanted very early, it will answer to lift clumps from the border in November and pot them.

SPARAXIS

See instructions under Ixia at page 338.

TIGRIDIA, or FERRARIA

The short-lived blossoms of the Tiger Flower are most gorgeously painted, and differ from everything else of the great family of Irids to which they belong. Much finer flowers are produced in the border than when grown in pots, and they present great variety, scarcely any two amongst hundreds showing flowers exactly alike. The usual time of planting outdoors is March or April, at a depth of three or four inches, and the flowers appear in June. Sandy loam and peaty soils are especially suitable. Although Tigridias are not quite hardy they will on a dry border pass the winter securely beneath a protection of litter. But where the soil is damp it is safer to lift them in October and store in the same manner as Gladioli. A bed of Tigridias makes an agreeable ornament in front of the window of a breakfast-room, as the flowers are in a brilliant state in the early hours of the day.

TRITELEIA UNIFLORA

This little gem belongs to the spring garden, and should be the companion of the Dog's-tooth Violet, the Crocus, and the Snowdrop. It will grow in any soil, and will produce an abundance of its violet-tinted white flowers, which, when handled, emit a faint odour of garlic. As a pot plant for the Alpine house it is first-rate. In the open, plant in October two inches deep.

TRITONIA

Tritonias are more showy than the Ixia or Sparaxis, but belong to the same group of South African Irids, and require the same treatment. They may be planted out in April, if prepared for that mode of cultivation by putting them in small pots in November or December. It is not advisable to tie them to sticks, for they are more elegant when allowed to fall over the edge of the pots, and suggest the 'negligence of Nature.'

TROPÆOLUM

T. tuberosum.—A few of the tuberous-rooted Tropæolums are hardy, but it is not wise to leave them in the ground, for damp may destroy them, if they are proof against frost. They are all graceful trailing plants, adapted for covering wire trellises, and may be flowered at any season if required, though their natural season is the summer. The compost in which they thrive best is a light rich loam, containing a large proportion of sand. The stems are usually trained on wires, but they may be allowed to fall down from a pot or basket with excellent effect, to form a most attractive tracery of leafage dotted with dazzling flowers. The sunniest part of the greenhouse should be devoted to the Tropæolums, and special care should be taken in potting them to secure ample drainage.

T. speciosum.—This showy variety is quite hardy, and is largely grown in Scotland where it may frequently be seen on cottage walls. The roots may be planted in either spring or autumn, and a moist, somewhat shaded position best suits the plant.

TUBEROSE

Polianthes tuberosa

This bulb is extensively grown in the South of France for the delicious perfume obtainable from its numerous pure white flowers. In this country it is widely known, but considering the beauty and exceeding fragrance of the blossoms it is astonishing that a greater number are not planted every season. Perhaps the fact that the bulbs are valueless after the first year may in a measure account for the comparatively limited culture. They are easily flowered as pot plants in a mixture of loam and leaf-mould, plunged in a bottom heat ranging between 60° and 70°. The growth is rather tall, and unless kept near the glass the stems become unsightly in length.

Culture in Pots.—When grown in pots, Tulips are treated in precisely the same manner as the Hyacinth, but several bulbs, according to their size and the purpose they are intended for, are placed in a pot. When required to fill épergnes and baskets, and other elegant receptacles, it is a good plan to grow them in shallow boxes, as recommended for Crocuses, and transfer them when in flower to the vases and baskets. This mode of procedure insures exactitude of height and colouring, whereas, when the bulbs are grown from the first in the ornamental vessels, they may not flower with sufficient uniformity to produce a satisfactory display. In common with the Hyacinth and Crocus, Tulips may be taken out of the soil in which they have been grown, and after washing the roots clean, they can be inserted in glasses for decorating an apartment. Early Tulips are often employed in this way to light up festive gatherings at Christmas and the early months of the year. But the pot culture of Tulips need not be restricted to the early varieties. The Darwin and May-flowering classes are also admirable when grown in this way, but it is important they should not be hurried into bloom. If placed in moderate heat and allowed ample time to develop, beautiful long-stemmed flowers may be had in March which will make a charming decoration for the drawing-room or the dinner-table.

Culture in Moss-fibre.—No bulb excels the Tulip in adaptability for bowl culture, given the treatment suggested for Narcissi and Daffodils on page 345, and particularly with respect to moisture.

Culture in the Open Ground.—For general usefulness the early Tulips are the most valuable of all, because of their peculiarly accommodating nature, their many and brilliant colours, and their suitability for the formation of rich masses in the flower garden. Any good soil will suit them, and they may be planted in quantities under trees if the position enjoys some amount of sunshine, because they will have finished their growth before the leafage of the trees shades them injuriously. If it is necessary to prepare or improve the soil for them, the aim should be to render it rich and sandy, and sufficiently drained to avoid a boggy character in winter. Plant in October or November, four or five inches deep, and six inches apart. The roots require no water and no supports, and may all be taken up and stored away in good time for the usual summer display of bedding plants. For geometric planting it is important to select the varieties with care, but a most interesting border may be made by planting clumps of all the best sorts of the several classes. The result will be a long-continued and splendid display, beginning with the 'Van Thols' (which are as hardy as any), following with the early class in almost endless variety, and finishing with the noble Darwin and May-flowering sections. The last named include a very large number of extremely handsome flowers, and their lasting beauty is of especial value at a season of the year when spring blooms are over and summer plants have scarcely begun to make a show.

As cut flowers Tulips are worthy of special attention. With very little care they not only maintain their full beauty in vases for a fortnight, but some of them actually increase in brilliancy of colouring. The May-flowering classes are perhaps the most appreciated for cutting, because of their great length of stem and the enduring character of the flowers. They are extremely beautiful in tall vases.

VALLOTA PURPUREA

This brilliant plant is nearly hardy in the Southern counties, and a cool greenhouse plant where it cannot be grown in the open border. To produce fine specimens a firm loamy soil is necessary, with abundance of water all the summer, and moderate supplies all the winter. The bulbs flower more freely when somewhat pot-bound. Therefore they should not be re-potted too often. Under these conditions feeding with clear liquid manure is necessary once a week from the time the flower-buds show until they begin to open. To dry off the bulb may weaken or kill it.

Those who cannot cultivate the Amaryllis will find the Vallota an excellent substitute.

VIOLET, DOG'S-TOOTH—*see page 327*

WINTER ACONITE

The Winter Aconite is the very 'firstling' of the year, for it blooms in advance of the Snowdrop, covering the ground with gilt spangles in the bleakest days of February. Any soil or situation will suit it, and it should be planted in large patches where a winter's walk in the garden affords pleasure. It should also be grown in quantity within view from the windows, for the benefit of those who, in the dreary season, cannot get out. The bulbs may be left in the ground for several years, or they may be taken up and stored after the leaves have perished.

ZEPHYRANTHES CANDIDA

Flower of the West Wind

A dwarf white Crocus-like flower, with foliage resembling the common Rush on a small scale. Plant in clumps from November to March in borders, and it will commence blooming about the end of July, and continue in flower until frost cuts it down. Any soil will suit this plant, and it thrives for several years if left undisturbed.

FLOWERS ALL THE YEAR ROUND FROM SEEDS AND ROOTS

Before proceeding to the duties which need attention in successive months of the year, it may be worth while to consider some of the points which constitute the alphabet of flower culture. To grow any plant in a pot is an artificial proceeding, and the conditions for its sustenance and health have to be provided. Among these conditions are temperature and accommodation. It is useless to attempt to grow flowers which require heat unless that necessity can be met. And it is equally useless to pot more plants than the space will accommodate when they attain their full size. A limited number, well grown, will produce a greater wealth of bloom, of finer quality, than a larger number which become feeble from deficiency of space for development. Nevertheless, there are many varieties raised in heat in the early months of the year which can be grown and flowered in the most satisfactory manner, without any kind of artificial aid, from sowings made in the open ground during April or May. The flowering will be somewhat later than from plants brought forward under glass; but as they receive no check from the very commencement, they will not be greatly behind their nursed relations; and they may even excel them in robust beauty, if they are treated intelligently and with a generous hand.

Good Soil for pot plants is not always obtainable at a reasonable cost, and sometimes the materials at hand must be made to serve the purpose. None the less is it true, that in proportion to the skill and experience of the cultivator will be his desire to secure a supply of loam, peat, and leaf-mould. Those who are capable of turning poor soil to the best account are precisely the men who will be most anxious to obtain the materials which are known to promote the luxuriant growth of pot plants.

The top spit of an old pasture makes capital potting soil. If taken from light land, it need only be stacked for one year before use. A heavy loam should be kept for at least two seasons, and in any case the heap should be turned and re-made several times. A slight sprinkling of soot between the layers of soil will be beneficial, and help to make it distasteful to grubs, wireworms, and other vermin. The frequent turning of the heap will not be wasted labour, for it equalises the quality, and tends to sweeten the whole by exposing new surfaces to the atmosphere; and this is a great aid to healthy growth.

Many plants thrive in peat, or in soil of which peat is a constituent, and some flowers cannot be grown without it. The peat may have to be purchased from a distance, but there is no difficulty in obtaining it.

A constant supply of decayed leaf-mould may possibly be arranged on the spot by sweeping up leaves and making a fresh heap every fall. In due time these leaves will decay and make useful potting soil. If this is out of the question, the requisite quantity must be purchased.

The preparation of soil for pot plants is frequently postponed until the day on which it is actually required. This is a bad practice, and results too often in the use of an improper proportion of the materials, and perhaps in their defective admixture. In this, as in all other operations connected with horticulture, the men who make all requisite arrangements in advance will achieve the highest results. In no pursuit of life is it more necessary to forecast coming wants than in the culture of flowers. We will suppose that three or four weeks hence many pots are to be filled with Primulas. The man who grows this flower with any degree of enthusiasm will not defer the preparation of the soil until the day arrives for potting the plants. He will determine in advance the proportions of loam, leaf-mould, and sand, have the whole thoroughly incorporated, and possibly sifted to remove stones. With these may come away some undecayed fibres, which make excellent material for laying over the crocks at the bottom of each pot. Forethought of this kind is certain of an ample reward.

Potting soil should also be in the right condition as to moisture. This is not easy to describe, but it must handle freely, and yet there should be no necessity for the immediate application of water after sowing seeds or planting bulbs. In the event of the compost being too dry, give it a soaking and allow it to rest for one or more days, according to the time of year and the state of the atmosphere.

Pots, new or old, should be soaked in water before use. They are very porous, and by absorbing moisture from the soil they may at once make it too dry, although in exactly the right condition before being placed in the pots. And old pots ought never to be used until they have been scrubbed quite clean. These may appear to be trivial matters, unworthy of attention. They have, however, an influence on the health of plants, and experienced growers know that a few apparent trifles make all the difference between success and failure. Pots which are dirty, or covered with green moss, prevent access of air, and tend to bring about a sickly growth. Cleanliness in horticulture is valuable for its own sake, and for the orderly routine it necessitates on the part of the cultivator.

Pots are known both by number and by size. They are sold by the 'cast,' and a cast always consists of the distinguishing number. The following are the numbers and sizes:—

Number in Cast				Inches
	72	Inside diameter across top		2-1/2
Small	60	"	"	2-3/4
Mid.	60	"	"	3
Large	60	"	"	3-1/2
Small	54	"	"	4
Large	54	"	"	4-1/4
Small	48	"	"	4-3/4
Large	48	"	"	5
	40	"	"	5-1/2
	32	"	"	6-1/4
	28	"	"	7
	24	"	"	7-1/2
	16	"	"	8-1/2
	12	"	"	9-1/2
	8	"	"	11
	6	"	"	12-1/2
	4	"	"	14
	2	"	"	15-1/2
	1	"	"	18

Watering is sometimes conducted on the principle that the usual time has arrived, and therefore the plants must have water. But do they need it? Press the fingers firmly on the surface; if particles of soil adhere it is too dry. Or tap the pots smartly with the knuckles or with a stick, when a clear and unmistakable answer will be obtained. Plants differ widely in their demand for water. Some are very thirsty, others require less frequent attention. The season of the year and the state of the atmosphere have also to be considered, as well as the fact that a heavy soil is more retentive of moisture than a lighter compost. A watchful eye and a willing hand will seldom err on this point. The water should always be of the same temperature as the house, otherwise the plants will be constantly checked. A tank in the greenhouse meets this requirement. In its absence, the watering-pots should be kept full under the stage, and they will be ready when wanted.

In the open ground, it is better to water a few plots thoroughly for two or three successive evenings, and then have an interval, rather than moisten the surface daily. The effect of constantly applying small quantities of water is to encourage the surface growth of roots. Then, if the sun shines fiercely on the soil, the first day of neglect results in immense mischief.

Drainage is easily managed. Into each pot put a crock almost the size of the bottom, with the convex side upwards. There need be no niggling to remove sharp angles, or to make the fragment shapely. Cover this with smaller crocks, and these with moss, or in some cases with small pieces of charcoal. If the compost has a proper admixture of sharp sand or grit, free drainage will be insured, and yet the soil cannot be washed through the pot. Silver sand is often employed, and there is nothing better for the purpose. But the sweepings from gravel walks, finely sifted, may be substituted. Road grit is often infested with weed seeds.

Ventilation is important, for a house full of plants cannot long be kept closed with impunity. The lights should be opened whenever the state of the weather may permit, and by doing this on the side opposite to the quarter whence the wind blows it is frequently safe to give air when it may be dangerous from other points of the compass; and it should be done early in the day, before the sun gets hot. Often the lights remain closed on a sunny morning until the atmosphere becomes stifling; and then perhaps plants which have been made sensitive by excess of heat are subjected to a killing draught.

In managing Temperature, there should be no violent alternations of heat and cold, for these bring speedy disaster; and, it is unwise to employ more heat than is actually necessary. Deviations from this rule are generally traceable to neglect. If the proper season for sowing seed of some important flower has been allowed to pass, an attempt is made to compensate for lost time by hurrying the growth in a forcing temperature.

Every needless degree of heat will be harmful, and result in attenuated growth, poverty of colour, or in the attack of some insect plague which the weakly plant seldom invites in vain. It is wise always to employ the lowest temperature in which plants will flourish. This necessitates the proper time for their full development, and will result in a sturdy growth capable of yielding a bountiful display of bloom. Occasionally it is requisite to force some special subject, such as bulbs for Christmas festivities. Even then it is advisable to augment the temperature very gradually, and to defer the employment of its highest power until the latest possible moment.

Plants are frequently taken straight from the forcing pit into a cold room, to their utter ruin. A moment's reflection will show the folly of such a proceeding. They should be prepared for the change by gradual transfer through lower temperatures; and if only a few hours are occupied in the process it will help them to pass the ordeal with less injury.

It should be an established custom to examine the seed-pans at least once every day, and morning is the best time for the task. If work has to be done, there is the whole day to arrange for its accomplishment. Whereas, if the visit is not made until evening, there may not remain sufficient daylight to do what is necessary. Just as seedlings are starting, a few hours' neglect will render them weak and leggy.

When transferring plants from seed-pans, it is usual to put them round the edges of pots. This is no mere caprice, but is founded on the well-ascertained fact that seedlings establish their roots with greater readiness near the edge of the pot than away from it.

In the following monthly notes, our principal object is to offer a series of reminders which will insure the sowing of various flower seeds and the planting of bulbs at their proper periods, and thus save the disappointment of losing some important display for a whole season. Those who have command of large resources will sow certain seeds a month earlier than we recommend, and their intimate knowledge and abundant facilities justify their practice. But we have especially in view the possibilities for an amateur, and of gardens moderate in extent, where appliances may not be of the most perfect kind.

When seeds are once sown or bulbs potted, the work is before the cultivator, and appeals mutely for attention. Therefore it is not our purpose to give detailed and continuous instructions month by month for every flower. Our remarks are limited to hints at the time for sowing or planting, and to some few points which may subsequently appear to demand notice.

For convenience of reference, the subjects are presented alphabetically under each month.

JANUARY.

In the open ground there is little or nothing of interest in the way of flowers, but the greenhouses and pits are full of promise. A constant watch must be kept on the barometer, and the materials for repelling frost or bleak winds should be at perfect command, so that there may be ample provision for saving plants from biting weather.

Achimenes are stove bulbs and cannot be grown without a sufficiency of heat. A warm greenhouse will answer for them, and some gardeners produce fair specimens in frames over hot-beds. The bulbs will lie dormant for a considerable time, so that it is easy to have a succession of flowers. A few should be started in January, employing sandy loam for the pots. Follow up with others at intervals.

Amaryllis may be sown in any month of the year, but the most satisfactory period is immediately after the seed is ripened, and it is advisable to put one seed only in each small pot. The slow and irregular germination of the finest new seed makes the separate system almost a necessity. A rich compost, well-drained pots, and a temperature of about 65° suit these plants.

Anemone.—See remarks under October.

Antirrhinums raised in heat now will flower from July onwards. Prick off the seedlings, and gradually harden for planting out in May. There are dwarf, medium, and tall varieties, of many beautiful colours.

Begonia, Tuberous-rooted.—The grace and beauty of this plant have placed it in the front rank of popular favourites. For the foliage alone it is worth growing, and the flowers are unique in both form and colour. Raising plants from seed is not only the least expensive process, but it possesses all the charm arising from the hope of some novelty which shall eclipse previously known varieties. As a matter of fact, new attractions either in colour or in habit are introduced almost every year. From a sowing made now plants should flower in July and August.

The seed is small, and requires careful handling. It is also slow and capricious in germinating, and many growers have their own pet methods of starting it. Good results are obtained by insuring free drainage, and partly filling the pots with rather rough fibrous compost, covered with a layer of fine sandy loam made even for a seed-bed. This is sprinkled with water, and the seed is sown very thinly. Some experienced growers make a rather loose surface, press the seed gently into it, and do not finish with a covering of soil. The majority, however, will find it safer to give a slight sifting of fine earth over the seed. Then comes a trial of patience, and as the seedlings appear at intervals, the wisdom of thin sowing will be apparent, for each one can be lifted and potted as it becomes ready, without wasting the remainder. An even temperature of about 65° is essential during germination.

Begonia bulbs which have been stored through the winter will need careful watching. Not until they start naturally should there be any attempt to induce growth, or in all probability it will result in the destruction of the bulb. Such as show signs of life should be potted in good soil, commencing with small pots, and shifting into larger sizes as the pots become full of roots. Until the final size is reached, remove all flowers. A warm humid atmosphere is favourable to them while growing, but when flowering begins moisture will be injurious.

Begonia, Fibrous-rooted, may also be sown at the end of this month or in February, and again early in March. Under similar treatment to that advised for Tuberous-rooted Begonias, the plants will be ready in June for transfer to beds or as an edging to borders.

Canna.—From the popular name of Indian Shot it will naturally be inferred that the seed is extremely hard and spherical. It needs soaking in water for about twenty-four hours before sowing. Even then it will probably be a considerable time in germinating, and there will also be variable intervals between the appearance of the seedlings. A high temperature is necessary to insure a start; but after the young plants are transferred to single pots, they should be kept steadily going in a more moderate heat until ready for the border or sub-tropical garden in June. Meanwhile they will need re-potting two or three times, and should have a rich and rather stiff compost.

Carnation.—Seed of the early-flowering class should be sown in heat during this month and again in February. With very little trouble, plants can be brought forward and transferred to the open ground, where they will give a splendid display in about six months from the date of sowing.

Chrysanthemums of the large-flowering perennial type can easily be raised from seed. If sown during this month or in February in a moderate heat, the plants will flower the first season. Pot the seedlings immediately they are ready, then harden, and put them out of doors as early as may be safe. This treatment will keep them dwarf and robust. Seedlings should not be stopped, but be allowed to grow quite naturally.

Cinerarias should have air whenever it is possible. Choose the middle of the day for watering, and do not slop the water about carelessly, or mildew may result. In houses which are not lighted all round, the plants should be turned regularly to prevent them from facing one way. Such specimens are worthless for the dinner-table, and will be diminished in value for decorating the drawing-room.

Cyclamens are still in the height of their beauty. The pots have become so full of roots that ordinary watering partially fails of its purpose. An occasional immersion of the pots for about half an hour will result in marked benefit to the plants. The flowers, when taken from the corm, should be lifted by a smart pull. If cut, the stems bleed and exhaust the root.

Where a succession of this flower is valued, a sowing should be made in this month. Dibble the seed, an inch apart and a quarter of an inch deep, in pots or pans firmly filled with rich porous soil; and place in heat of not less than 56° and not exceeding 70°; the less the temperature varies the better. Cyclamen seed is both slow and irregular in germinating, and sometimes proves a sore trial even to those who are blessed with patience. As the seedlings become ready transfer to small pots, and shift on as growth demands, always keeping the crown of the corm free from soil. The increasing power of the sun will render shading essential; yet a position near the glass is most advantageous to the plants.

Freesia.—This elegant and delicately perfumed flower is annually raised in large numbers from seed. From this month to March sowings may be made in heat, and as the roots are extremely brittle, re-potting is a delicate operation.

Gesnera.—Those who have once grown this handsome conservatory plant will not afterwards consent to be without it. The richly marked foliage contrasts admirably with the flowers. Sow in the manner advised for Gloxinia, and the two plants may be grown in the same house.

Gloxinia.—From two or three sowings, and by a little management, it is easy to have a supply of this magnificent flower in every month of the year. Sow thinly in new pots filled with a light porous compost, and see that the drainage is exceptionally good. Give the pots a warm moist position, and a light sprinkling of water daily will assist germination. The first seedlings that are ready should be lifted and pricked off without disturbing the remainder of the soil. Follow up the process until all are transferred. Although the leaves may rest on the surface, the hearts should never be covered. Pot off singly when large enough, and shift on until the 48-size is reached. For ordinary plants this is large enough, but extra fine specimens need more pot room, and so long as increased space is given the flowering will be deferred. Between the plants there must be a clear space or the leaves will decay through contact. While growing, a moist atmosphere, with a temperature of 60° or 65 °, will suit them; but immediately flowering commences, humidity is a source of mischief. The most forward plants from this month's sowing will, if well treated, begin to flower in June.

Grevillea robusta.—Seed of this exceedingly handsome shrub may be sown at any time of the year, and the pots containing it must be kept moist until the seedlings appear. How long it will be before they become visible we cannot tell. Germination may not occur until hope has died, and the pots have been contemptuously relegated to some obscure corner. But after the young plants are pricked off, they will give no trouble, except to re-pot them two or three times, and to take care that they do not perish for want of water.

Hollyhock.—This stately border flower is occasionally grown and flowered as an annual, and some gardeners succeed in producing satisfactory plants, carrying fine double blossoms, superb in colour and of noble proportions. Where this method is possible it is necessary to sow in the opening month of the year, and to use well-drained pots or seed-pans. Cover the seed with a sprinkling of fine soil, and place in a temperature of 65° or 700. In about a fortnight the seedlings will be ready for pricking off round the edges of 4 1/2-inch pots. But as a rule the finest spikes are obtained from a sowing in July or August.

Petunia.—About the third week of this month a sowing should be made to produce plants for indoor decoration. Late in February or early in March will be soon enough to prepare for bedding stuff. Sow thinly in good porous soil, and give the pots or pans a temperature of about 60°. They should have a little extra attention just as the seed is germinating, for that is a critical time with Petunias. Uniformity in temperature and moisture, with shade when necessary, and plenty of pot room, are the secrets of success in growing these plants.

Statice.—The Sea Lavenders make attractive border subjects, but the sprays of flowers are probably still more valued for cutting and, when dried, for the winter decoration of vases in association with Everlastings. Seed of the half-hardy varieties may be sown from January to March in gentle heat, transferring the plants to the open in due course.

Verbena.—This flower should be grown with as little artificial aid as possible. In fact, the more nearly it is treated as a hardy plant the more vigorous and free blooming will it be. A temperature of 60° is sufficient to raise the seed at this period of the year; and after the plants are established in pots, heat may be gradually dispensed with. Sow in pans or boxes filled with rich, mellow, and very sweet soil. Transfer to thumb pots when large enough, and give one more shift as growth demands, until the plants are ready for bedding out in May. There is a choice of distinct colours, which come true from seed. Green fly is very partial to the Verbena, especially while in pots; it must be kept down, or the seedlings will make no progress.

FEBRUARY

A Considerable number of important flowers should be sown during this month. The precise dates depend on the district, the character of the season, and the resources of the cultivator. Should the month open with frost, or with rough, wet weather, it will be wise to exercise a little patience. Where there are insufficient means for battling with sudden variations of temperature, choose the end rather than the beginning of the month for starting tender subjects. Govern the work by intelligent observation, instead of following hard and fast rules. But in no case should fear of the weather form an excuse for the postponement of necessary work.

Annuals and Biennials, Hardy.—It is one of the merits of hardy annuals and biennials sown in late summer for blooming in the following spring that they need very little attention. Still, they ought not to be entirely neglected. They should be kept scrupulously free from weeds, and it may be evident that a mulch of decayed manure is necessary to protect and strengthen them for a rich display of colour in the spring. Such varieties as have to be transplanted should be watched, and the first suitable opportunity seized for transferring them to flowering positions.

Abutilon is a flowering greenhouse shrub which answers well under the treatment of an annual. It does not need a forcing temperature at any stage, nor is the plant fastidious as to soil. The seed, which is both slow and irregular in germinating, may be sown in pots. As the young plants become ready they should be pricked off and kept steadily growing. When leaves drop, it indicates mismanagement, perhaps starvation. A well-grown specimen, when the buds show, will be two feet high, and bear examination all round.

Anemone.—Against the practice of planting roots of this elegant flower we have not a word to say. On the contrary, there is much to be advanced in its favour. Arrangements of colour can be secured which are impossible of attainment from seedlings. Still, there can be no doubt that the supposed necessity of depending alone on bulbs has proved a barrier to the growth of Anemones in many gardens, and on a large scale. The culture from seed is of the simplest character, no appliances whatever beyond those at the command of the cottager being needed. The prime requisite is a rich moist soil. Where this does not exist naturally, a liberal dressing of mellow cow-manure, and, in dry weather, a diligent employment of the water-can, will render it possible to grow superb flowers of brilliant colour. The best way of making the seed-bed is to open a trench, putting a layer of decayed manure at the bottom, and mingling a further quantity with the soil when it is returned. The addition of some light compost or sand to the surface may or may not be necessary to prepare it for the seed. We prefer sowing in rows and lightly scratching the seed in. Some growers only sift a little sand over, and the practice answers well. Weeds must be removed with care until the seedlings appear, and these are a long time in coming. Thinning to six inches apart, and keeping the bed clean and moist, constitute the whole remainder of the work of growing Anemones.

Aquilegia sown this month in a frame will produce plants which may flower later in the year, provided the season is favourable; but they will certainly pay for this early sowing in the succeeding spring. The plant is quite hardy, therefore seed may be sown later on in the open for a display in the following year.

Asparagus *(Greenhouse foliage varieties)*.—The finely feathered sprays of *A. plumosus* have become indispensable for bouquets, buttonholes, and general decorative purposes. *A. decumbens* and *A. Sprengeri* are most graceful plants in hanging-baskets. Seed of the three varieties should be sown in heat in either February or March.

Auricula.—The Show Auricula is one of the reigning beauties of the floral world, and, like the Rose, has its own special exhibitions. Although the flower merits all the admiration it receives, yet it must be confessed that some amateurs indulge in a great deal of needless coddling in the work of raising it. One quality there must be in the grower, and that is patience; for seed saved from a single plant in any given season, and sown at one time, will germinate in the most irregular manner. Months may elapse between the appearance of the first and the last plant. The lesson to sow thinly is obvious, so that the seedlings may be lifted as they become ready, without disturbing the surrounding soil. Both the Show and the Alpine varieties should be sown in pans filled with a mixture of sweet sandy loam and leaf-mould. They may be started in gentle heat, but this is quite optional. The Auricula is thoroughly hardy against cold, and glass is only employed as a protection against wind, heavy rain, and atmospheric deposits.

Begonia, Tuberous-rooted.—Seed may still be sown for a summer display. Transplant seedlings which are ready, and later on pot them singly.

Calceolaria, Shrubby.—Seeds sown in pans placed in a frame or a greenhouse of moderate temperature will insure plants for outdoor summer decoration. Transfer the seedlings to pots quite early.

Campanula.—By sowing seed in gentle heat during February many of the Campanulas will flower the same season. These hardy plants require but little heat, and they should be given as much light and air as possible. They may be grown on in pots for the decoration of rooms or the conservatory, or planted out on good ground in the open border. The half-hardy trailing variety, *C. fragilis*, is specially adapted for suspended baskets or large vases. Seed is generally sown in February or March; when ready the seedlings are transferred to pots.

Celosia plumosa.—Seed may be sown either now or in March, and the routine recommended for Cockscombs will develop splendid plumes. Re-pot in good time to prevent the roots from growing through the bottoms of the pots.

Cockscomb.—The ideal Cockscomb is a dwarf, well-furnished plant, with large, symmetrical, and intensely coloured combs. Seed of a first-class strain will produce a fair proportion of such plants in the hands of a man who understands their treatment. Sow in seed-pans filled with rich, sweet, friable loam, and place in a brisk temperature. Transfer the seedlings very early to small pots, and shift on until the size is reached in which they are to flower. Directly they become root-bound the combs will be formed.

Cosmea.—To prevent the disappointment which is sometimes experienced by growers of this attractive half-hardy annual, it is essential to sow a reliable early-flowering strain. Start the seed on a gentle hot-bed in February and plant out the seedlings in May or June when the danger from frost is past.

Dahlia.—Both the double and single classes can be grown and flowered from seed as half-hardy annuals. A sowing in this month will supply plants sufficiently forward to bloom at the usual time. Some growers begin in January, and provided they have room and the work can be followed up without risking a check at any stage, no objection can be raised to the practice. For most gardens, however, February is safer, and March will not be too late. Sow thinly in pots or pans filled with light rich soil, and finish with a very thin covering of fine leaf-mould. When the seedlings are about an inch high, pot them separately, taking special care of the weakly specimens, for these in point of colour may prove to be the gems of the collection. After transplanting, a little extra attention will help them to a fresh start.

Dianthus.—From sowings made this month or in January, all the varieties may be raised in about 55° or 60° of heat, but immediately the seed has germinated it is important to put the pots in a lower temperature, or the seedlings will become soft. They should also be transferred to seed-pans when large enough to handle.

Fuchsia.—It is now widely known that Fuchsias can be satisfactorily flowered from seed in six or seven months, and from a good strain there will be seedlings well worth growing. Sow thinly on a rich firm soil, and give the pots a temperature of about 70°. While quite small transfer the plants to the edges of well-drained pots, and later on pot them singly into a compost consisting chiefly of leaf-mould until the flowering size is reached, when a proportion of decayed cow-manure should be added. The Fuchsia is a gross feeder, and must have abundance of food and water. Aphis and thrips are persistent enemies of this plant, and will need constant attention.

Geranium seed may be sown at any time of the year, but there are good reasons why the months of February and August should be chosen. Seedlings raised now will make fine plants by the end of June, and begin to flower in August. They are robust in habit, and from a reliable strain there will be a considerable proportion of handsome specimens. Sow in pans filled with soil somewhat rough in texture, and the surface need not be very smooth. Lightly cover the seed with fine loam. To have plants ready for flowering in the summer it will be necessary to give the seed-pans a temperature of 60° or 70°, and follow the usual practice of pricking off and potting the seedlings.

Gladiolus.—It is not common to grow this noble flower from seed, but the task is simple, and seed good enough to be worth the experiment is obtainable. In large pots, well drained and filled with fibrous loam and leaf-mould, dibble the seeds separately an inch apart and half an inch deep. A temperature of 65° or 70° will bring them up, and when they reach an inch high the heat should be gradually reduced. The seedlings need not be transplanted, but may remain in the same pots until the grass dies down, and the corms are sifted out in September or October.

Gloxinia.—The directions under January are applicable, but it will be necessary to provide shade for the seedlings as the sun becomes hot, especially after they have been re-potted.

Kochia trichophylla.—A beautiful half-hardy ornamental annual shrub, symmetrical in form. From seed sown during this month or in March plants can easily be raised for indoor decoration or to furnish a supply for beds and borders. When well grown and allowed plenty of space from the beginning, each specimen forms a dense mass of bright green foliage which changes to russet-crimson in autumn.

Lobelias occupy a foremost place for bedding, and are sufficiently diversified to meet many requirements. Indeed, there is no other blue flower which can challenge its position. The compact class is specially adapted for edgings; the spreading varieties answer admirably in borders where a sharply defined line of colour is not essential; the *gracilis* strain has a charming effect in suspended baskets, window-boxes, and rustic work; and the *ramosa* section grows from nine to twelve inches high, producing large flowers. All these may be sown now as annuals, to produce plants for bedding out in May. Put the seed into sandy soil, and start the pans in a gentle heat.

Mimulus, if sown now and treated as a greenhouse annual, will flower in the first year. It is one of the thirstiest plants grown in this country, and must have unstinted supplies of water.

Nicotiana.—Where sub-tropical gardening is practised the Tobacco plant is indispensable. To develop its fine proportions there must be the utmost liberality of treatment from the commencement. Either in this month or early in March sow in rich soil, and place the pans in a warm house or pit. Put the seedlings early into small pots, and promote a rapid but sturdy growth, until the weather is warm enough for them in the open ground. The Nicotiana also makes an admirable pot plant for the conservatory or greenhouse, where it is especially valued for its delightful fragrance.

Pansy.—Although the Pansy will grow almost anywhere, a moist, rich soil, partially shaded from summer sun, is necessary to do the plant full justice. Many distinct colours are saved separately, and the quality of the seedlings is so good that propagation by cuttings is gradually declining. Sow thinly in pots or pans, and when the young plants have been pricked off, put them in a cool, safe corner until large enough for bedding out. The soil should be plentifully dressed with decayed cow-manure.

Pelargonium.—In raising seedling Pelargoniums, it is well to bear in mind that worthless seed takes just as much time and attention as does a first-class strain. The simplest greenhouse culture will suffice to bring the plants to perfection. A light sandy loam suits them, and the pots need not go beyond the 48-or at most the 32-size. Flowering will be deferred until re-potting ceases.

Petunia.—Towards the end of the month the seedlings raised in January for pot culture will be ready for transferring to seed-pans. It will also be time to sow for bedding plants, although the beginning of March is not too late.

Phlox Drummondii.—The attention devoted to this flower has made it one of the most varied and brilliant half-hardy annuals we possess. The *grandiflora* section includes numerous splendid bedding subjects which flower freely, and continue in bloom for a long period. These and others are also valuable as pot plants, and even in the greenhouse or conservatory they are conspicuous for their rich colours. All the varieties may be sown now in well-drained pans or shallow boxes. Press the seeds into good soil about an inch apart, and as a rule this will save transplanting; but if transplanting becomes necessary, take out alternate plants and put into other pans, or pot them separately. The remainder will then have room to grow until the time arrives for bedding out.

Polyanthus.—Either now or in March sow in pans filled with any fairly good potting soil, and do not be impatient about the germination of the seed. Many sowings of good seed have been thrown away because it was not known that the Polyanthus partakes of the slow and irregular characteristics of this class of plants. As the seedlings become ready, lift them carefully and transplant into pans or boxes, from which a little later they may be moved to any secluded corner of the border, until in September they are put into flowering quarters. While in the seed-pans they must be kept moist, although excessive watering is to be avoided. Should the summer prove dry, they will also need water when in the open ground.

Primroses of good colours are admirably adapted for indoor decoration, and there is no occasion to grow them in pots for the purpose. Lift the required number from the reserve border without exposing the roots; pot them, and place in a cool frame until established. Plenty of space, no more water than is absolutely essential, and progressive ventilation, comprise all the needful details of cultivation. Seed sown in this month or in March, in pans or boxes, will produce fine plants for flowering in the succeeding year.

Primula.—The elegant half-hardy varieties *P. obconica grandiflora* and *P. malacoides* may be sown any time from February to July, the earliest of which will commence flowering in the succeeding autumn and winter. The aim should be to keep the plants as hardy as possible, giving them air whenever conditions are favourable.

Ranunculus.—Although it is not usual to grow this flower from seed, it is both easy and interesting to do so. Sow in boxes containing from four to six inches of soil, and as there need be no transplanting, each seed should be put in separately, about an inch and a half apart. A cool greenhouse or frame will supply the requisite conditions for growing the seedlings. When the foliage has died down, sift out the roots, and store in dry peat or cocoa-nut fibre for the winter.

To secure an immediate display of Ranunculuses it is necessary to plant mature roots. The soil in which they especially thrive is an adhesive loam or clay. This happens to be unfavourable to their safety in the winter, and therefore it is wise to defer planting in such soils until this month. A very simple procedure will suffice to produce handsome, richly coloured flowers. If possible, choose for the bed a heavy soil in an open situation, and dress it liberally with decayed manure. Give the land a deep digging, and lay it up rough, that it may be benefited by frosts. In January and February fork it lightly over several times, with the double purpose of making it mellow and of enabling birds to clear it of vermin. Traps made of hollowed Potatoes will also assist the latter object. Not later than the third week of February the roots should be planted in drills drawn six inches apart and two inches deep. Put them at intervals of four inches in the rows, with the claws downwards, and cover with fine soil. Keep the bed free from weeds, and give abundant supplies of water in dry weather. When the foliage is dead, lift the roots and store for the next season.

The Turban Ranunculus is less delicate than the named varieties, and there need be less hesitation about autumn planting.

Ricinus.—The Castor-oil Plant is largely cultivated for its striking ornamental foliage, and under generous treatment it will attain from four to six feet in height. It is a half-hardy annual, and should be grown in the same manner as Nicotiana.

Salpiglossis merits its increasing popularity. A sowing at the end of this month or the beginning of March will insure plants in condition for the open ground in May. A moderate hot-bed is requisite now, but in

April the seed may be sown on prepared borders for a summer display of the veined and pencilled flowers.

Solanum.—The varieties which are grown for winter decoration are much prized when laden with their bright-coloured berries. Sow the several kinds in heat, and transfer the seedlings straight to single pots filled with very rich soil.

Stock, Intermediate.—To form a succession to the Summer-flowering, or Ten-week, varieties in July and August, seed of the Intermediate Stocks should be sown in gentle heat during February or March. The treatment accorded to Ten-week Stocks, described on page 379, will suit the Intermediate varieties also.

Sweet Peas have in recent years become such an important ornament to the garden and the flowers are so highly prized for household use that no effort is spared to insure a long-continued display. With this object in view seeds are sown in pots and the seedlings transplanted, as soon as weather permits, to the ground specially prepared in the preceding autumn. Those who did not sow in September should do so in the latter part of January or during February. A forcing temperature is injurious, and the plants thrive best when given practically hardy treatment.

Vallota purpurea.—This handsome bulbous plant is not quite hardy, but in several of the Southern counties it may be grown in the open ground, with only the shelter of dry litter or a mat. In pots the bulbs should not be allowed to go dry through the winter; and when growth commences in spring, water must be given freely. Good loam suits the Vallota, and it is desirable to avoid re-potting until the flowering period has passed: when a transfer becomes necessary, disturb the roots as little as possible.

Verbena, if not sown last month, should be got in promptly, for it is important not to hurry the growth of this plant by excessive heat.

Wigandia is a half-hardy perennial, grown exclusively for its noble tropical foliage. If started now, it will attain a large size as an annual. It is impossible to grow this plant too well. A lavish employment of manure and water will secure stately specimens. The instructions given for Ricinus apply equally to the Wigandia.

MARCH

The first duty is to ascertain that there are no arrears to make good or failures-to replace. If any sowing has gone wrong, do not waste time by repining over it, but sow again. Growing flowers under artificial conditions is a prolonged struggle with Nature, in which the most experienced and skilful gardener need not be ashamed of an occasional failure. But the cause of the failure should, if possible, be ascertained for future guidance. We say if possible, because the secret cannot always be discovered. There may have been every apparent condition of success, and yet, for some inexplicable reason, there has been disappointment. As a rule, however, the cause will be found by the man who is determined to make every failure the stepping-stone to future success.

The lengthening days and the growing power of the sun demand increased vigilance and activity. Danger of frost remains, and, worse still, there may come the withering influence of the north-east wind, which scorches delicate seedlings as with a breath of fire.

Annuals, Hardy, may be sown in the open from February to May. Perhaps a list of the principal flowers comprised under this denomination may aid the memory. Several of the following are not strictly hardy, but for practical ends they may be so regarded.

Abronia
Acroclinium
*Alyssum
*Asperula
Bartonia
*Cacalia
Calandrinia

Calendula
Candytuft
Centranthus
Chrysanthemum,
annual
Clarkia
Collinsia
Collomia
Convolvulus minor
Coreopsis
Cornflower
Dimorphotheca
Erysimum
Eschscholtzia
Eutoca
Gilia
Godetia
*Gypsophila
Hawkweed
Helichrysum
Hibiscus
Jacobea
Kaulfussia
*Larkspur
*Lavatera
Layia
*Leptosiphon
Leptosyne
Limnanthes
Linaria
Linum
Love-lies-bleeding

*Lupinus
Malope
Marigold
*Mathiola
*Mignonette
Nasturtium
Nemophila
Nigella
Phacelia
Platystemon
*Poppy
Prince's Feather
Rudbeckia
Salpiglossis
Sanvitalia
Saponaria
Silene
Sunflower
Swan River Daisy
Sweet Pea
Sweet Sultan
Venus' Looking-glass
Venus' Navel-wort
*Virginian Stock
Viscaria
Whitlavia
Xeranthemum

Hardy annuals are worth better treatment than they sometimes receive. They may be sown at once where they are intended to bloom, and for the varieties preceded by an asterisk this method is a necessity, because they do not well bear transplanting. In every case sow thinly, and afterwards thin boldly, for many of the flowers named will occupy a diameter of one or even two feet if the soil is in a condition to do them justice. Give the ground a deep digging and incorporate plenty of manure, except where Nasturtium is to be sown. A rather poor soil is necessary for this annual, or the flowers will be hidden by excessive foliage.

Abutilon.—There is yet time to raise plants for blooming in the current year. The seedlings must be potted on regularly to render them robust and free-flowering.

Aster.—Only those who are closely acquainted with the modern development of this handsome flower can have any conception of its varied forms and colours. There are dwarf, medium, and tall varieties in almost endless diversity, and nearly all of them will be a credit to any garden if well grown. Too often, however, flowers are seen which are a mere caricature of what Asters may become in the hands of men who understand their requirements. To grow them to perfection the ground should be trenched in the previous autumn, where the soil is deep enough to justify the operation. If not, the digging must be deep, and plenty of decayed manure should be worked in. Leave the ground roughly exposed to the disintegrating effects of winter frosts; and in spring it should be lightly forked over once or twice to produce a friable condition, in which the roots will ramify freely and go down to the buried manure for stimulating food. If by such means stiff land can be made mellow, it will grow Asters of magnificent size and colour.

In sowing it is not wise to rely on a single effort. We advise at least two sowings; and three are better, even if only a few plants are wanted. This diminishes the risk of failure and prolongs the flowering season. Prepare a compost of leaf-mould and loam, mixed with sharp sand to insure drainage. Towards the end of the month sow in pots or in seed-pans on an even surface; and we lay stress on a thin sowing, to avoid the danger of the seedlings damping off. Barely cover the seed with finely sifted soil, and place sheets of glass on the pans or pots to check rapid evaporation. If water must be given, immerse the pots for a sufficient time, instead of using the water-can. A cool greenhouse, vinery, or a half-spent hotbed is a good position for the pans, and a range of temperature from 55° to 65° should be regarded as the outside limits of variation.

Auricula.—Seed may still be sown; indeed, April will not be too late. Partially submerging the pans when water is needed saves many seeds from being washed out and wasted.

Balsam.—Although this flower comes from a tropical climate, it is not very tender; a gentle hot-bed is quite sufficient to bring up the seed. Two or three sowings are advisable to secure a succession of bloom, and for the first of them the middle of this month is the proper time. It is important that the soil for this plant should be light, rich, and very sweet. When the seedlings show their first rough leaves, lose no time in pricking them off, and they should afterwards be potted early enough to promote a dwarf habit.

Calceolaria.—Plants from last year's sowing will begin to move, and should be shifted into their final pots before the buds show. The eight-inch size ought to contain very fine specimens. The compost for them should be prepared with care several days before use. Put the plants in firmly, and place them in a light airy greenhouse. As soon as the pots are filled with roots an occasional dose of manure water will be beneficial until the flowers begin to show colour, when pure soft water alone will be required. Tie out the plants some time before the buds attain full size.

Clerodendron fallax.—A charming stove plant, producing large heads of bright scarlet flowers suitable for greenhouse decoration. From seed sown in March or April there should be a show of bloom in August or September following.

Coleus is strictly a stove perennial. But our short winter days do not maintain a rich colour, and it will in almost every instance give more satisfaction if treated as an annual, enjoying the beautiful and varied foliage during summer and autumn, and consigning the plants to the waste-heap as wintry days draw near. We do not advise the sowing of seed earlier than March, because a considerable amount of daylight is necessary to the development of rich tints and diversified markings in the foliage. The essentials for raising plants from seed are good drainage, a temperature which does not fall below 65°, the careful employment of water, and the early transfer of the seedlings. The green plants may be thrown away immediately they reveal their character, but those which show delicate tints in the small leaves will abundantly compensate for all the care bestowed upon them.

Dianthus.—Put the seedlings into single pots, and harden in readiness for transplanting to the open in May or June.

Dimorphotheca.—This valuable half-hardy annual, a native of South Africa, known also as the Star of the Veldt, may be flowered within six weeks from time of sowing. Plants may be raised by starting seed this month or in April, in pans of light soil given the protection of a frame. Transplant in May, in well-drained soil, choosing a warm sunny spot. In the open, seed may safely be sown in May or June. Plants potted on from the early sowing will make a most attractive show in the conservatory, or seed may be sown in pots and the seedlings thinned to three or four in each.

Gaillardia.—To secure a supply of plants for the open ground in May, seed of all the varieties may be sown during this month. Prick off early and keep them dwarf.

Geum.—From seed sown this month or in April, the popular double variety, Mrs. Bradshaw, may be brought into flower in the first year. The seedlings should be pricked off into boxes and gradually hardened for putting out in May or June.

Gladiolus.—This is one of the most stately and beautiful flowers grown in our gardens. Some of the varieties are strikingly brilliant; others are exceedingly delicate in tint and refined in their markings. The culture may be of the most primitive kind, or it may become one of the fine arts of horticulture. Simply put into the ground and left to fight their own battle, the corms sometimes produce splendid spikes of flower, although not so imposing as better culture might have made them. Under skilful care the flowers are magnificent in size and colour.

The main work of preparing the ground should be done in autumn. Now it is only necessary to give the soil two or three light forkings, and those not deep enough to bring the buried manure to the surface. This frequent stirring is beneficial in itself, and it promotes the destruction of the foes which prey upon Gladiolus roots. Small Potatoes, roughly hollowed out, or pieces of Carrot, may be used as traps for wireworm and other vermin. Planting is sometimes done at the end of this month, but as a rule it is better to wait until the beginning of April.

Gloxinia.—There is yet time to secure a brilliant summer display from seed. Bulbs which have been stored through the winter need attention. Where these flowers are wanted early, and there is plenty of room, a commencement will probably be made in February; but in the greater number of gardens March is soon enough. Assuming the bulbs to be sound, they should be potted in a mixture of loam, peat, and sand. Those which start first must be re-potted for a forward supply. While growing, manure water twice a week will help to produce fine flowers, intense in colour; but when the flowers open, the liquid manure must be abandoned, and pure soft water be given as often as necessary, for Gloxinias cannot endure drought. Shading is an important matter from the commencement, and particularly during the flowering period.

Hollyhock seedlings will be ready for putting into thumb pots. Directly they are established, begin to prepare them for planting out in May.

Impatiens.—Some growers find a little difficulty in raising this elegant flower from seed. Probably it arises from sowing too early. Where there is a command of sufficient heat no trouble should be experienced in March, and it is essential to sow very thinly for two reasons. Crowded seedlings are liable to damp off, particularly in dull, moist weather, and they are so fragile that it is well-nigh impossible to transfer them from the seed-pots until they are about an inch high.

Lavatera.—As the Mallows do not transplant well it is desirable to sow in the flowering positions. Good ground is necessary to insure fine specimens, and ample space must be allowed for the plants to develop. The seed may be sown from March to May.

Lobelia.—The perennial varieties make splendid border plants, and are easily grown from seed. Sow during February or March, in moderate heat, and in due time transfer to a deep rich loam. Their dark metallic foliage and brilliant flowers are most conspicuous, and admirably fit them for the back row of a ribbon border, or for groups in the mixed border.

Lupinus.—Seed of the annual varieties may be put in from March to May, and it is necessary to sow where required for flowering, as transplanting is not satisfactory. The perennial Lupines may also be flowered as annuals by sowing seed in March or April.

Marigold.—Both the African and French varieties are of importance late in the season, for they continue to bloom until cut down by frost. The former reaches the height of from eighteen to thirty inches, and the colour is limited to yellow in several shades, from pale lemon to deep orange. The latter is more varied in habit as well as in colour, and the Miniatures make excellent bedding plants. In hot dry seasons Marigolds entirely eclipse Calceolarias, because they can well endure drought and a short supply of food; whereas the Shrubby Calceolaria does not thrive under such conditions. All the varieties of Tagetes may be sown now on a moderate heat, and they should be pricked off into pans or boxes in readiness for transferring to the open ground in May.

Marvel of Peru.—The treatment prescribed for Balsam will suit this plant. In the first year it will grow to a considerable size, but will not, as a rule, attain to its full dimensions until the second season. It is a half-hardy perennial, and when saved through the winter will need protection from frost.

Mignonette finds a welcome in every English garden; and to add to its attractiveness there are now yellow, red, and white varieties, in addition to such forms as dwarf, pyramidal, and spiral. Mignonette can be grown without the least difficulty; indeed, it will reproduce itself from seed shed in the previous year. Nevertheless, it is true that in the majority of gardens justice is seldom done to this flower, for the simple reason that there is not sufficient faith in its capabilities. Each plant will cover a space of at least one foot, and we have seen specimens a yard across, bristling with flower-spikes which are delightfully fragrant. The soil for it should be made firm, just as an Onion bed is treated. Except for this one point, the culture of a hardy annual is all that is necessary. Mignonette does not transplant successfully, but otherwise it is very accommodating. The seedlings are frequently taken off by fly as fast as they appear above ground. Soot and wood-ashes applied in good time are the best preventives; but a second sowing may be necessary, and it should be made immediately the loss is discovered.

Nemesia.—For the earliest display of this beautiful annual the first sowing should be made in pots under glass during this month. In the open border seed may be sown in both May and June. Occasionally a little difficulty is experienced in raising plants under artificial conditions, but those who sow in beds or borders from the same packet of seed during the months named, will find that the culture is quite easy.

Pentstemon.—The treatment recommended for the perennial section of Lobelias will exactly suit this flower.

Phlox Drummondii.—There is still time to sow. Established seedlings should be gradually hardened by free access of air, until they are ready for the open ground.

Phlox, Perennial, may be raised from seed sown in shallow boxes in the early part of this month, and placed in moderate heat. Transplant the seedlings when ready, gradually harden, and plant out in rich soil one foot apart, or put them into vacant places in the shrubbery. Aid with water if necessary.

Poppy.—The annual varieties do not well bear transplanting, especially from light soils, and therefore, as a rule, it is advisable to sow where the plants are intended to bloom. They make conspicuous lines and clumps among shrubs; and this is especially the case with the huge flowers of the double class. Sow in March and April, and commence thinning the seedlings while they are small. They should ultimately be left about one foot apart. The perennial Poppies may also be flowered as annuals if sown in this month and transferred to open quarters when large enough.

Schizanthus.—Elegant half-hardy annuals, which can be grown as specimens for the conservatory, or in quantity for open borders. Sow in gentle heat, and pot on the seedlings.

Solanum.—For a succession of the varieties which are grown for their berries, sow again in heat, and make a sowing of the ornamental-foliaged kinds for sub-tropical gardening. The latter are rather more tender, and need a somewhat higher temperature than the former. They must all have liberal culture to bring out their fine qualities.

Statice.—The hardy annual varieties of Sea Lavender may be sown during March or April, and the best results are obtained by starting the seed in pans and planting out when the seedlings are far enough advanced in size. Seed of the hardy perennial kinds should be sown from April to July on light soil, and transplanted later on to flowering quarters.

Stock, Ten-week.—The increasing favour shown for Annual Stocks is in part no doubt attributable to the growing appreciation manifested for all kinds of flowers. But it is traceable in a still greater measure to the augmented purity, brilliance, and variety in colour of modern Ten-week Stocks, as well as to the enhanced reliability of seed in producing double flowers. We need say nothing of its perfume, for this is a quality which the most unobservant can scarcely fail to notice.

Although the Ten-week Stock is half-hardy, it must not receive the treatment of a tender annual; indeed, one of the most important points in growing it is to avoid any excess of artificial heat. A little assistance at the commencement it must have; but the aim should be to impart a hardy constitution from the moment the seedlings appear. We are not advocating reckless exposure to chill blasts, but the necessity of giving air freely whenever there may be a fair opportunity. The best of seed-beds can be made in pans or shallow boxes filled with sweet, sandy soil. In these sow thinly, so that the young plants may have abundant room. Even a little apparent wastefulness of space will be repaid by stout and vigorous growth. From the middle to the end of the month is a suitable time for sowing.

Sweet Pea.—This flower is so much in demand for decorative purposes that a prolonged display should be secured by successive sowings, commencing in this month and continuing until May, or even to June, where the soil and circumstances are specially favourable. The value of groups of Sweet Peas in borders and for enlivening shrubberies is now thoroughly appreciated, and it is not uncommon to see fine clumps among dwarf fruit trees.

Tigridia, or Ferraria.—Finer flowers are generally obtained from the open border than from pots, and the bulbs should be planted out three or four inches deep in March or April. Sandy loam and peat suit them admirably. On a dry border these bulbs will pass the winter safely, but in wet land it will be perilous to leave them out.

Verbena.—It is possible to raise Verbenas in the open from seed sown in drills on light soil, but the attempt is a little hazardous. There is, however, no danger at all in sowing in pans placed in a cool frame. The plants should be potted immediately they are large enough to handle. The flowering from this sowing will be rather late, but not too late for a good show of bloom.

Zinnia.—The double varieties are now grown almost to the exclusion of single flowers, and the former are so incomparably superior, that they are judged by the severe rules of the florist. With this plant it is useless to start too early. Towards the end of the month a commencement will be made by experienced growers, but the comparative novice will be wise to wait until the beginning of April. Sow in pots filled with a compost of leaf-mould, loam, and sand, and be quite sure there is effectual drainage. Plunge the pots in a temperature of about 60°.

Many half-hardy flowers, such as Acroclinium, *Convolvulus major*, *Linum rubrum*, Nemesia, Salpiglossis, Schizanthus, and others, which at an earlier period can only be sown with safety under protection, may now be consigned to the open ground without the least misgiving. A knowledge of this fact is of immense value to owners of gardens that are destitute of glass, for it enables them to grow a large number of flowers which would otherwise be impracticable. Of course, the flowering will be a little later than from plants raised earlier in heat.

Annuals, Hardy, which were not sown in March should be got in during this month and in May. A large number of beautiful subjects are available for the purpose, the most popular of which are named on page 373.

Aster.—When the seedlings attain the third leaf, they should be pricked off round the edges of 60-sized pots; later on put them singly into small pots, from which the transfer to the open ground will not cause a perceptible check. As the plants do not thrive in a close atmosphere, it is important to give air freely on every suitable occasion, or they cannot be maintained in a healthy growing condition. A second sowing should be made about the middle of the month, following the routine already advised. A sowing in drills on a carefully prepared bed in the open ground is also desirable, and in some seasons it may produce the most valuable plants of the year. Asters come so true from seed that the bed may be arranged in any desired pattern. Thin the plants early, and continue the process until they are far enough apart for flowering. A distance of eight inches is sufficient for the miniatures, ten inches for the dwarfs, and twelve or fifteen inches for the tall varieties.

Balsam.—About the middle of this month will be the time for a second sowing, and the seed may be raised in a frame without artificial heat.

Canterbury Bell.—Sow in good soil from April to July and transplant when ready. Under generous treatment these hardy biennials make a beautiful display in borders and the pure colours show with striking effect against the dark foliage of shrubs.

Carnation.—Any time from now until August will be suitable for sowing, and if the seed has been saved from a first-class strain, a good proportion of very fine flowers will be produced in the following year. For these plants florists have always considered it important that the potting soil should be prepared months before use, and there are good reasons for the practice. If this is impossible, see that the compost is sweet, friable, and, above all, free from that terrible scourge of Carnations, the wireworm. Even sifting will not rid the soil of its presence with certainty, but by spreading thin layers of the mould evenly upon a hard, level floor, and passing a heavy roller over it east and west, then north and south, the wireworm will be disposed of. Or dressing the soil with Vaporite two or three weeks in advance of potting will often prove effectual. Turfy loam three parts, leaf-mould one part, decayed cow-manure one part, with an addition of sharp sand, make a first-class compost. Sow in well-drained 48-sized pots, cover the seed very lightly, and place in a frame. Transplant the seedlings immediately they can be handled, when a cool, shaded pit will keep them in hard condition. After six or eight leaves are formed it will be time to plant them out. In the following spring the usual routine of staking and tying must be followed.

Chrysanthemum leucanthemum (Marguerite, or Ox-eye Daisy).—Seed of these well-known perennial varieties may be sown any time from April to July. There are several greatly improved forms of this popular flower which may now be had in bloom from May until early autumn. Start the seedlings on a bed of light soil, and when large enough transplant them to positions for flowering in the following year.

Cyclamen.—The bulbs which have been flowering in pots through the winter are now approaching their period of rest, and they must not be neglected if they are to make a satisfactory display next season. Water should be gradually diminished until the foliage dies off, and then the corms will require shade, or they will crack. Dry treatment generally results in an attack of thrips, and each root must be painted with some good insecticide to destroy the pest. Cyclamen should never be allowed to become actually dust-dry; but if the pots can be plunged in a shaded moist pit, watering will rarely be necessary. In June the pots may be buried to the rim in a shady spot until August, when it will be time to re-pot and start the bulbs into growth. The chief enemies of Cyclamen are aphis and thrips. Fumigation will settle the former; for the latter, dip the plants in a solution of tobacco-water and soft soap.

Dahlia, seedlings must have plenty of water, and be kept free from aphis while in pots. Instead of taking out the leading shoot, as is often done, give it the support of a neat stick. The plants should also be potted on as growth demands, the important point being to maintain steady progress without a check until they can be planted out. At the same time they must be hardened in readiness for removal to the open ground; and if the work is carried on with judgment, the plants will be dwarf, and possess a robust constitution capable of producing a brilliant display of flowers until frost appears.

Gladiolus.—Assuming that the beds have been properly prepared, we have now only to consider the question of planting, and no better time can be chosen than the beginning of April. Some eminent growers are at the trouble of taking out the soil with a trowel for each bulb. In the opening, a bed of sand and wood-ashes or powdered charcoal is made, on which the root is placed. Others lay them in deep drills, partly filled with a similar light mixture. Whichever method is adopted, the crown of the corm should be left about four inches beneath the surface. The distance between them may vary from twelve to eighteen inches, and the greater space is a distinct advantage when attending to the plants subsequently. The same rules apply to the planting of clumps.

Kochia trichophylla.—Sow seed where the plants are to stand, or in a prepared bed from which they can be transferred to make clumps, lines, or single specimens where the attractive foliage will be most effective.

Lobelia.—Early in the month transfer the seedlings to pans or boxes, but the latter are preferable. Not a single flower should be allowed to show until the plants are established in the open ground. Although Lobelias are very attractive in pots, they cannot be satisfactorily grown in them, with the exception of the *ramosa* varieties. But the object is easily attained by potting plants from a reserve bed after they have developed into good tufts. From a stiff soil they can be lifted and potted with facility; and a light soil will cause no difficulty if the bed be soaked a short time in advance. After potting, the plants will give no trouble, except to supply them with water.

Marigolds can be raised in a cold frame, and towards the end of the month there will be no risk in sowing in the open ground. The plants thrive in a sunny position, even in scorching seasons.

Marvel of Peru.—If not sown last month, there is no time to lose; and with a little care seed can now be germinated without artificial heat. When the plants come to be transferred to the open, put them, if possible, in sandy loam, exposed to full sunshine.

Mignonette.—Successional sowings may be made up to the end of June. Give each plant plenty of room. By removing the seed-pods as fast as they are formed flowering is greatly prolonged.

Nasturtium.—Both dwarf and tall varieties are usually treated as hardy annuals, with the exception of the date of sowing. None of the Nasturtiums are quite hardy, and if sown in March the plants are liable to destruction by late frosts. It is therefore usual to sow in April or May, according to the district, and the growth is so rapid that the plants are full of bloom before the summer has far advanced. Sow on poor soil always.

The *Tropæolum canariense* (Canary Creeper) may be raised in pans from a March sowing for planting out in May, or seed can be sown in the open during April.

Petunia.—Plants from the first sowing will be ready for small pots, and they must be kept going until the 48-or 32-size is reached. All Petunias rebel if root-bound, and the double varieties are especially impatient in this respect. After each transfer give them a sheltered, shady position and attention with water until they start again. Good drainage and careful ventilation are essential, or the foliage will lose colour. Seedlings intended for beds may be transferred direct from the seed-pans into 60-sized pots.

Picotee and Pink.—See the culture prescribed for Carnation.

Ricinus.—At quite the end of the month or the beginning of May, seed put into the open ground will produce splendid specimens if treated with a lavish hand. Take out the soil for a depth of eighteen inches or two feet, and fill the space to within three inches of the surface with a mixture of rich soil and well-decayed manure. Upon each bed thus made place three Ricinus beans in a triangle, and when they are up, thin to one plant at each station, and this, of course, the strongest. This mode of growing Ricinus will astonish those who have been accustomed to allow the plant to struggle through existence in the ordinary soil of a garden border. Plentiful supplies of water must be given in dry weather, and stakes will be necessary to save the specimens from injury by wind. It is too early for putting out those raised in heat.

Stock, Ten-week.—Where the requisite quantity of seed has not been sown, it must be done promptly. If there happens to be a cold frame on a spent hot-bed to spare, it will exactly suit the seedlings when they are ready for transferring. Make the surface fresh by adding a little rich soil, and put the plants in rows three or four inches apart, allowing three inches between them in the rows. In seed-pans, however, space cannot be afforded in this liberal fashion, but they will make a full return for rather more than the usual spacing. To maintain a dwarf habit, it is imperative that the plants should be kept near the glass.

Where there are no facilities for growing Stocks in the manner described seed may be sown at the end of the month in the open ground, and with a little care there will be a handsome show of bloom. The seedlings are subject to the attacks of turnip fly, which is a terrible foe to them in the seed-leaf stage; in fact, the plants are sometimes up and gone before danger is suspected. A light sprinkling of water, followed immediately by a dusting of wood-ashes, just as they are coming through, will save them, but it may be necessary to repeat the operation two or three times until they are out of peril. A rich and friable seed-bed is one remedy for the fly, for it promotes rapid growth, which speedily places the plant beyond the power of its insect adversary. But if open-ground culture exposes Stocks to one hazard, it saves them from another, as mildew does not attack them unless they have been transplanted. Stocks come so true from seed that it is easy to arrange a design in any desired colours. Sow in drills from nine to fifteen inches apart, according to the height of the variety, and cover the seed very lightly with fine soil. The bed must be protected from birds, and a dressing of soot will keep off slugs. Begin to thin the plants early, but do not forget that some single specimens will have to be taken out when the flowers show, and that is the time for the final thinning.

Sunflowers do not well bear transplanting, hence the seed should be sown where the plants are intended to flower. During its brief season of growth, the Sunflower taxes the soil very severely, and to develop its full proportions decayed manure must be freely employed to a good depth, and unstinted supplies of water will be necessary in dry weather.

Zinnia.—The first week of this month is as good a time as any to sow seed, and the conditions named under March should be followed. When the seedlings are an inch high, pot them separately, and place in a close, shaded frame until they are established. Then give air more and more freely while the plants are being trained to bear full exposure.

MAY

This is the chief month for bedding, and the crowded state of pits and houses creates a natural anxiety to push forward the work; yet the exercise of a little patience may save many a valuable lot of plants from being injured past recovery. Although the days are long, and perhaps sunny, the nights are often treacherous, especially in the early part of the month. The first business is to prepare the plants gradually for transfer to the open ground by free exposure whenever there is a favourable opportunity. Take off the lights on genial days, and by degrees open them at night, until they can be dispensed with altogether. About the second week of the month it will generally be safe to put the most hardy subjects on a bed of ashes, under the shelter of a hedge or wall, before planting them. Begin with Antirrhinum, Dianthus, Phlox Drummondii, Stock, and Verbena. A little later on, others which are rather more delicate, as, for instance, Balsam, Begonia, Dahlia, Petunia, Zinnia, &c, can be treated in the same way, until the great bulk of them are in final quarters. Sub-tropical plants, such as Nicotiana, Ricinus, Solanum, and Wigandia, had better be kept under control till the first or second week of June.

Annuals.—There is still an opportunity of sowing many varieties, and also to make further sowings of others that are already showing signs of promise. The practice of insuring a succession of all flowers much in demand for vases, of which Sweet Peas are an example, is on the increase, and deserves to be further extended. Another point is that many annuals which require heat in earlier months may with confidence be sown during May in the open ground.

Hardy Biennials and Perennials.—Seed of many favourite biennials and perennials may be safely sown in the open ground during May, June, and July, and as a general rule the finest plants for flowering in the following season are obtained from the earliest sowings. The bed for the seed should be prepared with care and a friable loam is the best for the purpose. Immediately the seedlings are large enough to handle, transplant to small rich nursery beds and shift to flowering positions in the autumn. A number of these subjects are dealt with individually in the calendars for the months named, and others which are suitable for the purpose are:

Anchusa italica
Aster sub-cæruleus
Aubrietia
Candytuft (Iberis)
Cheiranthus Allionii
Chrysanthemum leucanthemum
Coreopsis grandiflora
Cynoglossum
Digitalis
Gaillardia
Galega officinalis
Gaura Lindheimeri
Geum
Gypsophila paniculata
Heuchera
Lupinus
Cenothera
Poppy, perennial
Pyrethrum
Saxifrage
Thalictrum
Verbascum
Viola

Antirrhinum is admirably adapted for a dry and sunny position, in which it will thrive and flower freely.

Balsam.—Towards the middle of the month a final sowing may be made with safety in the open ground. Former seedlings will need potting on until they reach the eight-inch size, and at each transfer put the plants in rather deeper than before; this encourages the growth of roots from the stems. While increasing the pot-room not a bud will show; but immediately the roots are checked by the pots, flowering will commence. The old method of stopping and disbudding not only spoiled the plants, but robbed them of the finest flowers, which are invariably produced on the main stem. Since the natural method of growing Balsams has been in favour it is usual to see grand specimens covered with immense flowers.

Campanula.—The hardy perennial varieties may be sown in the open during the present month to provide seedlings for transplanting to flowering positions in autumn. Should there be any good reason for delay it will not be too late to sow in June or July, but the finest specimens are generally produced from May sowings. The best results can always be obtained by raising the required number annually and discarding the plants after they have flowered in the following season.

Cineraria.—Those who care to have Cinerarias in bloom during November and December may do so from a sowing made at the beginning of April, but it is not usual to start so early. Our own practice is to sow twice, during the present month and again in June, to insure a succession. From this month's sowings we look for our finest plants. The Cineraria is easy to raise and to grow, but it will by no means take care of itself. It has so many enemies that unusual vigilance is necessary to flower it to perfection. It thrives in a compost of turfy loam, with a little leaf-mould added; but the soil should not be over-rich, or there will be much foliage and few flowers. Still, as the plant is a rapid grower, it must not be starved, neither must it suffer for lack of water. Pots or pans may be employed for the seed; and as the young plants grow freely, they may go straight to thumb pots without the usual intermediate stage of pricking off.

Coleus should be finally shifted into 48-sized pots. If signs of decline become manifest, weak liquid manure water given occasionally will revive the plants and intensify their colours. During the summer any ordinary greenhouse or conservatory will suit them, provided they are shaded from fierce sunshine.

Cyclamen.—The strongest seedlings should now be ready for 60-sized pots. Abundant but judicious ventilation, plenty of water, and freedom from aphis, are the conditions to be secured.

Dahlia.—Make the ground on which this flower is to be planted thoroughly rich. It is a rapid grower, and cannot attain to fine proportions on a poor soil. If the plants are carefully prepared for the change by free exposure on genial days, and also during warm nights, they will scarcely feel the removal. When first put out, dress the surrounding soil with soot to prevent injury by slugs, which show a decided partiality for newly planted Dahlias. Give water freely when requisite, and in staking the plants take care that the ties do not cut the branches. These ties will require attention occasionally during the summer and autumn.

Delphinium.—Sow the perennial varieties on a prepared bed. Thin early, without removing all the weaker seedlings, and when sufficiently advanced to bear removal, transfer to borders where the plants are to flower.

Hollyhocks may be put into the borders when the weather is quite warm. Wait until the end of the month, or even the beginning of June, rather than have them nipped by an untimely frost. Like the Dahlia this plant must have unstinted supplies of water and abundance of manure. A tall stake, firmly fixed, will also be necessary for each plant.

Nicotiana.—Seed may be sown on an open, sunny border, but it is a waste of seed and labour to put it into poor soil. Prepare the ground beforehand by deep digging, and by incorporating plenty of manure. If the near presence of other plants renders this impossible, drive a bar into the soil and work a good-sized hole. Fill it with rich stuff to within a few inches of the surface, and finish with fine soil, on which sow the seed. This method can only be adopted for light land. In the event of a cutting east wind after the seedlings are up, improvise some kind of shelter until the danger is past.

Petunias are very sensitive under a frost or cold wind. Therefore be in no hurry to bed the plants until quite the end of the month or beginning of June, especially if the weather appears to be at all threatening. A good mellow soil, free of recent manure, suits them. If unduly rich, it will strengthen the foliage at the expense of the flowers, and will also postpone the blooming until late in the season.

Portulaca.—It is useless to sow until the temperature is summerlike. If necessary, wait until the close of the month, or longer, before putting in the seed. This flower will endure neither a moist atmosphere nor a retentive soil. Sow on raised beds of light soil, the more sandy the better; and in seasons which speedily burn the life out of other plants, Portulacas will display their beauty, no matter how fiercely the sun may beat upon them. Water will occasionally be necessary, but it should never be given until there is obvious need for it. Portulacas are easily grown in pots or window-boxes, and they will bloom profusely where many other flowers only wither and die.

Primula.—Almost every season witnesses the advent of some novelty in this flower, either in colour or in form. And the plant is now worth growing for the beauty and diversity of its foliage alone. The flowers range from pure white through all shades of tender rose up to a deep, rich crimson. After years of earnest effort, two beautiful blue flowers have been obtained. There are also several elegant double strains, and these possess a special value for bouquets, because of their enduring quality. All the varieties, including the popular Star Primulas, can be grown with ease in any soil which is fairly rich and friable. Equal parts of leaf-mould and loam, with a little sand, will suit them to perfection. Fill the pots firmly, taking precautions to insure effective drainage. A thin layer of silver sand sifted over the soil will aid an even sowing by showing up the seed. As a finish, shake over just enough fine soil to hide the sand. Thin sowing is important, because the most reliable new seed is almost certain to germinate at intervals, and the plants which come first can then be lifted without imperilling the remainder. Prick off as fast as ready round the edges of small pots, and shade until established. Then give air more and more freely.

Stock, Ten-week.—The preparation of the soil is the first business, and whether the Stocks are intended to be grown in small groups or alone in beds, the treatment should be the same in either case. With light land there is no difficulty; it is only needful to dig it well, and to incorporate a sufficient quantity of decayed manure. If disposed to incur a little extra trouble to give the plants a start, take out some soil with a trowel, and fill the hole with compost from the potting shed. This course is indispensable on heavy land; and assuming it to be rich enough, the quickest and most effectual way is to make drills six inches deep at the proper distances, and nearly fill them with prepared soil, in which the Stocks can be planted. For a short time afterwards provide shelter from the midday sun, but do not keep them covered a moment longer than is necessary. In planting it must not be forgotten that an uncertain proportion of single specimens will have to come out. On this account it is advisable to put them in small groups, and remove the surplus even if they are double,

Sweet William.—The introduction of several new varieties has created a fresh interest in this fine old garden favourite. This is one of the hardy biennials that will not be hustled. On a nicely prepared bed in the open sow thinly in drills either during this month or up to July. In due time transplant in rows, affording sufficient space for each specimen to become stocky, and in autumn transfer to flowering quarters.

Verbena.—Beds for Verbenas should be rich, mellow, and very sweet. A poor soil not only produces poor flowers, but it materially shortens the blooming period. Peg the plants down from the outset, and allow them to cross and recross each other until there is a sheet of glowing colour.

Wallflower.—This fragrant spring flower is not always grown as well as it might be. It is often sown too late to become established before winter sets in. Sow now in drills nine inches apart on friable loam. Thin to three inches apart, and transplant the thinnings. A little later repeat the operation, so as to leave the plants at a distance of six inches in the rows. Assist them with water if necessary.

Zinnia.—A sowing in the open ground about the middle of the month will provide plants in gardens where there are no means of raising them artificially at an earlier date. Even those who possess a stock will be wise to put a final sowing in the open. If possible, choose a sunny border sloping to the south, and make the soil rich, fine, and rather firm. Drop seeds in little groups of three or four at each spot, allowing fifteen or eighteen inches between the groups. Cover lightly, and eventually thin the plants to one at each station.

JUNE

The days are now at their longest, and plants in pits and houses should have the full benefit of it. By opening the lights early, and shading in good time, the flowering period will be greatly prolonged. Ply the syringe over plants infested with aphis until they are quite clean. In some instances, it may even be wise to pinch off young shoots which are covered with the fly.

Keep Verbenas, Petunias, and the taller varieties of Phlox Drummondii pegged down; this furnishes the beds and helps to check evaporation.

Rain and watering alike tend to harden the ground; and as this condition does not favour growth, the surface should be frequently broken with the hoe.

Anemone.—Those who grow this flower from seed should make another sowing now or in July, even if they have thrifty plants from the February sowing. By this arrangement the flowering period is prolonged, and the finer blossoms will probably come from this month's sowing.

Aquilegia seed will germinate now in the open ground, and the plants need no protection during winter.

Balsam.—As a rule, it is unwise to put Balsams into beds or borders before the first week of this month. The plant revels in warmth and light, and should have an open, sunny position. Its succulent nature will indicate the necessity of giving abundant supplies of water. For so fleshy and apparently fragile a plant, it is astonishing how well it stands in a strong wind. From good strains the separate colours come so true that the design of a bed can be accurately arranged. As pot plants Balsams need no support, provided they are kept dwarf and stout, and they make admirable decorative subjects. But for indoor use it is easy to grow them in the open ground, and when well advanced they can be lifted with care and potted. This procedure offers the advantages of a choice of colours even from mixed seed and a selection of the most robust plants.

Begonia, Tuberous-rooted.—This has proved to be one of the most elegant and refined bedding subjects we possess, and it appears to become more popular every year. The plant is also freely grown in the reserve border to produce flowers for cutting. Employ specimens that are large enough to make a show at once, and select plants of the short-jointed class for outdoor work. They must have unusually rich soil.

Calceolaria.—For wealth of bloom, combined with richness and intensity of colouring, the Herbaceous Calceolaria has no rival among biennials. A large greenhouse filled with fine specimens in their full splendour is a sight which will not soon be forgotten. One great source of interest lies in the annual changes in shades of colour, and the variations in the markings of individual flowers. From a first-class strain of seed, high expectation will not be disappointed. Indeed, the excellence of seedlings is so fully recognised, that there is not the smallest advantage in propagating the plant by the tedious method of cuttings. But Calceolarias will not be trifled with. They must have an even temperature and unremitting attention to maintain a thriving condition. Fill the seed-pans or pots firmly with a compost which is both rich and porous; the last point is of great consequence in helping to secure free drainage. Make the surface perfectly even, and whiten it with silver sand; this answers the double purpose of revealing the seed and afterwards of showing when it is sufficiently dusted over with fine soil. Whether or not this method be adopted, the sowing must be thin and even, and as the seed is exceedingly fine, the task is rather a delicate one. Sheets of glass placed over the pans and turned daily will check rapid evaporation. Place the pans in a moist, shady spot, where the temperature is constant, and germination will take place in from seven to nine days, when the glass must be promptly removed. Then comes a critical stage, and a little neglect may result in the loss of past labour, and necessitate a fresh start. Still keep the pans in some sheltered corner which can be thoroughly shaded from the sun. This question of shade needs much vigilance. So also does the supply of water, which must not be administered wholesale, but rather by frequent gentle sprinklings. On the appearance of the second leaf, promptly prick off the seedlings in carefully prepared pots, allowing about two inches between them. They will need dexterous manipulation because of their small size, but a skilful hand will transfer them without injury, and perhaps with a little soil adhering to the roots. As all the seedlings will not be ready at one time, it will probably require about three operations to clear the seed-pans, and the early removals should be so made as to avoid injuring the remainder. A pen, with the point firmly pressed into the holder, makes a small handy implement for the task. Retain the seedlings in a sheltered position, and continue the attention as to shade and watering. In about a month the plants will be ready for thumb pots.

Canna.—In the mixed border, and also in the sub-tropical garden, Cannas are much valued for the exceeding grace and beauty of their foliage. They should be put into very rich soil; and, like all other plants of rapid growth, they will need copious supplies of water in dry weather. In mild districts and on dry soils the plants may remain out all the winter, under the protection of a heap of ashes. But, as a rule, it will be necessary to store them in frames until spring; and they may be finer in the second than in the first season.

Cineraria.—To insure a succession, and where a sufficient stock is not already provided, another sowing should be made, following the method advised last month. The seedlings, when transferred to small pots, should be put into a close frame, and be sprinkled with water morning and evening until the roots take hold. At first it is desirable to keep them fairly warm, but in a fortnight the heat may be gradually reduced and more air be given until cool treatment is reached. The plants will need potting on up to November, when they should go into the final size; and, except for special purposes, 6-1/4-or 7-1/2-inch pots are large enough. Cinerarias are sought after by every pest which infests the greenhouse. We need only say that by fumigation, sulphur, or by syringing with a suitable insecticide, the plants must be kept clean, or they cannot be healthy.

Daisy, Double.—The finest blooms are obtained from seedlings raised annually, and the general practice is to sow in the open ground during this month or July. When large enough transplant to good ground for blooming in the following season. The new Giant forms of the Double Daisy are of superb size, closely resembling finely shaped Asters in form.

Dianthus.—For a display next summer, sow in drills drawn six inches apart in an open situation, and cover the seed lightly with fine soil. Shade the spot until the plants show.

Geranium.—Sometimes a difficulty is experienced in bringing Geranium seedlings into flower. They possess so much initial vigour that the production of wood continues to the very end of the season. Plants which show signs of excessive growth should be put into the border without removing the pots. This check to the roots will throw the plants into luxuriant bloom.

Gladioli are very liable to injury by high wind, and stakes should be put to them in good time. Each plant may have a separate support, and this is the most perfect treatment; or the stakes may be at intervals, or at the ends of rows, connected by lengths of strong, soft material, to which intervening stems can be secured. The work should be done carefully, and if the flowers are intended for exhibition they must also be shaded by some means. This may be a cheap or a costly proceeding; but in whatever manner it is carried out, security is essential, or the whole bed may be ruined.

Hollyhock.—A sowing in the open ground will produce plants for wintering in the cold frame; and if generously treated, they will make a fine show in the following year.

Myosotis.—During this month sow Sutton's Pot Myosotis and bring forward in a cold frame for winter decoration, for which purpose this plant is rapidly increasing in favour. Seed of the hardy varieties may also be sown now or in July, choosing a shady spot in the open ground. Transplant when large enough.

Nicotiana.—To expose Tobacco plants before warm weather is established will give them a check from which they may not recover until the summer is half over, if they recover at all. Spare frames with movable lights will prepare them admirably and save labour. The second week of this month is generally warm enough for the planting. The seedlings must have a very rich soil, and abundance of water in dry weather. A heavy mulch of decayed manure will supply them with food and check evaporation.

Pansy.—From the end of May to the end of July seedlings may be raised in the open ground. Thin and transplant when ready.

Polyanthus to be sown from May to July on a shaded border. Thin the seedlings boldly, and bed the thinnings. Those raised early will flower next spring, but the later seedlings cannot be depended on for blooming in the first year.

Portulaca.—The weather may have been too cold and wet for sowing in May, or seed then sown may have failed; happily, there is yet ample time for raising this flower, in either beds or pots.

Primrose.—This fine old favourite may be grown from seed in various tints of yellow and almost any shade of colour from white to deep crimson; an effective blue has also been achieved. Primroses make beautiful pot and border flowers. Seed may be sown from May to July. Seed-pans can be used, or the sowing may be made in drills in the open. In the latter case, a free dressing of soot must be employed to render the spot distasteful to slugs. When transplanting, give the plants a deep retentive loam if possible, and a shady position.

Primula.—To insure a succession of flowers next spring, make another sowing as advised under May. Seedlings which are ready should be got into small pots, and afterwards they must be re-potted when necessary; but never shift them until the pots are full of roots, and always put them in firmly up to the collar.

Solanum.—The berried varieties may be grown entirely in pots, or they can be put into beds for the summer, from which they will lift for potting again just as the handsome berries are turning colour. The spiny-leaved varieties are valuable for sub-tropical gardening. Small plants are of little worth, hence they should be put into very rich soil, with a thick layer of manure on the surface, and have copious supplies of water to induce free growth.

Stock, Spring-flowering.—This valuable section, which includes the popular Brompton strain, usually comes into bloom in May and June. Seed is sometimes sown where the plants are to flower, but a certain degree of risk attends this mode of procedure, and Spring-flowering Stocks are so valuable that they are worth more careful treatment. Either now or in July sow in pans, and place them under shelter until the plants are an inch high; then stand them in the open for a week before planting out.

Stock, Winter-flowering.—For their refreshing colours and delightful perfume Stocks are highly prized during the winter months. To have them in flower at Christmas, seed of Christmas Pink or Beauty of Nice should be sown in June. It is usual to grow three or more plants in a pot, according to size. At the fall of the year place them in the conservatory or a cool greenhouse, and give assistance in the form of weak liquid manure as soon as the buds appear. Other suitable varieties, of which there are a number, may be sown in July or August for flowering indoors through the winter and spring months.

Wallflower.—If no seed was sown in May the task ought not to be neglected this month.

Zinnia.—The first week of June is about the right time to bed Zinnias, and there are three facts to be borne in mind concerning them. They do not transplant well, and therefore a showery day should, if possible, be selected for moving them. In the absence of rain, be liberal with water. They are very brittle, and should have a position somewhat sheltered from the full force of the wind; and as they revel in sunshine, the more roasting the season the finer will be the flowers.

JULY

Antirrhinum.—A sowing in drills during the present month or August will supply plants for flowering next year. Transfer direct from the seed-bed to the positions where they are intended to bloom.

Calceolaria.—If more plants are wanted, sow again. Among the seedlings which we left last month just as they had been pricked off, it will soon be evident that there is a wide difference between the strength of the plants. As a rule, the most robust are those in which yellow largely predominates. These make bright and showy decorative plants, but the colours that are especially valued by florists will probably come from the seedlings which are weakly in the early stage. Hence these should be specially prized, and under skilful management they may be grown into grand specimens. The thumb pots for Calceolarias need careful preparation with crocks covered with clean moss or vegetable fibre, and they must be filled with rich porous compost. Transfer the plants with extreme care, and place them in a sheltered part of the greenhouse or in a shaded frame, allowing free access of air on the leeward side. If aphis has to be dealt with—and it is very partial to Calceolarias—fumigation is the best remedy. Choose a quiet evening for the operation; on the following day carefully water the plants and shade them from the sun.

Campanula.—The perennial varieties may still be sown, either in pans or in the open. Give them a good light soil, and do not stint the supply of water.

Cyclamens which are forward enough should be shifted into 48-sized pots. Follow up the process until all are re-potted.

Lobelia.—In pots or pans sow seed of the perennial varieties to provide plants for the borders next year. Pot off singly when ready, and protect in a cold frame through the winter.

Mimulus sown in the open ground will flower in the following spring. If possible, make the seed-bed in a moist retentive soil and in a shaded situation.

Primula.—To force the growth of this plant is to ruin it. The most satisfactory results are invariably obtained from specimens which have matured slowly, and have been treated as nearly hardy after the seedling stage. From this month up to the middle of September it will be quite safe to expose them freely, day and night, except in inclement weather. Even in the winter protection is only needed from frost, damp, and keen winds.

AUGUST

Annuals and Biennials, Hardy.—In the majority of English gardens the spring display of bulbous flowers is too often followed by a dreary blank, which is almost unredeemed by a touch of colour, except that afforded by the late Tulips and a few other flowers which are relatively unimportant. The brilliance of the Crocuses, Hyacinths, and early Tulips serves to throw into relief the comparative barrenness which follows. And the contrast is rendered all the more striking by the cheerful spring days. It is at this juncture that annuals and biennials from summer or early autumn sowings light up the garden with welcome masses and bands of fresh and vivid colouring. They are then so valuable that it is surprising they are not more commonly grown, especially as the cost of seed is very trifling. Even the transitory character of some of them is an element in their favour, for they do not interfere with the summer bedding arrangements. Such flowers as Pansy and Viola, however, produce a long-continued show of bloom.

The following list contains the varieties which are best adapted for the purpose:—

Alyssum, Sweet
Antirrhinum
Asperula azurea setosa
Calandrinia umbellata
Calendula officinalis fl. pl.
Candytuft
Cheiranthus Allionii
Chrysanthemum, Morning Star
Chrysanthemum, Evening Star
Chrysanthemum inodorum plenissimum
Chrysanthemum segetum gr.
Clarkia
Collinsia
Coreopsis
Cornflower
Erysimum
Eschscholtzia
Gilia tricolor
Godetia
Iceland Poppy
Larkspur, dwarf rocket
Leptosiphon
Limnanthes Douglasii

Linaria, pink
Nemophila
Nigella, Miss Jekyll
Papaver glaucum
Phacelia tanacetifolia
Poppy, Shirley
Saponaria calabrica
Scabious
Silene
Sweet Sultan
Venus' Looking-glass, purple
Virginian Stock
Viscaria
Whitlavia

Sow thinly, not later than the middle of the month in cold districts, but September will be early enough in the Southern counties. Drills are preferable to broadcasting, because the beds are more easily weeded and kept in order. Thin the rows early, so that the plants may become stout and hard before winter overtakes them. Early in the new year transplanting must be resorted to during open weather if the plants are to be flowered in heavy soil; but on light, rich land, sow where they are intended to bloom.

Annuals under Glass.—The flowers available for winter and spring blooming are naturally few in number compared with those which fill gardens and conservatories during the summer months. But it is not generally realised that several favourite outdoor annuals are as serviceable for flowering under glass in the short days of the year as they are for growing in the open ground in summer, and they are the more valuable for winter and spring use as no elaborate system of cultivation is needed. Any greenhouse or conservatory from which frost can be excluded will grow these annuals well. Seed should be sown in August or September, in pots or pans placed in a cool house or frame. When the seedlings have made some progress, prick them off into the pots in which they are wanted to flower, and grow steadily on, bearing in mind always that the most important point is to keep the plants as hardy as possible by giving air at every favourable opportunity. The following varieties are especially suitable for winter and spring flowering under glass:—Alonsoa; The Star and Dunnettii varieties of Annual Chrysanthemum; Clarkia elegans; Dimorphotheca; Gypsophila elegans; Linaria; Nemesia Suttoni; Nicotiana, Miniature White and N. affinis; Phlox, Purity; Salpiglossis; and Swan River Daisy.

Asters for indoor decoration should now be lifted from beds or borders and potted. It is worth a little trouble to accomplish the task with the least possible injury or disturbance to the roots. Light soils should have a good soaking of water on the previous evening, to prevent the mould from crumbling away.

Carnation.—Seed may still be sown as advised in April; but to carry the plants safely through the winter it is necessary to have them strong before cold weather sets in.

Chionodoxa can be forced with the same ease as Roman Hyacinths. A 48-sized pot will accommodate several bulbs.

Cinerarias are frequently placed in the open during this month and September, and as it tends to impart a hardy constitution, the practice is to be commended. A north border under a wall will suit them, but the proximity of a hedge should be avoided. Before the plants are put out see that they are quite clean, or it may be necessary to restore them to the house in order to rid them of some troublesome pest.

Clarkia.—The varieties of the Elegans class make very handsome pot plants, and to insure the requisite number seed must be sown in well-drained pots during this month or early in September.

Cyclamen.—Where Cyclamens are extensively grown it is usual to make the first sowing in August, and many gardeners regard this as the most important period for securing healthy young seedlings. A common mistake with beginners is to raise them in too high a temperature. On this and other points useful suggestions will be found in the article commencing on page 256.

Dianthus.—Either now or a little later transfer seedlings to flowering quarters, and if possible put them into sandy loam in a sunny spot.

Freesia.—Few and simple are the conditions necessary to the well-being of this beautiful and delicately scented flower. The fine specimens to be seen occasionally in cottagers' windows in the Isle of Wight attest the ease with which it can be grown in a congenial atmosphere. The bulbs are exceedingly small in proportion to the flowers, and the rootlets are so fragile that potting on is to be avoided. A 48-sized pot will hold five or six bulbs, and the soil should consist largely of decaying vegetable fibre, such as peat, leaf-mould, and turfy loam. The pots can be stood in any sheltered position out of doors, under a covering of cocoa-nut fibre or other light material, until the foliage begins to grow.

Geranium.—A sowing in August will supply plants for flowering next summer, and the directions given in February are suitable, save that heat can now be dispensed with. These late seedlings will need more care to carry them through the winter than plants raised earlier in the year.

Gerbera.—These charming flowers make admirable subjects for the greenhouse and conservatory, and an excellent display may also be obtained outdoors if a sunny well-drained part of the garden be selected for the plants. August is the best month for sowing seed. Plants required for indoor blooming should be potted on as may become necessary. Those for the open ground must be thoroughly hardened off for planting out in the early summer of the succeeding year.

Hyacinths, Italian and Roman.—Obtain the bulbs as early as possible, and pot them promptly. Place them in any spare corner of the open ground, where they can be covered with cocoa-nut fibre or leaf-mould until the roots are formed. A child can grow these flowers; and they should be largely employed for bouquets and for indoor decoration during the dark winter days.

Mignonette.—For winter flowering sow in 48-or 32-sized pots, filled with light rich soil. Put the seed in little groups, thin to three or five plants in each pot, and give them the benefit of full daylight close to the glass. When flowering commences do not allow seed to form. If the spikes which have passed the heyday of perfection are cut off, the plants will break again and flower a second time.

Narcissi.—The first potting of early varieties is made this month as soon as the bulbs can be obtained.

Pelargonium.—The remarks under Geranium apply to this flower also.

Picotee.—Follow the instructions given for Carnation.

Schizanthus.—To do full justice to this flower seed should be sown now for plants to be kept through the winter in any house which is sufficiently warm to exclude frost.

Scilla præcox, or sibirica.—The treatment which suits Roman Hyacinths will answer for this bulb also, when required for flowering indoors. The two form an admirable harmony in blue and white.

Silene.—All the most useful varieties of Catchfly are hardy against cold, but not entirely so against damp. They possess a special value for their sparkling appearance in spring. Sow in light sandy soil, in which they will pass the winter safely. On a heavy loam the transplanting system must be resorted to in February or March.

Stock, Intermediate.—This section is valuable for indoor decoration in spring. No artificial heat is necessary to raise the seed; in fact, it is not wise to employ it. Either in this month or early in September sow the required number of pots and plunge them in ashes in a frame until March. Thin the seedlings to three in each pot. Before flowering, a rich top-dressing will be beneficial; and manure water—weak at first, but stronger by degrees—will intensify the colours.

Stock, Spring-flowering.—A bed prepared under trees or shrubs will afford some shelter from winter frost. Make it thoroughly rich, and in it plant the seedlings. Should the growth be very rapid in September, the plants will probably become too succulent to endure the stress of winter. If so, lift them and plant again on the same spot.

Sweet Pea.—The modern culture of this delightful flower includes deep trenching and the liberal use of manure. Those who intend to sow during September in the open must get the trenched ground into perfect order early in the present month. The details are important and are fully described in the article commencing on page 303.

SEPTEMBER

Agapanthus taxes the soil severely, and must have ample nourishment in pots. It is also one of the thirstiest bulbs known, but is quite hardy, and will thrive in the open if planted in a deep rich loam at any time from September until March.

Alstroemeria.—Although related to the Ixia, this bulb may be trusted to the open ground in all but the coldest districts of the country. It is not suitable for pot culture, but in a dry border it may be allowed to remain undisturbed for years. Plant quite nine inches deep.

Amaryllis.—The proper time to commence operations with these superb flowers is during their season of rest, which ranges from September to March. Pot them in firm loam, enriched with leaf-mould, and containing a fair proportion of sand. Very little water is required until growth begins, and then it must be increased with the progress of the plant. Start them by plunging the pots in a temperature of about 65°, and when they are coming into bloom, remove to a warm greenhouse or conservatory. After the flowers have faded, allow the plants to complete their growth, and then slowly reduce them to a resting condition without permitting the bulbs at any time to become quite dry.

Anemone.—The tuberous varieties are valuable as pot plants, not only for their flowers, but also for the distinctive character of the foliage. The roots may be potted from now up to the end of the year, so that a succession of flowers can be easily insured. When plunged in a pit or frame to preserve them from frost, watering is all the attention they will need, but of this there must be plenty, particularly when the plants begin to flower. Pot the roots between one and two inches deep, in rich soil, and with the eyes upwards. A large pot will accommodate several roots.

Babiana.—Treat in the same manner as the Ixia.

Begonia, Tuberous-rooted.—Lift the plants which are in the open ground, and pot them to complete their season in the greenhouse; but if they are not wanted for this purpose, they may remain in the beds until October. When the stems fall, still retain the bulbs in their own pots, and store them in a dry cellar or shed, under a layer of cocoa-nut fibre. They need protection from both damp and cold. Neither hurry the drying off of the roots, nor attempt to force the growth in spring, but wait until they start naturally.

Calceolarias ought now to be in large 60-pots, placed close to the glass to insure a dwarf habit. During sharp weather they may be taken down, but should be restored immediately the danger is past. Much heat in winter will be injurious; a range of 45° to 55° should be considered the limits of variation in temperature. Pot the plants on as growth demands.

Crocus.—For indoor decoration, two or three separate lots should be potted at intervals of a fortnight; and the named varieties are worth this mode of treatment, both for the size of their flowers and for the exceptional brightness and diversity of their colours. Use a light rich soil, and put six to eight corms in a 48-sized pot. They may also be grown in quantity in large seed-pans or in shallow boxes. When coming into flower, the roots may be freed from soil to facilitate the packing into ornamental baskets or vases.

Crown Imperial.—This bulb requires a rich loamy soil and an open position to bring it to perfection. Still, it will flower satisfactorily in a shrubbery, or under the shade of trees; and, so far as the roots are concerned, there is no occasion to divide them more than once in three seasons. Plant during this month, and on to the beginning of November.

Cyclamens in pots will pay for an occasional dose of weak manure water. Shut the plants up in good time on chilly evenings. If a sowing of seed was not made last month it should be put in without delay.

The hardy varieties, such as *C. europæum* and *C. Coum*, are cultivated out of doors; and in some of the warmer districts of the South of England the Persian varieties can also be successfully grown in the open. They are suitable for rockwork, or for little nooks and sheltered corners, in which some gardens abound. For their success good drainage, a warm position, and plenty of water in dry weather are essential. September and October are the best months for planting out.

Dog's-tooth Violet.—For small beds, or in front of a rockery, these compact and interesting little plants are valuable for spring flowering, and are worth cultivating for their foliage alone. They also succeed in pots, and thrive in peat, or in sandy loam and leaf-mould. A 48-sized pot will accommodate five bulbs.

Freesia.—Towards the end of the month these bulbs will be ready for removal to a cool greenhouse or cold pit. No heat is required—merely protection from frost and excessive moisture. The stems are so slender that support must be given early. As the plants do not bear re-potting, the danger of exhausted soil can be met by administering weak manure water occasionally.

Fritillarias belong to the same order as the Crown Imperial, and the conditions which suit that plant will answer for all the Fritillarias. The bulbs thrive in a deep loam, and they are quite hardy.

Gladiolus.—The potting of the early-flowering varieties should be commenced this month and continued according to requirements. As the corms of these Gladioli are small, several may be placed in a 32-sized pot. No great amount of heat is wanted for these flowers, a temperature of about 55° being quite sufficient for them.

Gloxinia.—As the season of rest approaches, place the plants in any airy position, and gradually reduce the supply of water until the leaves fall off. The bulbs may be stored for the winter in peat or in dry moss.

The majority of growers, however, never store a bulb, but rely entirely on seedlings raised annually.

Hyacinth.—To grow this flower successfully in glasses demands no horticultural skill, for children often produce very creditable specimens. It only requires the intelligent application of certain well-understood principles. Like all other bulbs, the Hyacinth should form its roots before top-growth begins. The flower is cultivated in water for two reasons: the pleasure derived from seeing the entire plant, and the decorative value insured by this mode of treating it. As darkness retards top-growth, but does not delay the production of roots, it is usual to place the glasses in a cool cellar; and if this happens to be airy as well as cool and dark, there is no better place in which to start the bulbs. Still, it must be admitted that darkness is not essential for the development of roots. But darkness and coolness alike tend to delay the growth of foliage until roots are formed. Therefore, if the cultivator resolves to have the plants in view from the commencement, he must place them in a low and uniform temperature. The water should always be pure and bright, although it must not quite touch the bulb, or the latter will rot. Wires to support the flowers are necessary, and those which are manufactured expressly for the purpose are both neat and effective. A rather low temperature, and free access of pure air, should be regarded as necessary conditions of health in all stages of growth. Hence it will be obvious that a mantelpiece, with its fluctuations of heat and cold, is a most unsuitable position for the glasses. We should like to add, that notwithstanding the high qualities of the Hyacinth, it is quite a cottager's flower.

For pot culture the Hyacinth is a grand subject. Prepare the pots carefully as to drainage, and fill them with a light, rich, porous compost. Remove a little soil from the central surface, and into this hollow lightly press the bulb, and press the soil somewhat firmly round it, leaving about half the bulb visible. If too much power is employed, the soil will be so compact that when the roots begin to grow, instead of penetrating, they will lift the bulb out of its proper position. There is always some risk of this, and it accounts for the practice of heaping over the pots a considerable weight of ashes. Of course this covering serves a second purpose in checking leaf-growth until the roots are established. Any cool and safe position will answer for storing the pots at this stage. For the earliest supply of flowers select single varieties, as these naturally come into bloom somewhat in advance of the doubles. When the tops begin to grow, remove the pots to a greenhouse or frame, and subdue the light for a brief period until the natural colour is gained. Thence transfer to the forcing-pit as requirements demand; and they will need a week or ten days to prepare them for use. It is easy to secure a continuous supply of Hyacinths from Christmas onwards by forcing successive batches of roots until the final display will come into flower without artificial assistance. To augment the beauty of the flowers employ as little heat as may be necessary, and defer the finishing temperature until the latest moment possible. For general decorative purposes, small pots will be found extremely convenient when a brilliant display is wanted in a limited compass; good specimens can be grown in the 48-size, but for exhibition the 32-size must be resorted to. Neither in pots nor in glasses should the bulbs be allowed to send up leaves from between the outer scales; these rob the central growth, and they should be carefully removed with a sharp knife.

Hyacinths, Italian and Roman, should be potted in successive batches to provide a continuous supply. When the roots are formed the pots may be removed to a pit or frame, and to the forcing temperature as the buds show. If they have been brought on gradually, a very few days in a warm pit or house will throw them into bloom. It is a source of astonishment to us that these flowers are not more extensively grown in private gardens. Immense numbers are annually consigned to the London markets, and find a ready sale for bouquets and table decoration. Of course these Hyacinths will not bear comparison with the splendid named varieties which come later, but the Italian and Roman classes are ready at a time when flowers are scarce and valuable. Like other bulbs of the same class, they may be shaken out of their own pots and transferred to ornamental contrivances.

Iris.—The tuberous varieties are all perfectly hardy, and may be planted at any time from August to December. Put into light soil three inches deep and nine inches apart they will give no trouble, except to lift and divide them every second or third season.

Ixia.—Babianas, Ixias, and Sparaxis may all be treated in precisely the same manner. In sheltered districts in the Southern counties they can be grown in the open ground; but otherwise the culture must be in pots under the shelter of a frame or greenhouse. A 48-sized pot will hold four or five bulbs, and they will thrive in any soil which contains a large proportion of sand. In spring they may be transferred to a sandy border, or they can be kept in pots for a couple of years when well managed.

Jonquil.—The treatment recommended for Narcissus will suit this highly perfumed flower, both for forcing and in the open ground.

Narcissus.—It is undesirable to hold these bulbs in a dry condition longer than is necessary, and those intended for pot culture should be got in promptly. A low temperature must be relied on for keeping back such as are intended to flower late. The Double Roman and the Paper White naturally come into bloom in advance of other sorts, and these should be selected for the earliest display. Give them a rich porous soil, and pot them rather firmly, but not so firmly as to render it impossible for the roots to penetrate, or the bulb will be raised above the soil. Place them in a cool spot, covered with suitable material to keep the bulbs in their places, and to prevent the foliage from starting prematurely. When top-growth commences, the pots must go into some house or frame where they can progress slowly until the moment arrives for forcing them. If the buds just show, about a week in a bottom heat of 65° will suffice to bring them to perfection. A succession can be brought forward at intervals by the same means, until the final lot will flower without artificial aid. And for the comfort of those who do not possess heating apparatus, we may add that the flowers grown naturally will probably be finer than those which have been forced.

Narcissus may also be grown in glasses in the manner recommended for Hyacinths, or in bowls and other suitable receptacles filled with moss-fibre.

In the open ground Narcissus should be planted in quantity, especially in spots where it appears to be naturally at home, and one of the most charming effects is obtained by putting them in the rough grass adjoining shrubbery borders. Instead of cutting the grass, it must be allowed to throw up flower-heads, and this affords the bulbs time to mature in readiness for the following season. The many forms of Double and Single Daffodil are effective border flowers, and the numerous varieties of Narcissus should be grown in clumps and patches in every spot which is suitable and vacant. In the reserve border of many gardens large numbers of Pheasant's Eye and other Narcissus are planted to supply flowers for cutting. They are peculiarly valuable for the purpose, and if cut when scarcely ready they will develop in water, and last for many days. In planting, be guided as to distance by the size of the bulb, allowing four or five inches between small sorts, and six to nine inches for large varieties; depth, six to nine inches.

Oxalis.—Except in a few sheltered districts, it will be necessary to cultivate this exceedingly pretty flower in frames, or in a sunny, airy greenhouse. It may also be forced in the stove with success. Put several bulbs in a pot, and give them a light soil with plenty of sand in it.

Snowdrop.—It does not improve the roots of this exquisite little favourite to keep them out of the ground, and they should, if possible, be planted early.

Sparaxis needs the same treatment as advised for the Ixia.

Sweet Pea.—Exhibitors of Sweet Peas and those who endeavour to secure the finest sprays for decorative purposes, commence the preparation of the ground during the present month and incur whatever expense may be necessary to insure a deep bed of rich friable loam in which the roots can ramify freely. It is also the practice to sow seeds about the middle of September in order to provide sturdy well-rooted plants in readiness for transfer to the prepared plots in early spring. Either pots or boxes may be used, and a frame is sufficient to bring the seedlings safely through the winter. The method is dealt with in detail on page 305.

From mid-September to the end of October, according to the locality, is an excellent time for sowing Sweet Peas outdoors where the soil is light and the situation fairly warm. Plants from autumn-sown seed are generally more robust and produce finer flowers than those raised from seed sown in the open in spring.

Tropæolum tuberosum.—In potting the tuberous varieties, insure efficient drainage, and use a compost of rich light loam mixed with sand. The foliage will trail over the sides of wire baskets with graceful effect, but it may be trained around balloon-shaped wires specially made for these flowers. The bulbs remain dormant all through the winter, and may be started at any time from September to March.

Tulip.—The early class of Tulips is of great value for forcing because of their brilliant colours and elegant forms. They take kindly to a high temperature, but forcing should not be commenced too early, nor should the heat be allowed to exceed 65° at the finish. Plunging is the most satisfactory method. Several bulbs may be put into one pot, but it is more convenient to grow them singly, so that flowers in exactly the same stage of development may be selected for use at one time. A continuous supply may be secured by potting batches at short intervals. When in bloom the roots can be washed free from soil for placing in vases. Decayed turf, with decomposed cow-manure and a proportion of sand, make an excellent potting soil for Tulips, and it will be all the more suitable if laid up in a heap for twelve months after being mixed.

OCTOBER

Anemone.—The tuberous-rooted Anemones may be planted in the open at any time from September to March, and from successive plantings a continuous display will be obtained from February until far into spring. For the choice named varieties it is customary for specialists to make elaborate preparations, into which we need not enter here. Splendid flowers can be grown in clumps and beds in ordinary gardens by deep digging, and the employment of a liberal dressing of decayed cow-manure. Plant the roots from four to six inches apart, and at a uniform depth of about three inches. In a heavy, retentive soil it is not advisable to risk a collection of named Anemones until January, unless a deep layer of light compost can be placed in the drills where the roots are to be planted.

Annuals, Hardy.—On light soils it will be safe to transplant these now; but on heavy land the risk is too great, and we advise waiting until February or March. Lift the plants with as much soil attached to the roots as possible.

Crocus.—Several flowers bloom in advance of, or as early as, the Crocus; but no other bulb of its own period can compare with it for brightness and effective colouring. Plant during this month and November, in groups and patterns wherever there is a vacant plot and bulbs can be found to fill it. Put them in at a uniform depth of about three inches. Drills are easy to draw, and are better for the bulbs than the objectionable plan of dibbling.

Cyclamen seed may be sown again this month. If properly grown, seedlings raised now will bloom splendidly next autumn.

Ferraria.—See Tigridia, page 379.

Gladiolus.—By the end of the month lift roots which have flowered, even if the stems are still green. Label them, and hang in an airy place to dry. A little later remove the foliage with a sharp knife. Then lay out the roots for about a fortnight, and when ready store them in paper bags or boxes placed on a dry shelf, secure from vermin.

Hollyhock.—In favoured districts and in light soil it will be safe to winter this plant in the open ground with merely the protection of a little dry litter. But in damp adhesive land it is perilous, and a cold frame will afford the requisite protection until May returns.

Hyacinth.—Considering the magnificent appearance of this flower, its culture is most simple. Any fairly good garden soil which is not too damp in winter will grow it; and the bulbs may be planted in clumps or beds in any design or arrangement of colour that taste may dictate. At six inches apart there will be a brilliant display, but the distance is quite optional. The crowns of the bulbs should not be less than four or more than six inches below the surface; the greater depth will slightly retard the flowering. When planted they will give no more trouble until the time arrives for lifting them to make room for other occupants.

Hyacinth, Feather, is an exceedingly beautiful border flower during May and early in June. The stems are from nine to fifteen inches high, and carry flowers whose petals are cut into slender filaments. It will grow in pots and in the open, in any soil which suits Hyacinths. Plant a good number in each group.

Hyacinth, Grape.—An interesting dark blue flower, which should be freely grown in mixed borders to bloom in April. Singly it is useless; plant good-sized clumps in soil which answers for bulbs.

Hyacinths, Miniature, are the delight of children, in whose honour many of the varieties are named. Except for their diminutive size, they are in all respects equal to their larger relations. The culture in pots, glasses, and beds is similar to that advised for the full-sized roots, save that the planting in open ground need not be quite so deep, three inches of soil over the crowns being sufficient.

Hyacinths, Italian and Roman.—Uncover the pots containing the earliest planting, and at first place them in a dimly lighted position. The application of heat will depend on the time the flowers are wanted; but when the plants are forward enough, plunge them in a temperature of 65°, and in about a week they will be ready for use.

Lachenalias rarely attain the proportions they are capable of for want of water in their growing state. They thrive in peat, and may be forced into flower at almost any season. Except in warm and sheltered gardens, they must not be planted in the open. Yet only sufficient warmth is required to keep frost at bay.

Leucojums are perfectly hardy bulbs which will grow in any garden. The flowers resemble Snowdrops, but are much larger. Plant in dense groups.

Narcissus.—From the natural characteristics of this bulb it is desirable that it should be planted early. Sometimes, however, it is impossible, consistently with other arrangements, either to pot or to plant Narcissus before October or November. In such cases it is consoling to know that from sound, well-ripened roots good flowers may be confidently anticipated, even from late plantings.

Ornithogalum.—In the open this bulb must have some protection during winter, to save its large fleshy roots from injury by frost. A heap of light manure or dry litter will answer the purpose. Plant six inches deep.

Scilla præcox can be grown almost anywhere, and in a light rich soil it blooms profusely. The bulbs will safely pass the severest winter in the open ground, and flower in February or March. The exact time depends on the climate and position. In sheltered spots and mild districts they will naturally bloom earlier than in bleak and exposed quarters. Plant in masses or lines, and the bulbs may remain undisturbed for years. A dense row makes an exceedingly beautiful background to Snowdrops. The other Scillas are equally hardy and valuable, and they all flower with great freedom.

Triteleia uniflora is a handsome white-flowering hardy bulb, which will grow freely in any garden. It is adapted for the company of any of the dwarf-growing bulbs, and may be employed in either lines or clumps. Plant the roots three inches apart and two inches deep.

Tuberoses are valued for the purity of their white flowers, and for the agreeable perfume they exhale. The bulbs may be potted singly or three in a pot. They thrive in a compost of loam and leaf-mould, and need a bottom heat ranging between 60° and 70° to bring them to perfection. The African bulbs are generally ready in September and the importations from America arrive in December and January.

Tulips may be planted in the open ground at any time during the month. We shall say nothing as to the arrangement of colours, nor as to the form of the beds, for both points admit of endless diversity. The mixed border may be enlivened with groups of many varieties, and if they are judiciously selected, there will be a succession of flowers for several weeks in the spring.

Wallflower.—After the summer bedding plants are cleared, Wallflowers may be usefully employed to fill beds with green foliage all the winter. They will flower freely in spring, when their colour and fragrance will be especially welcome, and they can be removed in time to make way for a different display for the summer.

Winter Aconite is not dismayed by frost or snow, but will put forth its golden blossoms in the dreariest days of February, and after the flowers have passed away the foliage will remain as an ornament. To put in single roots is useless; it is far better to plant a few large patches than to fritter away the flower in a number of small and inconspicuous groups.

NOVEMBER

Cyclamen.—Where there is a large demand for this flower, another sowing may be made this month, unless it was done in October. With so important a subject it is not wise to depend on a single venture. The seedlings will afford a valuable succession to those started in August.

Gladiolus.—The soil which answers best for the autumn-flowering section is a medium friable loam, with a cool rich subsoil. A light loam can be made suitable by trenching, and putting a thick layer of cow-manure at the bottom of each trench. And a heavy soil may be reduced to the proper condition by the free admixture of light loam or sand. Autumn is the proper time for doing this work, and the ground should be left rough, so that it may benefit by winter frosts. Wireworms are deadly enemies to the Gladiolus corms, and an effort should be made to clear them out. Happily, they will flock to traps such as Potatoes and Rape cake, and their destruction is a mere question of daily attention. Planting must, of course, be deferred until spring.

Hyacinthus candicans is generally grown in the company of other flowers which attain to something like its own imposing proportions. In good soil the spikes grow three feet high. It may be planted from this time until March.

Lilies are an ornament to the cottage garden, and they grace the grandest conservatory. Many of the most superb varieties, including the king of all the race, *L. auratum*, can be magnificently flowered in the open border; and we have seen fine specimens of the *Lancifolium* varieties grown in pots without the aid of pit or frame. It is therefore obvious that there are no difficulties in the culture of Lilies. In borders the best soil for them is a deep, rich, moist loam. Peat and leaf-mould also answer; but a stiff clay will not do unless it has been cultivated and mixed with lighter stuff. Plant the roots at least six inches deep, at any time they are in a dormant state, or can be obtained in pots. Their position in the border should be clearly marked, or the roots may sustain injury when the soil is forked over.

The noble appearance of *L. auratum* will always command for it a prominent place in the conservatory or greenhouse. It will grow in sandy peat, or in a mixture of loam, leaf-mould, and sand. The bulb should be put into a small pot at first. When this is full of roots, transfer to a larger size, and shift occasionally until the flower-buds appear, when re-potting must cease. A cool house will bring the plant to perfection, although it will bear a high temperature if wanted early. During growth water must be given freely and be gradually reduced when the flowering season is over.

The *Lancifolium* varieties require the same treatment, but it is usual to put several in one large pot. After the flowering is ended, instead of allowing the bulbs to become quite dry, keep them moist enough to prevent the fibrous roots from perishing, and they will start with all the greater vigour when the time arrives for repotting next season.

Lily of the Valley.—The forcing of this favourite flower generally begins in November, and it is important to secure roots which are thoroughly matured for the purpose. They must be finished in a high temperature, and if managed with judgment there will be plenty of foliage to set off the long spikes of charming white bells. When planted in the open ground a shaded spot should be chosen, which must be freely enriched with leaf-mould, and the plants will not need to be lifted for four or five years.

Ranunculus.—On a light dry soil, where there is no danger of the roots sustaining injury during winter, this is a suitable time for planting all the varieties. To do them justice the land must be liberally dressed with decayed manure, and the longer the bed can be made ready before planting, the better will it answer. Put the roots in drills drawn six inches apart and two inches deep and cover with fine soil. For retentive land it is advisable to defer planting until February.

Tritonia.—Perhaps the best way of treating this flower is to pot the bulbs now or in December, and keep them in frames until April, when they may be transferred to the open ground. A dry soil and a sunny spot should be found for them.

Tulip.—There is no better time for planting Tulips in beds than the first half of this month. The bulbs should be covered with four or five inches of soil according to size, and it is important that each kind should be put in at a uniform depth to insure a simultaneous display. On a heavy soil draw deep drills, and partially fill them with light compost, on which the roots should be planted. The late single varieties are the Tulips which were formerly so highly prized by florists. For these bulbs it was the custom to prepare the soil with extraordinary care when the Tulip craze was at its height. After the amazing folly of paying 300l. for a single bulb, the minor folly of extravagance in preparing the soil may be readily pardoned. Happily that phase of the business has passed away, and handsome Tulips are now grown without such a prodigal expenditure of money and labour. The site for this flower should be sunny, the soil fairly rich, and the drainage good. With these conditions insured, and roots which are sound and dense, it is easy to obtain a magnificent show of Tulips.

Zephyranthes Candida can be grown in any soil, and if possible the bulbs should be planted in some spot where they may remain unmolested through several seasons. The flowers appear about the end of July, resembling a White Crocus in form, and the blooming continues until cold weather sets in. Planting may be done between November and March.

DECEMBER

Only the idle or the half-hearted gardener will complain that he has no work to do in the short dark days of this month. Although there may be little or nothing to plant or sow, and few flowers need repotting, yet there are soils to obtain and store for future use; former heaps to turn over and remake; dead leaves to remove from plants in pits and houses; stakes and neat sticks to prepare for subjects which will need support by-and-by; beds and borders to enrich, and many other duties to perform. In the evenings, too, there are new combinations and fresh harmonies in colour to be designed for beds and groups in borders; the requirements for the coming season to consider while experience gained during the closing year is still fresh in the memory; the position of plants in pits and frames and houses to forecast, so that the plan of the summer campaign may be clearly understood, and all the resources of the garden be under intelligent control. The fluctuations of the thermometer have also to be watched, and means adopted to save plants from injury by a sudden fall of temperature. Altogether, there are abundant sources of profitable employment for those who have a mind to work.

Bulbs, such as Hyacinths, Tulips, Crocuses, &c., which have not been planted, will have commenced growing, notwithstanding the precautions taken to prevent it, thus showing that they ought to be in the ground. The growth has been made at the expense of the bulb itself, for there are no fibrous roots from which to draw support. Therefore it can scarcely be expected that the flowers from very late plantings will be quite so good as the same bulbs would have produced had they been put in at an earlier period. Still there are cases when the delay is unavoidable, and it is reassuring to know that sound bulbs carefully set at the proper depth will produce flowers only in a degree inferior to those from earlier plantings.

Bulbs in store, such as Begonia, Dahlia, Gladiolus, and Gloxinia, should be passed in review. Examination will almost certainly reveal some unsound specimens, and their removal may save valuable companions from their contaminating influence. This practice should be followed up about once a fortnight until all are eventually planted.

THE PESTS OF GARDEN PLANTS

The life-history of plant pests and ground vermin, with the best means of saving various crops from their ravages, are dealt with in a series of valuable leaflets issued by the Ministry of Agriculture and Fisheries. These leaflets embrace a very large number of subjects, several of which belong to the farm and the orchard and are beyond the scope of the present volume. Others are rarely met with, but concerning those which are common to the majority of gardens we offer information which will, we hope, enable readers to safeguard their crops from disaster.

When adverse weather operates injuriously on vegetation the plagues that infest garden plants usually acquire increased power in proportion to the degree of debility to which vegetation is reduced. This circumstance perfectly accords with the general law of Nature, and is full of instruction as to the means of saving plants from serious injury by vermin. The keen, dry east wind that so often jeopardises fruit crops is usually followed by visitations of fly and maggot, and in this case the cause is beyond human power or forethought. But neglect of watering and air-giving to pot plants can be avoided. Good cultivation not only insures fine specimens, but is often the means of preventing the plants from failing under the attacks of Aphis, Mealy Bug, and other enemies against which the gardener has to fight an unceasing battle.

Insects are among the frailest of living creatures and they perish at a touch. As they breathe through the pores of the skin, water alone—the promoter of life and cleanliness—is death to them; and they are still more subject to sure destruction when to the water is added an active poison, such as tobacco, or a substance that adheres to them and stops the process of breathing, such as glue, clay, sulphur, soft soap, and the numerous preparations that are specially made to annihilate insect hosts.

The various stages through which the larger insects pass place them within our power at some period of their existence. The butterfly may float beyond the reach of harm, but in the caterpillar or the chrysalis state it can be dealt with effectually. Again, we may be powerless to destroy the Chafer grubs as they feed or hibernate beneath turf, but in their perfect state as Cockchafers or Rose Chafers many may be beaten down during quiet evenings, and others can be shaken from Roses at dawn or sunset. A knowledge of the life-history of injurious insects will suggest what is to be done and the right time for doing it, so that often by simple treatment they may be destroyed.

The expense of preparing mixtures and washes may be in some degree lessened by economy of application. A drenching-board fitted on a firm frame, should be provided in every place where plant-growing is carried on to any extent. The board should slope from a resting ridge at the base. The plant in its pot may be laid on the board, with the bottom of the pot against the resting ridge, and a pail should be put to catch the liquid used as it drains from the plant after syringing. Every general washing or fumigating should be followed by another at an interval of from a week to a fortnight, because, although the first operation may kill every insect, there will be many living eggs left, and these renew the race, and very soon bring the plants into as bad a state as ever, unless consigned to a happy despatch as their parents were. In some cases it will be more economical to feed than to destroy the vermin; and, as a rule, feeding vermin does not add to their numbers, in the same or any future season, for insect life is so strangely dependent on certain conditions of temperature, &c, that if the season is not favourable to a particular kind it will be scarce, no matter how plentiful it may have been in a previous year. In the case of the Turnip Fly, feeding is frequently the cheapest and surest way of saving the crop. It is customary with Dahlia-growers, and, indeed, with the growers of florists' flowers generally, to sow Lettuces where the flowers are to be planted, for so long as Lettuces are on the spot Slugs and Snails will prefer them to other food. As the Lettuces themselves serve the purpose of traps, the Snails and Slugs congregated about them may, towards evening, be caught and destroyed.

In using a mixture for the first time, it is advisable to try it on one plant only, and that, of course, the worst in the collection affected. If the preparation is too strong, the truth will be declared by the state of the plant within twenty-four hours; thus a little caution may prevent a great loss. Another good rule is to employ the several remedies in a rather weak state until experience has been gained, for not only has the strength of the medicine to be considered, but the management of the patient before and after it is administered. It is above all things important to be thorough in the cleansing of plants, because they succumb rapidly to the attacks of insects, and should be effectually and promptly cleaned or consigned to the fire. If left in a foul state they spread the infection to all around. In the space at our command it is only possible to notice a few of the garden pests, and we begin with one of the most frequent and troublesome of plant foes.

Aphis in some form or other is the most persistent and perplexing of plant pests. The Green Fly is the enemy of the softer kinds of vegetation, and the Blue and the Black Fly are common plagues of the Peach-house and the orchard. The tender body of the Aphis is instantly affected by conditions unfavourable to its life, and it is therefore easily killed; but its marvellous power of reproduction renders its extinction impossible, for in every instance a few escape, and very soon re-establish their race. Two methods for the destruction of Aphis are in vogue. One is fumigation by tobacco, either pure or in some of the numerous preparations offered, including several popular insecticides which have nicotine as a basis. These are both clean and effective. When a houseful of plants is infested no time should be lost, and the evening is most suitable for dealing with the pests. The plants ought to be quite dry and the house closely shut. A dense cloud of smoke without flame is required. Allow the smoke to do its deadly work during the night. Early next morning syringe the plants freely, and in the course of an hour or so give air. The other remedy is to use one of the many liquids which are inimical to the life of Aphis and other insect pests. To economise the liquid it is advisable to fill a pail or tub and immerse the plants individually. Take one in the right hand and spread the fingers of the left hand over the surface of the soil to prevent an accident; then turn the plant over and plunge the foliage in the liquid, moving it up and down briskly two or three times. If this is not practicable syringe the plants, taking care to wet the leaves on both sides. On the following day syringe with pure soft water.

Rose trees may generally be cleansed of fly by means of the garden engine and pure water only, the essential point being to direct the water on the trees with some amount of force for several evenings in succession whenever the fly threatens to obtain the mastery.

Soft soap dissolved in water makes a cheap and effectual wash for exterminating all kinds of Aphis, and to these ingredients quassia may with advantage be added. One pound of soft soap will suffice for ten gallons of water, into which stir the extract obtained by boiling one pound of quassia chips in water. Pot plants can be dipped in it as already advised, or the solution may be applied by means of the syringe. On the following day the plants should be cleansed with pure soft water.

The Bean Aphis, also known as the Bean Plant Louse, or Black Dolphin *(Aphis rumicis)*. Our illustration shows the wingless female and pupa natural size and magnified. The pupa is black with greyish white mottlings, while the female is deep greenish black in colour. This insect commonly attacks the young shoots and tops of Broad Beans. It is well to cut off the infected tops and burn them. Should the attack be repeated spray the Beans with a solution of soft soap and quassia.

The Pea Siphon-Aphis *(Siphonophora pisi*, Kalt).—Among the aphides peculiar to vegetables this is one of the most common.

Our illustration shows the natural size and an enlarged figure of the greenish-winged and green-tinted wingless females, as produced, not from eggs, but alive and developed. This insect is occasionally very destructive to Pea crops.

American Blight, or Woolly Aphis, generally appears first on trees grafted on dwarfing stocks, particularly the bad forms of the Paradise Apple. Rapidly the mischief spreads, healthy trees become infested, and unless checked an orchard is speedily ruined. Andrew Murray says that in bad cases of American Blight it is sometimes necessary to root up and burn all the trees, and let the ground remain unplanted for a year or two. Fruit trees should be examined periodically for this pest, and immediately the woolly spots are detected small tainted boughs should be pruned away, and from the mainstems and large branches diseased spots can be pared off. The operation may need a bold and vigorous hand if the trees are to be saved, and it is important that every scrap should be burned. There is almost certain to be a further appearance of the Blight, which should be destroyed by one of the many remedies known to be effectual. Fir Tree Oil Insecticide has proved to be an excellent remedy. Gishurst Compound, in the proportion of eight ounces to a gallon of water, with sufficient clay added to render it adhesive, makes a capital winter paint for Apple trees. But there is no cheap remedy equal to soft soap for smothering American Blight in the crannies of the bark. The soap may be rubbed into the diseased spots, or as a wash it can be brushed into the boughs.

Our illustration shows a piece of Apple twig with the aphides and their woolly material natural size. The enlarged figures represent the winged female and the wingless larva of the Apple Blight Aphis *(Schizoneura lanigera)*. The insect is deep purplish brown in colour, and the well-known bluish white cottony material naturally exudes from it.

The Carrot Fly *(Psila rosæ*, Fab.), with its larva, pupa, and perfect insect, is illustrated natural size and enlarged. The ochreous shining larvæ live upon the tap-roots of the Carrot, and by eating into them cause them to rot. In colour the body of the fly is an intensely dark greenish black, with a rusty ochreous head. The presence of the larvæ in the root is made known by the change in the colour of the leaves from green to yellow, and the attacked plants should be promptly forked out entire and burned.

It is well to dig the ground in autumn, so that the earth may be exposed to the frosts of winter and the pupæ to the attention of birds. After sowing, spray the Carrot bed with paraffin emulsion. Spray again after germination, and a third time when thinning is finished. The emulsion to be made by dissolving half a pound of soft soap in a gallon of boiling water. While still boiling, pour the liquid into two gallons of paraffin and churn thoroughly until a buttery mass results. This will keep for a long time in tins. Before use, dilute with twenty times the quantity of water—soft water if possible. This is an excellent preventive. After the work of thinning, the fly may also be kept off the plants by scattering over them ashes, sand, or earth, impregnated with paraffin. Carbolic powder and soot are both disagreeable to the insect. It has been observed that when singling the disturbance of the soil is favourable to the operations of the Carrot Fly. A copious watering when the task is ended will firm the earth round the remaining roots, and prevent the fly from easily getting down to deposit eggs.

Carrots and Parsnips are often attacked by the larva of a Carrot Moth *(Depressaria cicutella)*, which spins webs for security while feeding, and sometimes works havoc among the foliage. A simple remedy is to shake the caterpillars from the leaves of the plants, when they can be destroyed by the use of lime.

Celery Fly.—The apparent blisters in Celery leaves are spots deficient of leaf-green, which the larva of the Celery Fly has eaten. Dusting newly-planted Celery with lime or soot may do something to prevent the fly from laying its eggs, but the most certain preventive is to boil half a pound of coal tar in one gallon of water for twenty minutes, add fifty gallons of clear water, and syringe the plants about noon once or twice from the middle to the end of June. When once the grub has made a home, it should be crushed by pinching the leaf between the finger and thumb, or the injured portions of the leaves should be cut out and burned. In doing this it must always be remembered that the leaves are as much needed by the plant as the roots, and every leaf removed tends to diminish the vigour of the plant. Our illustration shows the Celery Fly (formerly known as *Tephritis onopordinis*, but now called *Acidia heraclei*) natural size and magnified. This fly is also destructive to the leaves of Parsnips, and is named *onopordinis* from its habit of frequenting the Cotton Thistle *(Onopordon Acanthium)*. The larva is white to very pale green, the fly is shining tawny. An Ichneumon Fly detects the larva of the Celery Fly in the Celery and Parsnip leaves, and lays its eggs in the body of the larva. These parasites, named *Alysia apii*, assist in reducing the numbers of the Celery Fly.

All Celery refuse should be destroyed by fire. Infested ground may, if suitable, be trenched, bringing the subsoil to the surface and burying the top soil containing the pupæ. Frequent rough digging and the exposure of fresh surfaces to be searched by birds will also do something to abate the number of this pest. But in bad cases it will be necessary to resort to gas-lime, which poisons the pupæ and eventually benefits the soil, although in the season immediately following its use crops may be less satisfactory than usual.

Onion Fly.—Onions are frequently attacked by the larvæ of the Onion Fly, and in some instances the entire crop is destroyed. Our illustration shows the natural size of the fly and maggot, with magnified representations of both. The fly lays six to eight eggs on an Onion plant, generally just above the ground. These eggs hatch in from five to seven days, according to the temperature, and the maggots at once burrow into the Onion. The result is soon visible in the discoloration of the leaves which turn yellow and begin to decay. Several generations of the insect, the scientific name of which is *Phorbia cepetorum*, appear in the course of a single season. A close ally is the Cabbage Root Fly *(P. brassicæ)*, the destroyer of Cabbage roots.

Among the numerous methods of preventing attack and of destroying the grubs the following are worth attention:—

Where this pest proves very troublesome it may be desirable to transfer Onion growing to new ground until the infested land has been purged of the pupæ. Instead of throwing useless Onion material on the waste heap to afford the fly a home for its eggs, every scrap should be burned. As the preparation of an Onion bed approaches completion, powdered lime well mixed with soot, in the proportion of two bushels of the former to one of the latter, may be sown evenly over the surface and raked in. Sand impregnated with paraffin sown along the drills has answered as a preventive. Vaporite is a destroyer of the pupæ; this preparation has proved deadly to ground vermin generally. Earthing up the Onions was proved by Miss Ormerod's experiment to be effective. The objection to this procedure is the probability of enlarged necks which are not wanted. An emulsion, composed of one pint of paraffin, one pound of soft soap mixed with ten gallons of water, thoroughly churned by a hand syringe and sprayed over the young plants in a fine mist, is a valuable preventive. The dose may be repeated after rainfall, if necessary. The quantities named suffice for a small plot only. Soapsuds are destructive to the maggots, disagreeable to the fly, and beneficial to the young plants. The suds should be sprayed over the bed from a watering can on the first appearance of a yellow colour in the grass. As a final suggestion reference may be made to a singular fact which we do not profess to explain, viz. that transplanted Onions are very seldom touched by grub. The modern practice of raising seedlings under glass in January or February, and planting out in open beds in April, offers the advantage of a long season of growth combined with comparative immunity from attack by the Onion Fly.

Turnip Fly, or Flea, is well known to the gardener, and is the most troublesome of all the aërial pests of the farm, and one with which it is most difficult to cope, not only because of its general diffusion and numbers, but because it produces a succession of broods throughout the summer, and is therefore always in force, ready to devour the crop immediately it appears. The so-called 'Fly' is a small beetle named *Haltica (Phyllotreta) nemorum*, strongly made, and decidedly voracious. The larvæ are not to be feared, except that, of course, they in due time become beetles. In the perfect state this winged jumping insect makes havoc of the rising plant of Turnips, but the crop is only in danger while in the seed-leaf stage. It is in the spring and early summer chiefly that the ravages of these insects occasion perplexity, for they awaken from their winter torpor active and hungry, and have a ready appetite for almost any cruciferous plant. Hence we see the leaves of Radishes pierced by them, and all such weeds as Charlock, Cuckoo Flower, Hedge Garlic, and Water Cress serve them for food until the Turnip crops are on the move, when they will travel miles, even against the wind, to wreck the farmer's hopes. The Cabbage Flea *(Haltica oleracea)* in some districts is equally troublesome, if not more so. Whole Cabbages may be destroyed by this pest, and even Hops are often ruined by it.

Preventive and remedial measures that can easily be carried out in a garden may be impracticable on a farm. We propose to enumerate them briefly as they occur to us, leaving the ultimate choice of weapons to those who may unfortunately find occasion to use them.

One precaution is to insure a quick germination of the seed and strong growth of the plant in its seed-leaf stage. The cotyledons are tender and tasty, perhaps sugary from Nature's process of malting; and while the seed-leaf is assailable the *Haltica* makes the best of the shining hour. The seed sown should be all of one age, and the newest possible, because of the need for a quick and strong growth. When a powerful artificial is sown with the seed, the quantity of seed must be increased, as a proportion may be killed by the manure. It is important always to drill Turnip seed; broadcasting seems to invite the Fly—at all events, a drilled crop is generally safer. Before sowing, the seed may be soaked in paraffin or turpentine. Of the two the latter appears to be the more successful in keeping the insects at bay.

Rolling an infested plant disturbs and weakens the insects and stimulates the young plant.

The sprinkling of slaked lime over the seedlings is at once a safe and an efficient process, and possesses the additional advantage of being beneficial to the plant. We are aware that it does not always succeed, but we are inclined to attribute the failure to a bad quality of the lime, or a careless method of employing it. There should be enough put on to make the plants white, and they will be none the worse for the whitening. Dustings of fine ashes or soot are scarcely less effective, but salt must not be used, for it injures the plants and does not hurt the beetle. All such dustings should be done in the early morning, while the plants are wet with dew. To apply a dusting at midday, when the sun shines gaily, is to waste time, and probably many of the recorded failures might be explained if we knew at what hour and in what sort of weather the work was done. Nets and sticking boards have been tried and found effectual, and yet such things are rarely used. A board thickly covered with white paint, drawn over the plot on a still, sunny day, soon becomes a black board by the myriads of *Halticas* that jump at and remain attached to it, the victims of their extravagant love of light. Old sacks soaked in paraffin and drawn over the drills impart a disagreeable flavour to the leaves, and a very fine spray of paraffin distributed by a machine specially constructed for the purpose has proved effective.

Finally, this, in common with all other insects in the winged state, needs a dry air and some degree of warmth for its health and happiness. Many kinds of larvæ need moisture, but no winged insect can abide moisture long, and herein is a clue to the eradication of Turnip Fly. By the simple process of spraying the plant three or four times a day, until it is out of the seed-leaf, and the danger is over, it is possible in the garden to wash out the *Haltica*; and any kind of insecticide or flavouring, such as quassia, may be mingled with the water to render the plants distasteful to the insects.

The illustration on page 422 shows the Turnip Fly in its three stages, and in each case of the natural size and magnified seven diameters.

Daddy Longlegs, or Crane Fly, in its perfect form of a fly *(Tipula oleracea)* does no harm, but the grubs, known by the familiar name of 'leather-jackets' owing to the toughness of their skins, are terribly destructive. During late summer and autumn the female fly deposits its eggs in large numbers in turf, in garden soil and amongst garden refuse. The eggs are hatched in a fortnight or so and the dark grubs lie in the ground through the winter, inflicting their maximum, amount of injury to young crops in spring and early summer. Where song birds are scarce the Tipula is capable of utterly destroying grass and of seriously ravaging the Kitchen Garden; but cultivation, aided by the robins, thrushes, nightingales, and other birds, will keep the insect within bounds, even after a hot summer favourable to its increase. Where this pest is known to exist, an application of Vaporite at the time of preparing ground for sowing or planting will destroy many of the grubs. The regular use of the hoe is also to be recommended, for by the disturbance of the soil the enemy is exposed to the sharp eye of the robin and other feathered gardeners.

Root-knot Eelworm.—One of the worst pests that a Cucumber-grower has to deal with manifests itself by the presence of minute warts or nodosities, chiefly on the rootlets. These warts, which are caused by the action of innumerable small thread-like worms named *Heterodera radicicola*, range from the size of a pin's head to that of a pea, and when they are present in large numbers the total failure of the Cucumber crop is the invariable result. The eelworms are probably introduced to Cucumber-houses in infected water. Each worm is about one-seventyfifth of an inch in length and is at first coiled up inside a transparent egg. At maturity the eggs crack open, and the worms on emerging bore into the most tender rootlets, and there lay their eggs. These eggs speedily hatch inside the plant and new eelworms are produced, which traverse the rootlets in every direction.

These *Heterodera* are by no means peculiar to the Cucumber; they attack the roots of Tomatoes and Melons, and the roots, stems, and foliage of many other plants. Our illustration shows some very small Cucumber rootlets, natural size, with the eelworms in the eggs, and also emerging from and free of the empty eggshell (enlarged eighty diameters).

Immediately symptoms of the pest are apparent from the wilting of the foliage and stems, all infected plants should be removed and burned. The soil must also be cleared out and the interior of the house thoroughly washed with a solution of carbolic acid in water:—one part of the former to eight parts of the latter. To purify the infected soil, use a solution of carbolic acid (one part) and water (twenty parts) and saturate three times, at intervals of a fortnight. Another remedy is to mix weathered gas-lime freely with the soil. In either case the soil will be unfit for use for at least six weeks after treatment. When the house has been well cleansed, fresh compost should be used, to which the addition of lime and soot, mixed with the soil, will be beneficial.

Mealy Bug.—This plague is by no means confined to plants under glass. In the case of a lot of stove plants badly affected, the desperate course of committing the whole to the fire, and then repairing and painting the house, is often the cheapest in the end. We have known a Pine-grower compelled to destroy a houseful of plants that have been infested by the introduction of a plant from a buggy collection. Mealy Bug may be known by its mealy, floury, or cottony appearance. It has a great fancy for Grape vines. One of the best remedies is Gishurst Compound, prepared at the rate of eight ounces to a gallon of water, with clay added to give it the consistence of paint. Miscellaneous stove plants may be cleansed by washing with a brush and soft soap. Our illustration shows a group of Mealy Bugs natural size, with one insect magnified.

Red Spider is present in almost every vinery, however well managed. A moist atmosphere is a great, though not a certain preventive; but it is not possible, without injury to the vines, to keep the air of the house always so humid that the Spider is unable to obtain a lodgment. Syringing promotes a moist atmosphere, and is unfavourable to the Red Spider, which thrives best in heat and dryness. But the most decided repellent of Spider is the use of sulphur on the hot-water pipes. This may be managed by sprinkling dry sulphur on the pipes, or by making a thick solution of sulphur, clay, and water, with which the pipes should be painted. Be careful not to raise the heat at the same time, for if the pipes are hotter than the hand can bear fumes destructive to vegetation will be given off. Melons and Cucumbers may generally be kept clear of Spider by means of the syringe only; but when Melons are ripening they must be kept rather dry, and it is very difficult indeed to finish a crop without having the plants attacked by Red Spider. Gishurst Compound answers admirably to remove Spider from house plants. The mixture should consist of one and a half or two ounces to one gallon of water, and should be applied with a sponge. The scientific name of the Red Spider is *Tetranychus telarius*. Our illustration shows one of these destructive red mites natural size, and two individuals greatly magnified.

Scale.—A very common species, found on many kinds of stove and other plants, is the *Lecanium hibernaculorum*, here illustrated on a twig, natural size, and magnified. It is brown, tumid, and commonly somewhat more than hemispherical in shape. Besides this species there is the *L. filicum* of Ferns, the *L. hemisphoericum* of Dracænas, the *L. rotundum* of the Peach, and the common *L. hesperidum*, or Orange-tree Bug, which is one of the flat species, and it spreads to a great variety of plants. The Scale insect sucks the sap from plants, and in some instances the ground beneath the foliage is wet and soddened by the falling sap. Spirit of turpentine applied with a soft brush is considered to be a good remedy for Scale. It is, however, advisable (as in other remedies) to test this on a small number of plants at first. A near relative, a large brown *Coccus*, infests pomaceous trees, and is especially partial to the Pyracantha, which it often kills outright. The Scale of the Vine is *Pulvinaria* or *Coccus vitis*. Careful washing with soap and water, and the destruction of each separate Scale as soon as seen, can be recommended for the extirpation of this pest.

Thrips may pursue their mischief to a great extent before they are discovered by the novice, for their minute size and their habit render them inconspicuous. But the black deposit they make reveals their existence to the experienced eye, and the debilitated condition of the plants they have attacked would soon compel attention were there no such deposit to tell the tale. The Indian Azaleas are apt to be beset by Thrips, as the Grape-vine is by Scale, the Pineapple by Mealy Bug, and the Rose by Green Aphis. Atmospheric humidity is a powerful preventive, as is also the promotion of vigorous growth by a plentiful supply of water to the roots of the plants; in fact, starvation and a dry, hot air will soon bring an attack of Thrips. Generally speaking, the best remedy is fumigation with tobacco. Or tobacco water and a solution of soft soap, together or separately, if carefully applied, speedily make an end of this troublesome pest. A special preparation may be made as follows: Take six pounds of soft soap, and dissolve in twelve gallons of water, add half a gallon of strong tobacco water, and dip the plants in the mixture. Before they become dry, dip again in pure rainwater to remove the mixture. If too large to dip, apply the mixture with the syringe, and in the course of a quarter of an hour or so syringe with pure rainwater. Our illustration shows the Thrips in the larval and winged state, natural size and greatly magnified.

Ants.—These extremely interesting insects are frequently troublesome in gardens, and in the spring of the year the small red species mars the appearance of lawns by throwing up numerous heaps of fine soil. It is easy to destroy them by dropping a mixture of Paris Green and sugar near their runs. But as Paris Green is a poison, animal life must be considered. We recommend a simple remedy which entails no danger, but it must be followed up persistently. Purchase a few common sponges, as large as a man's fist. Dissolve one pound of Demerara sugar in two quarts of warm water. Immerse the sponges, wring out nearly all the liquid, and place them near the ant runs. Twice daily throw the sponges into hot water, and repeat the process until the ants are cleared. Nests located under walls can be destroyed by boiling water.

Caterpillars cannot often be treated in a wholesale way without injury to the plant. Hence it is usual to rely on hand-picking, and, tedious as this may be, a little perseverance will accomplish wonders. We have seen a fruit garden, literally hideous with clusters of Caterpillars in spring, completely cleared by a few days' steady work, costing but a trifle, and only needing to be conducted so that in removing the vermin there should be no harm done to the crops. In the same way the Gooseberry grub should be disposed of. Precautions cannot be taken against Caterpillars, but the careful cultivator will in good time look for patches of eggs and clusters of young Caterpillars on the under sides of leaves, and will carefully nip off the leaves on which the colonies are feeding, and make an end of them. This enemy cannot be raked in rank and file, but must be taken in detail, as in guerilla warfare.

Earwigs are the dread of the florist, for they spoil his best Dahlias and Hollyhocks, and are too partial to Chrysanthemums. They are readily trapped, as they like to go up to a high, dry, dark retreat; hence a bit of dry moss in a small flower-pot, inverted on a stake, will entice them into your hands; and if you are determined to keep down Earwigs, this way is sure, though, perhaps, not easy, because it must be followed up morning and evening from the beginning of June onwards. The hollow stems of the Bean make good traps, as indeed do hollow stems of any kind, for Earwigs love to creep into close, dark shelters after their nocturnal meal; and the cultivator who has resolved that he will not be eaten up by them needs only to persevere, and he may depend on trapping every Earwig within the boundaries. Unfortunately, they use their wings freely, and so travel from the sluggard's garden to find 'fresh woods and pastures new.'

Slugs are serious plagues to the gardener, and they sometimes appear in large numbers so suddenly as to suggest the idea that the little Slugs have come down in showers. Young crops are especially liable to injury from these vermin, and it is not easy, even in well-kept gardens, to keep them down. Constant attention is necessary, particularly in wet seasons. But here, as in the case of many other kinds of vermin, means may be adopted that will accomplish the double purpose of destroying the plague and benefiting the land; for lime, salt, soot, and nitrate of soda are certain Slug-killers, and will usually pay for their employment by their enrichment of the ground. The nice point always is to employ them advantageously. It should further be borne in mind that a Slug slightly touched by lime or salt has the power of throwing it off by means of the slimy exudation with which the creature is endowed. But if again quickly assailed in a similar manner death is certain to follow. Land made ready for sowing may be pretty well cleared of Slugs by broadcasting it with salt. Unfortunately, these destroyers are only effective in fine weather. In rainy seasons, or when a crop is rising, it is necessary to resort to trapping, and many kinds of vegetable refuse make tempting baits for Slugs. Pieces of Orange peel, suitably placed, are soon covered with the vermin, especially in the winter during intervals of frost. Cabbage leaves, sliced Turnips and Potatoes, or almost any waste vegetable may be used. The traps should be scattered about at dusk, and be gathered up in the morning, and buried in pits, or destroyed by fire.

Gas-lime is highly destructive to Slugs, but when first applied it is poisonous to plant life. An excellent method of using it is to dress the surface in autumn at the rate of from four to six cwt. per acre, and to dig the ground deeply four weeks later.

Rows of Peas are easily protected by a covering of barley sweepings, or by charcoal broken very small and flavoured with paraffin. Slaked lime, carefully used, is also employed with satisfactory results.

Snails.—In their methods of attacking garden vegetation, and in the extent of damage they cause, Snails may be placed in the same category as Slugs. During the day the Snail usually remains in hiding, emerging from rockeries and creeper-covered walls in the evening or after a shower of rain. They may be trapped by one of the methods suggested for Slugs, and preference should be given to the use of Cabbage leaves. It will, however, be safer to protect young plants by giving heavy dressings of lime or soot. Hand picking is the surest means of dealing with them, and in the winter months large numbers may be collected from among box edgings, the base of ivy-covered walls and similar shelters. Birds, especially thrushes, show a marked partiality for Snails.

Wasps are a terrible scourge in some gardens. They spoil a large quantity of fruit, and jeopardise the remainder by forcing the harvest before the crops are ready for gathering. When the localities of the Wasps' nests are known, it is a simple task to dispose of them. Turpentine and gunpowder were formerly in vogue, especially among the younger members of the community, to whom a spice of danger is always an attractive element in the fun. But these are clumsy methods of destruction and will not compare with the far easier remedy of poisoning the colonies by means of cyanide of potassium. Dissolve one ounce of the drug in a quarter of a pint of water. This will be sufficient to destroy several nests, but it is a deadly poison, and must be kept in a place of safety. Soak a piece of rag in the fluid, and lay it over the entrance to the nest. There is no occasion to run away; not a Wasp will venture out, and those which return from foraging will not lose their tempers and find yours, but at each successive attempt to enter their home they will become feebler, until they fall near or beneath the drugged rag. After an hour or two the nest may be dug out, when every insect, including queen and pupæ, will be found dead.

If the colonies lie beyond your frontier, or their positions cannot be ascertained, the enemy must be disposed of by stratagem and in detail. One of the best modes of trapping them is to put some injured fruit beneath one of the trees, and over it a hand-light raised about three inches above the ground by stones or pieces of wood placed at the four corners. This light must have a rather large hole at the top. Upon it should rest another light from which egress is prevented, except through the apex of the lower light. After the Wasps have visited the fruit, they will rise into the first light, and gradually find their way through the opening into the one above, from which not one insect in a hundred will escape. In a trap of this kind we have seen an enormous number of Wasps and Hornets which had been lured to death within a few hours.

Another simple and effective method of destroying these pests is to pour a small quantity of ale mixed with sugar into glass jars and suspend them from branches of Pear or Plum trees. The vessels must be emptied every few days and the liquid renewed.

Wireworm is the most persistent and destructive of all the ground vermin. There are fully a dozen species of beetles the larvæ of which are known as 'Wireworms,' and of these the 'Spring-Jacks,' 'Click-Beetles,' and 'Blacksmiths'—*Elater obscurus, E. lineatus*, and *E. ruficaudis*—are the most prevalent. The female beetle deposits her eggs in the earth in the height of the summer, and in due time the worms emerge and commence their depredations. These worms have a tenure of three to five years in their subterranean homes, during which time they feed voraciously, and are not very particular as to what they eat. Their muscular power renders them expert in burrowing, and they are well protected by their horny jackets. When their term of feeding is completed, they descend to a considerable depth and change into the chrysalis state, from which they come forth as jumping beetles in the course of July and August, a certain proportion remaining in the ground to complete their final change in spring. Their power of destruction is then at an end. They resort to flowers, lead a merry life for a short time, and when they pass away leave plenty of eggs to continue the race of Wireworms.

For practical purposes the Wireworm may be regarded as inhabiting every kind of soil and consuming every kind of crop. The crops it is most partial to are Grass, Potatoes, Turnips, and the juicy stems of all kinds of cereals. The larvæ may be trapped by burying in the ground pieces of Potato, or better still thick slices of Beet root; the spots to be marked, and the traps examined every few days, when the Wireworms can be destroyed. Superphosphate sown along the drills with seed has saved spring-sown crops from destruction; and Vaporite, a proprietary article, has also been used with marked success. The latter gives off a gas smelling of naphthalene which kills the Wireworms. Soot is a well-known remedy, and by its use the crops are also benefited.

Woodlice are very destructive but easily caught, and they may be completely eradicated by perseverance. When a frame or pit is infested, they can be destroyed wholesale by pouring boiling water down next the brickwork or the woodwork in the middle of the day. If this procedure does not make a clearance, recourse must be had to trapping. In common with Earwigs, they love dryness, darkness, and a snug retreat; but while a mere home suffices for Earwigs, a home with food is demanded by Woodlice. Take a thumb pot, quite dry and clean. In it place a fresh-cut slice of Potato or Apple, fill up with dry moss, and turn the whole thing over on a bed in a frame or pit. Thus you have devised a Woodlouse trap, and next morning you may knock the vermin out of it into a vessel full of hot water, or adopt any other mode of killing that may be convenient. Fifty traps may be prepared in a hundred minutes; and those who are determined to get rid of Woodlice may soon make an end of them.

Rats and Mice.—Traps are efficient while they are new, and almost any reasonably good contrivance will answer for a time, but will fail at last, or at least for a season. To keep down Rats and Mice effectually there must be invented a succession of new modes of action, for these creatures—Rats especially—are so clever that they soon see through our devices, which then fail of effect. Generally speaking, two rules may be prescribed. In the first place it is imprudent to fill up their holes or stop their runs; let them have their way. If you stop them, they will make new thoroughfares, to the further injury of the foundation; and, besides, when you are acquainted with their runs, you know where to put traps and poison for the vermin. As to the best poison, there is nothing so effectual as arsenic; but it should be employed with great care, and before it is brought on the premises the question of safe storage must be considered. A fat bloater split down and well rubbed with common white arsenic will kill a score of Rats, provided only that they will eat it.

Cut it into four parts, and place these in or near their runs, and cover with tiles or boards to prevent dogs and cats obtaining them. If this fails, try bread and butter dressed with oil of rhodium and phosphorus. The oil of rhodium seems to possess an irresistible attraction for these vermin. When dry food is preferred, there is nothing so good as oatmeal; and it is a golden rule to feed the Rats for a few days with pure oatmeal, and then to mix about a fourth part of arsenic with it. Several proprietary articles are offered for the destruction of Rats. Before resorting to these means of annihilating vermin it is necessary to take steps to prevent the bodies from proving a nuisance after death. A good fox-terrier will keep a large garden free from Rats and Mice.

THE FUNGUS PESTS OF CERTAIN GARDEN PLANTS

Many of our garden plants are liable to the attacks of fungi. Cures are in most instances unknown, but in some cases preventives—which are better—have been adopted with partial or entire success. Plants raised from robust stocks, grown in suitable soil and under favourable conditions, are known to be less liable to disease than seedlings from feeble parents, or those which have been rendered weakly by deficiencies in the soil or faulty cultivation. Whether weakness is hereditary, or is attributable to a bad system, the fact remains that disease generally begins with unhealthy specimens, and these form centres of contamination from which the mischief spreads. It is, therefore, important that seed from healthy stocks should be sown, and that a vigorous constitution should be developed by good cultivation.

Anbury, Club, or Finger-and-toe.—The disease known by these various names is common in the roots of cultivated cruciferous plants such as Cabbages, Kohl Rabi, Radishes, Swedes, Turnips, &c., and also in many cruciferous weeds, including Charlock and Shepherd's Purse. The cause of this disease is an extremely minute fungus, which may lie dormant in the soil for several years for want of a comfortable home, and when a cruciferous plant becomes available the fungus fastens on the fine roots, multiplies rapidly in the tissues, and produces malformation and decay. After the disease has made some progress insect agency frequently augments the mischief, so that on cutting open a large decaying root it is not unusual to find the interior packed with millipedes, weevils, wireworms, and other ground vermin.

Unlike the Potato disease, which spreads from plant to plant through the atmosphere, the fungus of Finger-and-toe infects the ground, and from the first spot attacked the disease spreads rapidly in all directions and in various ways. It may be carried by the soil adhering to implements or the boots of labourers. And each patch becomes a new centre of infection which is spread by digging or raking. Every scrap of infected soil, or of diseased fibre which may be added to the manure-heap, distributes the virus over a wider area, so that Finger-and-toe may suddenly appear in parts of the garden which have hitherto been free from this troublesome pest. A very simple experiment will prove the certainty and ease with which the spores may be introduced to fresh land. Macerate the tissue of old Finger-and-toe in water; use this on young isolated plants of Cabbage or Turnip and in a short time the plants will be infected.

The fungus which produces Finger-and-toe is known as *Plasmodiophora brassicæ*, and it belongs to the *Myxomycetes*, or 'slime-fungi,' which, as a rule, live upon decaying vegetable material. The protoplasm of the fungus ramifies among and within the tissues of the roots of attacked plants, and eventually produces an amazing number of spores so small that more than thirty millions would be required to cover a superficial inch. A microscope of great power is necessary to reveal them to human vision.

The spores are capable of resting in a state of vitality for a long time, and can easily withstand the frosts of winter. The illustration shows at A the fungus in its protoplasmic condition, and at B its ultimate sporiferous or 'seed'-producing stage, after the protoplasm has changed to a mass of minute spores (enlarged five hundred and twenty diameters). When a spore in due course germinates, its protoplasmic contents escape through a small aperture in its wall and begin moving about of their own accord in a slow writhing manner. The movement is so much like that of the microscopic animal organism found in ponds, and called *Amœba*, that this tiny mass of moving protoplasm is called *Myxamœba*, to denote that it is an amœba-like form produced by one of the *Myxomycetes*. Each myxamœba is drawn out at one spot into a fine delicate tail or cilium, as at C, D, E, and is capable of a creeping motion in moisture. When quite free from the spores, transparent expansions or limbs extend from the bodies of the myxamœbæ, as at F, G, and when these organisms, after existing in the soil for a longer or shorter time, reach the roots of cruciferous plants, which they apparently enter through the root-hairs, they again assume the protoplasmic condition shown at A, and live within the cells, at the expense of the nurse-plant. Other cruciferous plants are less seriously damaged by the pest than are Turnips and Cabbages; but it is evident that if diseased Charlock is near Turnips, the latter are very likely to fall a prey to the disease. We advise the sowing of the best seeds, the eradication of cruciferous weeds, and the destruction by fire of all decaying Finger-and-toe material, for it is in this material that the spores of the disease rest ready for continuing the disease in the following season. It is also desirable that cruciferous plants should not be continuously grown in the same quarter—in other words, it would be prudent after an attack of Anbury not to repeat a cruciferous crop on the same ground, but to follow on with a crop of some other class.

Numerous experiments have shown that slaked lime can be relied on to destroy the spores of Finger-and-toe in infested land. An application of from fourteen to twenty-eight pounds per pole may suffice in the case of light soils, but fifty-six pounds per pole will not be too much on heavy land, and the dressing should be given either six or eighteen months before a Cabbage or Turnip crop is sown; the longer period is the more certain in its effect. Preference should be given to stone or rock lime over chalk lime. The former is much more powerful and efficient. It may be necessary to repeat the dressing twelve months after the first application. As regards the occurrence of Anbury in seed-beds, frequent transplantation is a very effectual mode of stopping its progress, for the little galls can be pinched off by the workman, and burned as he proceeds; and the plant, being invigorated by change of soil, will soon grow away from the affection. In transplanting Cabbages it is a good plan to discard and burn such plants as are obviously affected with Anbury. It is worthy of remark that in market-gardens this disease is by no means so prevalent as to interfere with the routine of cultivation, although the Cabbages, Broccoli, and Cauliflowers grown in these grounds are, under other circumstances, especially liable to attack. By 'other circumstances' we mean that market-gardens are generally kept under high cultivation, the land being perpetually turned and heavily manured; and these measures appear to be a preventive of Anbury, while they result in heavy crops. But on land less energetically tilled Anbury may prevail to such an extent as to interfere seriously with the order of cropping. Another very important mode of keeping down the pest consists in burning instead of burying the stumps and all other refuse of the crop that cannot be turned to account.

Confusion may be prevented if we point out that Club-root, Anbury, or Finger-and-toe—whichever name may be used—is quite distinct from an apparently similar malformation of the root which is sometimes induced by certain characteristics of soil, seed, or manure, and is in fact a case of reversion to the original wild type. Instead of a shapely, solid Turnip, the bulb is divided into a number of coarse, worthless tap-roots, caused by either poverty of the soil, careless cultivation, or a degenerated stock of seed. Those who save their own seed continuously for years are almost certain to become well acquainted with this malady. They will find a change of seed necessary, and at the same time an alteration in the routine of culture. A healthy, vigorous plant, derived from a pure seed-stock, does not easily make Finger-and-toe, but a sound root that stands for food and money.

'Grub.'—The wart-like growths formed upon the roots of Turnip and Cabbage by the little hard beetle known as the Turnip-gall Weevil, *Ceutorhynchus pleurostigma*, are also quite distinct from Finger-and-toe. By cutting across a malformed root of Turnip or Cabbage it is usually not difficult to determine the cause of the mischief. If it is Finger-and-toe the root will be found filled with decaying matter; in the case of Weevil attack the small legless maggots, commonly called 'Grub,' will be brought into view; and if it is merely an instance of reversion the cut root will appear to be healthy.

Potato Disease.—The fungus which causes the Potato Disease, or 'Blight' as it is sometimes called, was formerly known as *Peronospora infestans*; now it is recognised by scientific authorities as *Phytophthora infestans*. The mark of its pestilent touch on the foliage, and its destructive effect on the tubers, are unfortunately too familiar in gardens and on farms. In dry seasons its energies are restricted, but the scourge is never absent, and during wet summers the parasite may do its deadly work on such a vast scale as to cause a Potato famine. Moisture is a necessity of its existence, and in rotting haulm, decayed tubers, and damp soil the spores remain in a resting condition until they are afforded an opportunity of multiplying with the marvellous rapidity that invests the disease with its terrible power. A series of six illustrations, five of which are highly magnified, will enable the reader to follow the development of *Phytophthora infestans*.[1]

Spraying Potato plants twice or thrice with Bordeaux mixture has proved effective in warding off the attack of *Phytophthora infestans*, and the practice is now freely adopted, especially in humid districts. The first application should be given towards the end of June or early in July, immediately the haulm is sufficiently developed. The Bordeaux mixture is made in the proportion of four pounds of pure copper sulphate and two pounds of quicklime to forty gallons of water. The foregoing quantities will give what is known as the *one per cent.* mixture. For the *two per cent.* mixture the quantities of copper sulphate and quicklime must be doubled, but the amount of water should remain at forty gallons. In its effect on the fungus, however, little difference is to be found between the two solutions. The copper sulphate is stirred into a few gallons of hot water

1 For permission to reproduce the engravings numbered 1, 3, 4, and 5 from Professor Marshall Ward's 'Diseases of Plants,' we gladly acknowledge our indebtedness to the Society for Promoting Christian Knowledge. Professor W. Carruthers has kindly allowed us to use the illustrations numbered 2 and 6.

placed in a wooden tub or earthenware vessel. When quite dissolved, add twenty or thirty gallons of cold water. The lime, which must be freshly burnt quicklime, is then slaked in another vessel and thoroughly stirred with two or three gallons of water until it is of the consistency of thin cream. As soon as the liquid is quite cold, filter it through coarse sacking into the copper sulphate solution and add water to make a total of forty gallons. To be effective, Bordeaux mixture must be applied in the form of a fine spray, and not with a coarse-holed syringe.

The Burgundy mixture, the use of which is preferred by some, acts in a very similar manner to the Bordeaux mixture, and is made in the same way as the latter, except that washing soda (five pounds) is substituted for quicklime.

Those who leave Potatoes to rot in the ground because the crop is not worth digging, or who bury diseased haulm and tubers in a shallow trench, under the impression that it is a safe way of getting rid of worthless vegetation, are simply storing *Phytophthora* for another attack in the event of Potatoes being planted in the same land again. If buried at all, it must be at a considerable depth, but the effectual method is to destroy all Potato refuse by fire.

Wart Disease (Black Scab) of Potatoes *(Synchytrium endobioticum,* Percival).—This extremely infectious and destructive disease of the Potato has been given a variety of names in different parts of the country, but it is now generally known as the Wart or Cauliflower Disease, the latter term being attributable to the Cauliflower-like appearance of the outgrowth of the fungus. This outgrowth first shows in the eyes of the young Potato in the form of small wrinkled warts. These multiply and combine, thus creating a dark spongy scab which eventually decomposes. Where the disease is very rife it attacks haulm as well as tubers, and a yellowish-green mass may sometimes be found just above or just below the surface of the soil. As a rule, however, no outward indication of its existence is to be seen in the crop during the early stages of growth, but towards the end of the season the haulm of badly diseased plants often retains a fresh green appearance when the foliage of others, which are healthy or only slightly attacked, is dying off.

Infection is perhaps most commonly spread by the planting of diseased tubers. Another frequent means of dissemination is caused by consigning infected haulm to the waste heap instead of to the fire. The spores may also be introduced in manure from animals fed on diseased Potatoes in a raw state, and they may even be carried from one plot to another on garden implements or the boots of those who walk across infected ground. Immediately any sign of the disease is observed it should be dealt with promptly and in no uncertain manner. Every particle of the infected material must be carefully collected and burned. Dig out the soil around all diseased plants and burn this also. On infected land it is important that some crop other than Potatoes be taken in the season following the outbreak, and, if possible, such land should not be used for Potatoes for at least five or six years. But where garden space is limited, a contaminated plot may have to be requisitioned for Potatoes within two or three years. In such cases it is an excellent plan to dust the sets freely with sulphur at the time of planting and to repeat the application before earthing up.

Although for some years the unremitting labour of experts has been devoted to the investigation of Wart Disease, and innumerable experiments have been undertaken, no effectual remedy has yet been discovered. It has been found, however, that certain Potatoes are resistant to the disease, and by order of the Ministry of Agriculture and Fisheries none but 'immune' varieties may be planted in districts scheduled as infected areas. A notification of the existence of Wart Disease must be made to the Ministry immediately it is observed.

Leaf Spot of Celery.—This disease, which is caused by a minute fungus *(Septoria apii*, Chester), is capable of inflicting serious damage to the Celery crop unless prompt measures are taken to exterminate it. The first sign of its appearance is to be found in the leaves in the form of small brown patches. These are, however, quite distinct from the spots deficient of leaf-green due to the attack of the Celery Fly larvæ, and on close examination may be recognised by the presence of a number of very small black points. From the leaves the fungus quickly spreads over the leaf-stalks and finally to the heart of the plant, ending in its total collapse. So rapid is the multiplication of the spores, especially in moist weather, that a few diseased plants are capable of infecting a large plot within two or three weeks. Immediately discoloration of a leaf is noticed the affected portion of the plant should be picked off. If the stage of the disease is so far advanced that the outer leaf-stalks have become decayed, the entire plant should be removed and destroyed. It is of the utmost importance that every particle of diseased material be consigned to the fire and not to the waste heap. Spraying three or four times with Bordeaux mixture at intervals of two or three weeks may be helpful in the case of a light attack, but the safest course always is to remove and destroy any plant on which the fungus is found. One of the most frequent means of introducing Leaf Spot of Celery is through the use of infected seed, and therefore only seed which has been treated for the destruction of the fungus should be sown.

Lettuce Mildew.—This fungus is named *Bremia lactucæ*, formerly known as *Peronospora ganglioniformis*, and is sometimes of the most destructive character. It covers Lettuce leaves with a fine white bloom, which decomposes the leaves, and makes them adhere together in one putrescent mass. It should be looked for in its earliest stages, and be hand-picked and burned. Old Lettuce stumps should likewise be pulled and burned, otherwise they may harbour the disease.

Onion Mildew is caused by the fungus *Peronospora Schleideni*, which is occasionally disastrous in its effects, more especially in cold, wet seasons. It occurs at uncertain intervals of time with extraordinary virulence, and then utterly destroys the crops. Autumn sowing is considered a good preventive by many growers, as the disease is frequently fatal to spring seedlings. In its early stages the mildew may be successfully dealt with by freely dusting the plants with flowers of sulphur when wet with dew, or by the application of sulphide of potassium in the proportion of one ounce to a gallon of water. Otherwise all diseased material should be removed and burned.

Pea Disease.—Although garden Peas often suffer badly from the attacks of *Peronospora viciæ*, which is the cause of Pea Mould, yet the most deadly foe to Peas, especially late Peas, is a fungus of a totally different character. To such an extent does the Pea Blight sometimes devastate the later Peas, particularly in dry summers, that the whole crop is in some gardens completely annihilated. The name of the fungus of the Pea Blight or Mildew is *Erysiphe Martii.* Its attack is often made suddenly; the leaves then lose their natural green colour, and become yellowish and densely coated with a fine white bloom; this bloom becomes at length dusted over with innumerable minute black bodies, which look, under a lens, like tiny spiders'-eggs in the web. These little black bodies are filled with extremely small transparent vessels, and each vessel contains from four to eight spores or seeds. Our illustration shows this *Erysiphe* enlarged one hundred diameters, with two of the vessels containing the spores removed from the globular spots and further enlarged. The only safe way of dealing with infested Pea plants is to burn them. Many other species of fungi belonging to the same genus attack fruit trees, vegetables, and garden flowers. It is, however, unnecessary to illustrate them, as they more or less resemble the fungus of Pea Blight. They all arise from an *Oïdium* condition, similar to the *Oïdium* or Mildew of the Vine, and it is in this condition alone, as in the case of the Vine, that they can be reached by any fungicide.

Tomato Diseases.—The Tomato, like its near relative, the Potato, is subject to a number of destructive diseases which spread rapidly if allowed to become established. The most serious of these epidemics are found among crops cultivated under glass, where the forcing treatment which they often receive, and the soil and atmospheric conditions, render the plants abnormally susceptible to the attacks of fungi and insect pests. Perhaps the most virulent forms of disease with which the Tomato-grower is troubled arise from the attacks of parasitic fungi and bacteria, among which the following are most frequently met with:—

SLEEPY DISEASE, or TOMATO WILT.—In its outward symptoms and effects this disease somewhat resembles an attack of Root-knot Eelworm, but the swellings are absent from the root. The plants for a time appear quite vigorous and healthy, but when full-grown they suddenly wilt and die within a few days. The malady is caused by the fungus *Fusarium lycopersici*, which first invades the roots and ultimately eats its way through the substance of the collar or stem near the surface of the soil, in consequence of which the supply of water taken up by the roots is cut off from the leaves above ground and the plant collapses. There is no remedy for the Sleepy Disease of Tomato, and plants which bear evidence of infection should be carefully dug up and burned.

TOMATO 'STRIPE.'—This disease of the Tomato is comparatively common, and although the attacks are sometimes slight its ravages may be disastrous when conditions are favourable for its development. The presence of Tomato Stripe is usually first noticed about the time fruit is forming. The stems of the diseased plants then exhibit dark spots and elongated sunken stripes of a brown tint, and yellow patches, which turn brown later, appear on the leaves. Brown pits or depressions develop on the fruits and spoil their appearance. The disease has been traced to the action of a bacterium which closely resembles, or is identical with, that causing Stripe among Sweet Peas. This organism probably resides in the soil, and the signs of its attack are often visible in young plants. In severe cases the soil of the house should be removed and replaced with fresh loam. But when only slight traces of the disease are apparent, partial sterilisation of the soil by means of carbolic acid, as recommended for Root-knot Eelworm on page 425, may be adopted. One of the surest means of guarding against losses by Stripe disease, is to promote robust healthy growth, and to avoid extreme forcing conditions, particularly by the excessive use of nitrogenous manures. Where, however, forcing manures may have been employed in too large a quantity, an application of potash (in the form of kainit or sulphate of potash) and phosphatic fertilisers should be given to counteract the effect of the nitrogen. Immediately any trace of the disease is found, remove the affected part of the plant, if it is possible to do so without serious injury, but otherwise the entire plant should be uprooted and destroyed by fire. It should be remembered that the organism can be carried on the fingers and on tools, and therefore knives with which affected plants have been trimmed should be sterilised with lysol or some other antiseptic solution before being used on healthy plants.

TOMATO-LEAF RUST.—The leaves of the plant attacked by this disease rapidly become covered with a dull brownish velvety mould, or fungus, known as *Cladosporium fulvum*. From the mouldy spots and patches thousands of spores are readily carried by a slight current of air to the surrounding healthy crop, and unless prompt measures are taken to check the pest the whole house is rapidly involved. Excessive atmospheric moisture encourages the mould, and it is spread extensively if diseased plants are sprayed with water in the presence of healthy ones. Judicious management in air-giving, which is one of the fundamental principles of successful Tomato culture, will do much to prevent the attack of *Cladosporium fulvum*. Under regular examination the presence of the disease will be revealed before considerable damage can be inflicted, and when only a few leaves are affected, carefully remove and consign them to the fire. Spraying with the Bordeaux mixture at half the usual strength is recommended when the disease is first noticed. When the plants are bearing flowers or fruit, fungicides containing copper must not be used, but a solution of liver of sulphur, one ounce dissolved in six gallons of water, employed instead.

ROOT-KNOT EELWORM.—A dangerous insect pest which frequently attacks the Tomato, in common with the Cucumber and Melon, is the Root-knot Eelworm *(Heterodera radicicola)*. The root on which the swollen pea-like knots develop do not carry on their ordinary functions, and the leaves droop, the stem becomes limp, and the whole plant soon collapses and dies if the trouble is severe. The treatment suggested on page 425 should be adopted.

Sometimes the outdoor Tomato crop is attacked by *Phytophthora infestans*, the fungus responsible for the Potato Disease: Bordeaux mixture should be used to check it.

Directions for preparing the Bordeaux mixture are given on page 440.

Another useful preparation which checks many fungus diseases may be made by dissolving one ounce of potassium sulphide (liver of sulphur) in three or four gallons of water, to which should be added an ounce or two of soft soap. The last named greatly assists in the complete and uniform wetting of all parts of the foliage.

THE FUNGUS PESTS OF CERTAIN FLOWERS

Cineraria and Senecio Disease.—*Senecio pulcher*, soon after its introduction into England, was attacked, and in some gardens completely destroyed, by a fungus named *Puccinia glomerata*, or rather the *Uredo* stage of this fungus with simple, not compound, spores. The fungus is well known, being closely allied to that which causes the rust or mildew of corn crops. It is very common on the wild species of Groundsel in England, being especially frequent and virulent on the Ragwort Groundsel, *Senecio Jacobea*, from August to October. The leaves of infected plants are covered with rust-coloured dusty pustules, the *Uredo* condition of the fungus, and known in this stage as *Uredo senecionis*, sometimes termed *Trichobasis senecionis*. The fungus has a *Puccinia* stage of growth very similar to that of the Hollyhock fungus, *Puccinia malvacearum*.

At A is illustrated a fragment of a leaf of *Senecio pulcher*, natural size, and covered with the orange-coloured fungus; at B a small part of a *Uredo* pustule as seen bursting through the cuticle of the Senecio leaf.

No remedial measures for the extirpation of this fungus are known, but as garden Senecios and Cinerarias are infected by diseased plants of Wild Groundsel, it is desirable that plants of the latter (especially when diseased) should be destroyed. Weeds in and about gardens are a common cause of disease in cultivated plants. It often happens that a weed, being sturdy, is only slightly inconvenienced when attacked, whilst a cultivated plant will speedily succumb if attacked by the same fungus. This is the case in the *Sempervivum* disease. In this country the common House Leek is the nurse-plant, and is seldom much injured; but if the disease *Endophyllum sempervivi* gets among greenhouse species, every plant may be utterly destroyed.

Gladiolus, Crocus, Narcissus, and Lily Diseases.—In certain soils and situations where the ground is heavy and the atmosphere inclined to be humid the Gladiolus is very subject to a destructive fungoid disease. This is especially the case during unusually wet summers. The disease attacks the corm, and corrodes and decomposes the tissues, so that on cutting open a corm the whole interior, or such parts as are diseased, will be found permeated with a deep, foxy colour. It is believed by some persons that one stage of this disease is identical with the disease named 'Tacon' by the French, and in this country known as 'Copper Web,' *Rhizoctonia crocorum*. This *Rhizoctonia* is a mere spawn or mycelium, a mass of rusty-brown material like a thick coating of spider's web of a red tint. This parasite attacks the Crocus (especially *C. sativus*), the Narcissus, Asparagus, Potato, and other plants. Immersed in the softer and damper portions of the red substance of the corm may frequently be found great numbers of large compound spores, as illustrated at A (enlarged two hundred and fifty diameters). These bodies belong to the fungus named *Urocystis gladioli*; but whether they really belong to the spawn named *Rhizoctonia* there is no conclusive evidence, as the spores have never been seen on the threads or upon any spawn. The spores are very ornamental objects, consisting of from three to six compacted inner brown bodies, surrounded by an indefinite number of transparent cells. At maturity these spores break up as at B, and are the means of reproducing the fungus.

The Colchicum is attacked by a closely allied but different species of *Urocystis*—viz. *U. colchici*. The Ranunculaceæ are attacked by another ally in *U. pompholygodes* and Rye is attacked by a third in *U. occulta*. No method of cure has yet been published for this pest; it is, however, desirable that only sound and good corms should be planted, for if infected corms are placed in the ground it is one certain means of propagating the disease. The bars shown across the illustration of this disease are magnificent crystals, very common in Gladiolus corms.

Lilies are very subject to a disease in early summer: the leaves get spotted and damp, and rot off; the flower buds speedily follow, and leave the bare stalk. The disease of Lilies is caused by a fungus closely allied to the fungus of the Potato disease, and named *Ovularia elliptica*, known also as *Botrytis elliptica* (see illustration C). The spores are large, and produce zoospores, or spores with hair-like tails (cilia), capable of swimming about in water or upon moist places. This pest attacks a large number of species of *Lilium*, both before and after flowering. *Hyacinthus candicans* and some Tulips suffer from a very similar, if not the same, organism. This fungus has been described as a true *Peronospora*. Bulbs are subject to many fungus growths as *Volutella hyacinthorum*, *Didymium Sowerbei*, &c.; many fungi follow the decay of the bulb, others undoubtedly produce or greatly accelerate decay. No remedy is known, but we advise the purchase of the soundest and best bulbs. Good drainage and sufficient air are indispensable. All infected foliage and stems should be burned.

Disease of Hollyhocks and Malvaceous Plants.—In some parts of England the cultivation of the Hollyhock had at one time quite ceased owing to the attacks of a microscopic fungus named *Puccinia malvacearum*. In gardens and nurseries, where years ago Hollyhocks were one of the chief ornaments of the place, it became impossible to grow a single plant. The disease is not confined to the Hollyhock, for it attacks many malvaceous plants, notably the Mallows of our hedgesides. We have seen plants of the white variety of the Musk Mallow (*Malva moschata*) totally destroyed by this parasite. The home of the Hollyhock fungus is Chili, whence the Potato fungus reached us. The Hollyhock fungus first attacked the malvaceous plants of Australia, and then reached England in 1873 by the continent of Europe. The best and cleanest seeds of the Hollyhock should be purchased.

A fragment of a Hollyhock leaf is illustrated at A, dotted with the characteristic brown pustules; these pustules cover the stems as well as the leaves. At B is shown the edge of a pustule enlarged one hundred diameters and seen in section; to show the whole of a pustule in section from six inches to a foot of space would be required. Bursting through the skin of the plant may be seen a dense forest of threads, each thread bearing a spore with a joint across the middle. One pustule alone will produce thousands of these double spores. At C some of the threads and spores are still further enlarged to two hundred diameters, and at D one ripe spore is shown falling from the thread and breaking asunder—each piece is a reproductive body or spore. When mature, these minute spores or 'seeds' are carried in the air by millions. At E one of the compound spores is enlarged to four hundred diameters. As this disease is seated within the tissues of the plant, remedies are difficult of application, and in many cases attempts at cure have failed. No doubt the fungus is nursed by malvaceous weeds. Infected Hollyhock plants and allied weeds should be destroyed by fire or by deep burying.

Poppy Disease.—Garden Poppies are often attacked by a fungus pest closely allied to the fungus of the Potato disease, and named *Peronospora arborescens*. It grows sometimes in abundance on the common Red Poppy of cornfields *(Papaver Rhoeas)*, and it badly attacks *P. somniferum* and all its garden varieties. The fungus grows within the leaves, and emerges with a tree-like growth through the organs of transpiration (the stomates) on the under side of the leaves. Like the fungus of the Potato disease, it speedily sets up decomposition, and destroys the host-plant.

At A is illustrated one of the stems of the Poppy *Peronospora* emerging from the leaf, enlarged seventy-five diameters. The fungus of the Poppy is very much more branched than that of the Potato, and every minute branchlet carries a spore. To save confusion, a large number of spores are omitted from the branchlets in the illustration, and the branches growing from the stem both before and behind are for the same reason left out. At B a tip of a single branch is shown further enlarged to four hundred diameters. The spores in the Poppy fungus are unusually large and numerous: an infected plant will throw off many millions of such spores. All the putrefactive spawn of this fungus is inside the host-plant; cure, therefore, is difficult. This disease, like every other plant disease, is always at its worst in ill-kept places where red field Poppies are abundant. Field Poppies are often sown with unclean corn. As prevention is better than cure, all we can advise is, buy the best and cleanest garden and field seeds, cultivate in the best way, and look out for and burn, or deeply bury as soon as detected, all disease-stricken plants, whether wild or cultivated. When diseased plants of any sort are left to decay on the refuse-heap, it is the most certain way of propagating a plant disease for the next year.

Diseases of Violets.—Violets are subject to fungoid diseases, both in spring and autumn. The disease of autumn is caused by the brown *Puccinia violæ*, allied to the *P. graminis* of Corn and to the *P. malvacearum* of Hollyhocks and various malvaceous plants. The *Puccinia* of Violets has its yellowish or orange-coloured stage; it is then known as *Trichobasis*, or *Uredo violarum*. In spring and early summer Violets are often badly affected by a fungus named *Æcidium violæ*, which is apparently identical, however, with *Puccinia violæ*. This disease attacks leaves, stems, and sepals, and it is best examined on the leaves. In this position it is seen to consist of a considerable number of minute yellow pustules, each pustule less in size than a pin's head, and all congregated into one flat circular mass of about a quarter of an inch in diameter. This pest is very frequent on the Dog Violet, but it is perhaps equally common on the Sweet Violets of our gardens in early spring, and it not infrequently spreads to other species of *Viola*. One of the most destructive pests of Violas is found in *Æcidium depauperans*, so called because its effect is first to starve and attenuate, and then to totally destroy, plants of *Viola cornuta*. It is a close ally of *Ae. violæ*, but it differs in having its minute cups or pustules irregularly distributed all over the green parts of the host-plant instead of being congregated in circular patches, as in *Ae. violæ*. Our illustration shows, at A, a small portion of the stem of *Viola cornuta* attacked by *Æcidium depauperans*. The minute pustules are seen (natural size) distributed all over the stem, leaf-stalks, and ruined leaves; the effect of the fungus growth is to decompose the tissues of the plant. At B, a transverse section through the stem is illustrated and magnified twenty diameters. The section cuts through several of the abscess-like pustules, and it is seen how completely embedded they are in the flesh of the plant. At C, a pustule is seen in section, enlarged sixty diameters to show more clearly the innumerable spores, or 'seeds,' disposed in necklace-like fashion, which are destined to reproduce the pest in future seasons. Another disease of Violets in autumn is caused by a fungus named *Urocystis violæ*. This fungus causes gouty swellings to form on the stalks and principal veins. These swellings at length burst, exhibit black patches, and discharge sooty spores. The fungoid disease named *Phyllosticta violæ* is frequently common on Violet leaves in June. In this the spots are whitish. No cure is known, and it is always well to burn or deeply bury all infected leaves or plants.

Lightning Source UK Ltd.
Milton Keynes UK
10 September 2009

143533UK00001BB/3/A